T0076116

From Extraterrestrials to Animal Minds
Advance Praise

"An elegant and informative account of many popular mistakes about our evolutionary history, with fascinating details about creatures from surprisingly complex protists, through trilobites and almost-birds, all the way to our own peculiar species. Simon Conway Morris marshals his arguments and information to good effect, with an engaging openness to really peculiar theories and possible counterexamples."

—**Stephen R. L. Clark**, DPhil, emeritus professor of philosophy, University of Liverpool

"If, somehow, the Christian visionary poet-artist William Blake were a world-class Cambridge paleontologist, evolutionary biologist, and astrobiologist, drank gin and tonics, and was very, very funny, you would have a human marvel approaching Simon Conway Morris. You would also have a world at once physical and mythical—a true cosmic story come alive, become real, that we are writing even as we are being written by it. You would have a 'universe built on imagination,' which is to say, 'on consciousness.' You would have what appears to be so."

—**Jeffrey J. Kripal**, PhD, J. Newton Rayzor Professor of Religion, and associate dean of the School of Humanities, Rice University

"A classic Simon Conway Morris book. The point is not whether you agree with him (I rarely do) but whether you are intrigued, challenged, having fun with the play of ideas—as I always am. Equally, as always, I am deeply impressed by the profound understanding of evolutionary processes, as the author takes us on a dazzling tour through such topics as randomness and extinctions, to supposed missing links (Conway

Morris the paleontologist is very good here) and then on to animal minds and extraterrestrials. A worthy successor to *Life's Solution*. Highly recommended."

—**Michael Ruse**, PhD, professor emeritus at the University of Guelph, Canada, and author of *The Gaia Hypothesis*

"This book opens a fresh perspective on the evolutionary process, a very welcome change from the neo-Darwinian orthodoxy that has predominated for so long. Conway Morris shows convincingly that long term trends stretch over many millions of years, developmental patterns occur again and again in many kinds of convergent evolution, and all this take place in a universe built on imagination. Altogether surprising and liberating."

—**Rupert Sheldrake**, PhD, author of *The Science Delusion*

From

EXTRATERRESTRIALS

to

ANIMAL MINDS

This is as strange a maze as e'er men trod
And there is in this business more than nature
Was ever conduct of: some oracle
Must rectify our knowledge

—William Shakespeare, *The Tempest*, Act 5, Scene 1

I have dipped into Darwin. It's heavy going. The prose thick, grey and formidable, like porridge. . . . On the big question, the God question, he seems to have maintained—this one-time candidate for [holy] orders—a careful reticence, a curiously bland open-mindedness, an obtuse bewilderment. . . . There is a passage in [his autobiography] where the author laments the gradual loss of all taste for poetry, likewise, virtually, for music, painting, and fine scenery, and speculates (ever the man of science) on what has caused the atrophy of the relevant parts of the brain. Is he trying to tell us something?

—Graham Swift, *Ever After* (1992)

From EXTRATERRESTRIALS *to* ANIMAL MINDS

SIX MYTHS OF EVOLUTION

SIMON CONWAY MORRIS

TEMPLETON PRESS

TEMPLETON PRESS
300 Conshohocken State Road, Suite 500
West Conshohocken, PA 19428
www.templetonpress.org

Set in Minion Pro and Acumin by Westchester Publishing Services.

This paper meets the requirements of ANSI/NISO Z39.48-1992 (Permanence of Paper).

ISBN: 978-1-59947-528-8 (cloth)
ISBN: 978-1-59947-529-5 (ebook)

Library of Congress Control Number: 2021944404

A catalogue record for this book is available from the Library of Congress.

Printed in the United States of America.

22 23 24 25 26 10 9 8 7 6 5 4 3 2 1

For Felix, Nils, and Alexander

CONTENTS

INTRODUCTION

The heat in Venice was insufferable, the narrow *calle* like ovens. Even as the boat pulled away from the Fondamente Nove, despite the breeze picking up the heat still followed us across the Lagoon. On arrival I ate alone, so far as that is ever possible in an Italian restaurant, and not long after I was rowed by Gianni's son, Marco, across to San Francesco del Deserto. And there Mortimer was, standing at the same spot as we had said farewell so many years ago, grinning hugely, scarcely aged, among the lengthening shadows. Our meeting was characteristically English—understated, ironic, and amusing in turns. Later, after the evening meal, we walked around the cloisters.

Mortimer did almost all the talking, not loquacious, but as ever weaving his webs of imagination. Here, I can only offer an inadequate paraphrase. "Well, yes, I heard the reports, and every time I learnt from one reliable source or another that you were going off the rails, I was enormously encouraged. Certainly you gave convergent evolution a good run for its money, but as ever you are spot on, becoming far too popular these days: plenty of capable hands to take *that* story forward. But evolutionary myths, tut! tut!, splendid opportunity to rattle the cage, stir those slumbering guardians of received wisdom. Wouldn't be so bad if they were trying to slip by watchful dragons, but that's another area where they show a distressing weakness. But, as you say, not fairy tales but areas of received wisdom that are long overdue for careful reexamination—or as I prefer to think, a really good kicking. Hardly surprising that your discipline is quietly drifting in circles, stranded in an intellectual Sargasso Sea. One almost envies the physicists. Mislaying 90 percent of the visible universe is splendid; now they have the chance of making some real progress. Mind you, and I mean that literally, when they get the answer I don't think they are going to like it one little bit.

But your myths! In one way or another they all point in much the same direction, don't they? Obvious enough that evolution is strongly constrained. Convergence shows that well enough, and so too do those so-called missing links. No wonder those cladists get in such a tizz, caught in a web of their own making—or if you prefer, trapped in endless corridors made of mirrors. In the meantime evolution has better things to do. So it gets on with the straightforward business of combinatorics, shuffling and reshuffling the kaleidoscope of evolutionary possibilities. And the net result? Not only are the many permutations tested, but oddly enough the end result is pretty well guaranteed. It is the same story wherever you care to look. Just think of those squadrons of theropod dinosaurs taking to the air, those platoons of sarcopterygian fish marching in the same direction.

Helpful also, isn't it, for those mass extinctions weighing in on the side of the angels rather than serving as the dark forces of destruction? Those groups that got it in the neck were doomed in any case; for them, the writing was on the wall long, long before the final chop. And what about their successors, the lucky few that by the merest fluke slipped under the wire? Come off it! There they are, already lined up under starters' orders, indeed some already in the race. Now comes the day of doom: *Kapow!* No fun at all. But the dust soon settles, and who do we see emerging from the wreckage? The self-same groups that in the fullness of time would have taken over the world in any case. Mass extinctions accelerate the inevitable—far from the cliché of being destructive paradoxically, they are gratifyingly creative, giving evolution a 50 million year leg up, gratis."

Silence fell, and then Mortimer glanced up at Venus, shining high above the colonnades. Pointing skyward at our shimmering companion, he continued, "Extraterrestrials? On Venus? Outside chance of life, high up in the clouds, but pretty parlous place to be, I must say. Certainly wouldn't put any money on aliens there, or anywhere else mind you. Most amusing—convergent evolution tells you what all those extraterrestrials are going to look like, but blow me down, nobody up there ready to take the call. But then our materialist chums never wanted to know what the universe is really like, nor how very odd it is that we can make sense of at least some of it. Music of the spheres, as that wise old bird Evelyn Cheesman said."

As if on cue, a nightingale began to sing. Enraptured, we sat. "Glorious sound, but don't tell me that those liquid notes have anything to do with our music." Mortimer gestured again. "By the time Saint Francis stayed here in 1220, he had long been familiar with unseen but absolutely real worlds. So have many others, but, mind you, few with such playfulness. As I said when we parted all those years ago, the story doesn't stop this side of the grave. Even I cannot see all the connections, but I am certain you are on the right track, just like old Teilhard, a much neglected figure. Evolution gets us to where we need to be, but our very uniqueness, the gulf that now separates us from all other animals, provides a splendid spectator sport. Watch all those forlorn attempts to paper over the cracks, to fool ourselves that the mental world of a chimpanzee is just a dilute version of our minds, or rather a Mind.

As I said, to open a door and to enter into a narrative not of our making, to become mythopoeic beings, able to intuit ever deeper meanings, *Homo sapiens* transforms into *Homo narrans* and fifty other species. So we discover meanings, and being granted access to orthogonal realities are now capable of infinite explorations. Maybe that's where our extraterrestrials really are, but if so—as you yourself pointed out all those years ago—we'll only be meeting ourselves. Now then what are your plans? Ah yes, I suspected as much. More unfinished business, but tell me when you return." Mortimer laughed uproariously. "Keep up the good work! Only by going off the rails will you keep to the straight and narrow . . ."

From EXTRATERRESTRIALS *to* ANIMAL MINDS

1

The Myth of No Limits

"Once there were bacteria, now there is New York."[1] No fish swam above the Archaean stromatolites, no rainforests swathed the tropics of the Cambrian, no cities lined the seaboards of the Cretaceous Tethyan ocean. Through time the biosphere reveals a story of ever-greater complexity. So too, as Darwin outlined, we are taught to see this as a step-by-step process. Undeniable, but things are not quite so simple, either literally or metaphorically. First, New York does indeed stand where once only microbial mats flourished, but the bacteria are still with us (which is just as well).[2] More importantly, however deep we choose to climb down the phylogenetic ladder, true simplicity is strangely elusive. Second, the implication of the unfolding tapestry of evolution is that it is effectively without limits or boundaries. A closer examination suggests that—with one crucial exception—this is not the case.

INVENTING THE EUKARYOTE: FIRST SIMPLE, THEN . . . ?

What circumscribes life is an oddly neglected topic, but perhaps this is less surprising when there is so much to try to understand in the grand narrative of evolution. Central to this task is the identification

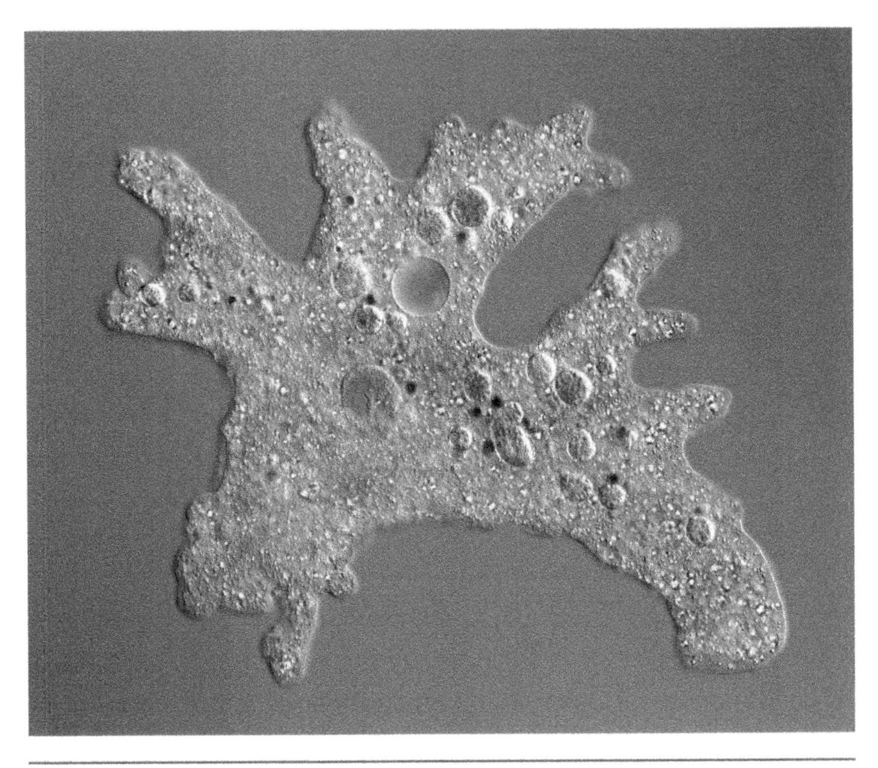

FIGURE 1.1 An iconic single-celled eukaryote, *Amoeba*.

of major transitions in the history of life,[3] epochal rearrangements that may build on existing diversity but usher in new worlds. Language, or perhaps more generally the human mind, is one such example; the evolution of multicellularity is perhaps another. But assuredly neither of these breakthroughs would have happened without the evolution of the eukaryotes, which were originally single-celled organisms, as we still see in the living *Amoeba* (figure 1.1). The subsequent diversifications of the eukaryotes have been staggering, populating the planet with mushrooms, sequoias, sperm whales, and for good measure giant kelp. If the starting point was a single cell then so too surely it was gratifyingly simple? Correspondingly we can then sit back to enjoy a story of unfolding complexity? Such would be the intuitive assumption. Given a rudimentary fossil record that scarcely preserves any cellular details and involves events that were probably underway more than two billion years ago, one might enquire how we would ever know one way or the other. The answer is relatively straightforward inasmuch as every living organism has an evolutionary footprint that will betray its origins.

Not that tracing these various spoor is a simple exercise. Form may be altered almost beyond recognition, and evolutionary convergence can fool even the seasoned observer (chapter 2). Genomes may be scrambled, bloated, or stripped down, and foreign genes can be parachuted into the cell in the process known as horizontal gene transfer. Neither, despite enormous advances in the technologies of evolutionary analysis, is it possible (or even necessary) to analyze every species. So we take representatives to serve as proxies for each of the major groups of eukaryotes. These, by general consensus, number five or six. Should you wish to know, you are a unikont, which among other things explains why my sperm have a single flagellum. To be slightly more specific we are opisthokonts (the flagellum is located at the posterior end of the sperm), thus placing us humans relatively close to the toadstools. These very major groups are, of course, further subdivided so that in turn we are also animals (metazoans), vertebrates, primates, and a very peculiar sort of ape (chapter 5).

In any event, all these groups, including the unikonts, stem from the ancestral eukaryote. We can then infer the relative complexity of this ur-eukaryote on the reasonable premise that if all the major groups share particular genes, molecules, or cellular structures, then we can be pretty confident that so too did the ancestral cell—which usually shelters under the acronym LECA, for "last eukaryotic common ancestor." It is now clear, for example, that LECA came equipped with mitochondria and these were essential for its (and ours) respiration. Mitochondria also serve as a canonical example of so-called endosymbiosis. That is, these tiny organelles were once free-living bacteria (specifically α-proteobacteria) that surrendered their freedom (or brokered some other agreement) to become permanently associated

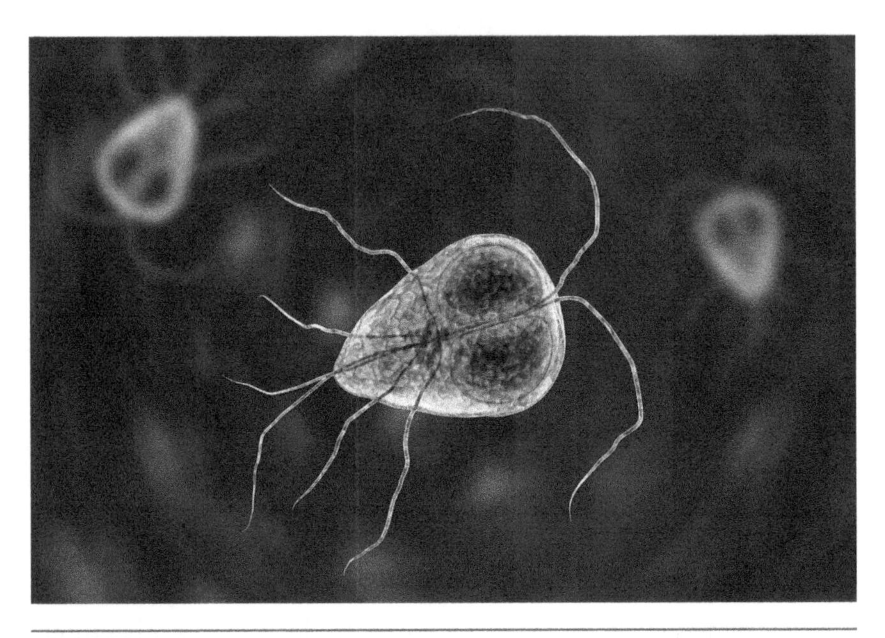

FIGURE 1.2 *Giardia*, a eukaryote to be avoided.

with the eukaryotes. Mitochondria are almost universal among eukaryotes. However, parasitic forms, such as that little parasitic horror *Giardia* (figure 1.2), lack mitochondria. In the heady earlier days of these studies, when the outlines of eukaryotic phylogeny were emerging, such forms were thought to give a glimpse of the very first stages of their evolution. But this is not so: the mitochondria are not so much lost as converted into a rudiment labeled the mitosome (consistent with *Giardia*'s effectively anaerobic existence).

Yet even if mitochondria were in place at the very earliest steps in eukaryotic history (or perhaps were even a pre-eukaryotic acquisition), we should still expect to see stories of unfolding complexity as each of the major groups elaborated in their different ways the elementary scaffolding provided by LECA. The exact reverse turns out to be the case: far from being simple, this common ancestor was astonishingly complex. To explain just how complex is almost as challenging as understanding the innumerable intricacies of the eukaryotic cell itself.

Without being overreductionist, one can consider the cell as a submillimetric factory, where master plans are executed, messages sent and received both within the cell and with the outside world, routine maintenance is undertaken, and disaster squads are always poised for action. It may sound somewhat routine, but what is truly dazzling is the cellular integration of function and efficiency in an immense series of complex cycles and networks.

Given that these arrangements did not fall out of the sky,[4] what then can we infer about the cellular situation of LECA? Let us start with the genome, where three items command our immediate attention. That the earliest eukaryotes engaged in cell division (mitosis) is unremarkable; much more surprising is that they were also capable of sexual reproduction (meiosis), even though this calls on sophisticated processes for the all-important exchange of genetic material between "male" and "female."[5] Second, eukaryotic DNA must, of course, issue the instructions; the sections that are involved with the actual coding (the exons) are, however, almost always interspersed with other sections of DNA—the introns—which ultimately have to be snipped out. Typically single-celled eukaryotes have relatively few introns whereas multicellular groups (notably plants and animals) are intron-rich. The default assumption would be that LECA was intron-poor and as time elapsed more and more introns were acquired, but the exact reverse turns out to be the case: LECA was extraordinarily intron-rich,[6] with introns perhaps accounting for two-thirds of the genome. What advantage this may have conferred is much less clear, but is another pointer to the complexity of this "primitive" genome. This is echoed by the mechanisms involved with the transcription of genes (where the RNA is read off the DNA) and the associated regulation (such as the employment of homeodomains). Once again, far from being simple trial attempts, the regulatory arrangements inferred in LECA match the complexity of living eukaryotes.[7] Indeed, wherever one looks in terms of early eukaryote genome management it is the same story of early complexity.[8]

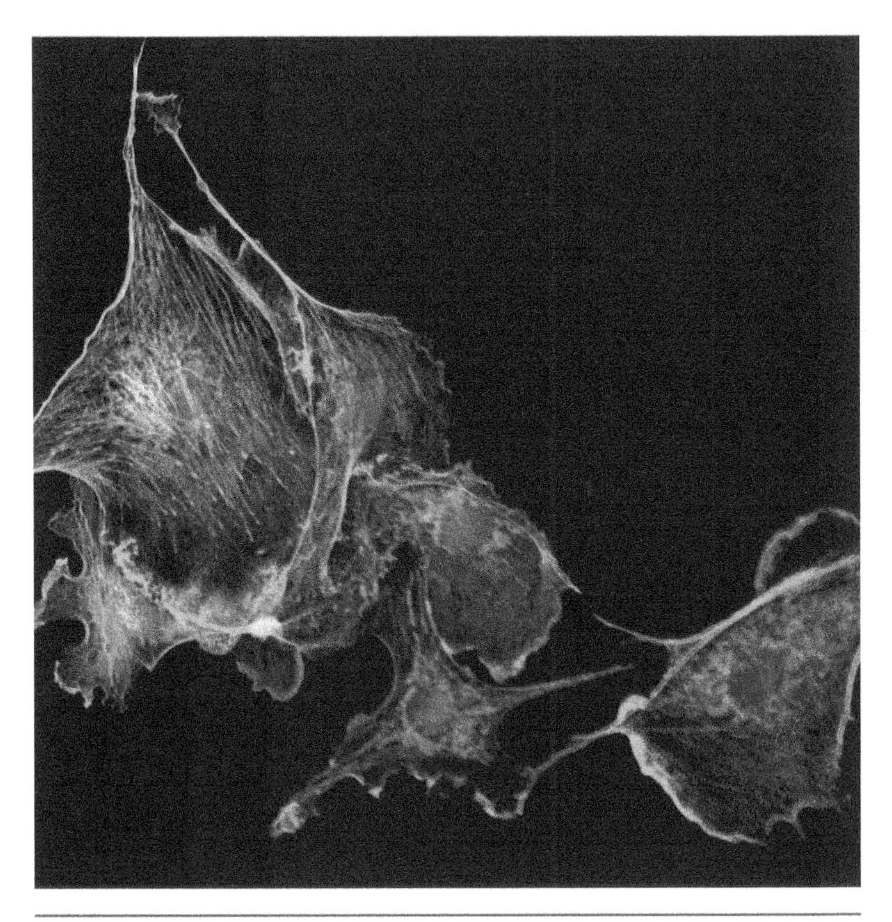

FIGURE 1.3 The cytoskeleton, a hallmark of eukaryotic complexity.

Beyond this hive of industry located in the nucleus, the rest of the cell contains the intricate scaffolding of the cytoskeleton (figure 1.3) (including filaments and microtubules) and an array of endomembranes such as the Golgi apparatus (which, once again, has a very deep ancestry[9]). In this regard, the eukaryotic cell is exceptionally dynamic, not least with its sophisticated molecular motors and ceaseless trafficking of various compounds. The point about these systems is not only are they very sophisticated nanomachines,[10] but most of the different protein families were already up and running in LECA.[11] Neither can any cell operate unless instructions can be dispatched and acted on and, just as importantly, incoming information from the outside world can be interpreted in a process known as transduction. Photons triggering an electrical response in the retina is a familiar example.

One important ingredient in these sorts of processes are small proteins known as ubiquitins. These "talk" to all sorts of other proteins and thereby control what they can do. As the name of the proteins and the process (ubiquination) suggest, these activities in the cell are all pervasive and central to eukaryotic existence. So does LECA possess a prototype ubiquination, and does the subsequent history of the eukaryotes demonstrate a series of subsequent elaborations? On the contrary: the ubiquination tool kit possessed by LECA was as complex as found today.[12]

What about the activities linked to signal transduction? Key to these processes are proteins known as guanine protein-coupled receptors (GPCRs). In us, for example, our capacity to taste, smell, and see all depends on GPCRs. The overall diversity and range of functions of these GPCRs, however, extend far beyond their involvement in sensory systems. Once again, one might infer a story of steadily unfolding complexity in the evolution of the eukaryotes. But when we turn to LECA, it is evident that the majority of GPCR families were already in place.[13]

So LECA was the exact reverse of simple,[14] but its unexpected complexity begs two questions. First, LECA obviously did not step out of thin air—they must have emerged from among the prokaryotes. They too can hardly be described as simple, so perhaps some of the complexity of LECA is inherited? Second, sheep and palm trees are among the innumerable descendants of LECA, but why then is this ur-eukaryote so much more complex than Darwinian orthodoxy would proclaim? With regards to the first point, the ancestry of the eukaryotes increasingly points to the archaeal bacteria playing a central role,[15] with the

FIGURE 1.4 Near boiling point; these volcanic waters are host to a diversity of bacteria.

Lokiarchaeota being pinpointed as prime suspects.[16] More generally these bacteria are referred to the Asgard group, with a penchant for scalding-hot hydrothermal vents (figure 1.4). A propensity for extreme environments is something of a hallmark of the Archaea and perhaps may also be a useful fingerpost toward the further reaches of extraterrestrial life (chapter 6).

In any event, like all the prokaryotes these bacteria are very sophisticated. So too it is apparent that a good part of the machinery that the eukaryotes rely upon is already present in the prokaryotes.[17] It would be surprising if it were not. There are, however, two important qualifications. First, the uses to which these proteins are put are often quite different in either group. Just as important, the molecular systems in the prokaryotes seldom show the range and diversity that is already evident in LECA. Obviously there was evolutionary continuity between the last prokaryote and first eukaryote, but the arrival of the eukaryotic cell really was a quantum leap in evolution.

BIOLOGICAL BIG BANGS

What then of evolution subsequent to LECA, those multitude of roads leading to, among other things, those sheep and palm trees, not to mention puffins and carnivorous plants? A point too often overlooked in discussions on evolution is how much is inherent in ancient forms. As George Gaylord Simpson remarked, "All the essential problems of living organism are already solved in the one-celled . . . protozoa, and these are only elaborated in man or the other multicellular animals,"[18] although later he also noted, "[Man] . . . is also a fundamentally new sort of animal."[19] In other words some aspects of subsequent eukaryotic success are inherent in LECA.[20] Its possession of homeodomains, for example, effectively prefigure the potential for multicellularity. This in turn was another momentous step because it ushered in the potential for differentiated tissues, not least nervous systems. Correspondingly, various molecular systems in single-celled eukaryotes will ultimately be employed by animals in the construction of nerves and synapses. And just as these configurations are inherent in more primitive eukaryotes, so with the appearance of a nervous system there is the potential for mind—or perhaps, more accurately, the capacity to access Mind (chapter 5).

The other striking feature of the post-LECA history is that in each of the eukaryotic groups there is a dramatic streamlining of the genomes, in a sense seeing a move from early complexity to relative simplicity.[21] This does not only involve "special" cases, such as among mitochondria and other organelles as well as endosymbiotic bacteria; this sort of streamlining is very widespread, and it also appears to involve an inherent cyclicity.[22] Rather remarkably, this shift toward relative simplicity turns out to be a more general feature of evolution.[23]

In any event, this story of complexity followed by simplification reveals a sort of biphasic evolution. In other words, we see short-lived times of radical reorganization succeeded by a mundane and much more extended interval of branching across ecospaces. This concept

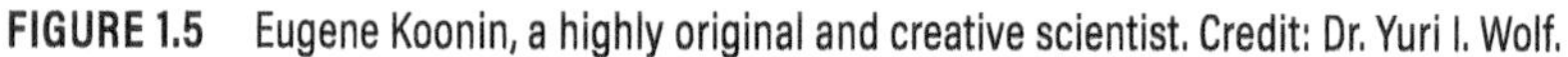

FIGURE 1.5 Eugene Koonin, a highly original and creative scientist. Credit: Dr. Yuri I. Wolf.

largely owes its formulation to Eugene Koonin[24] (figure 1.5). By drawing close comparisons to the Big Bang history of universes, Koonin has proposed an analogy between the rapid emergence of complexity and those post–Big Bang episodes of a dizzying cosmic inflation. This leitmotif of "unexpected" complexity early in a given history seems to be not only a recurrent feature but one that can be traced to the very dawn of life.

Long before LECA, there was the last universal common ancestor (LUCA), concerning which Nicolas Glansdorff and colleagues pithily observed, "Life was born complex."[25] Here among the key developments was the evolution of biological catalysts, the enzymes, without which the chemistry of life would be impossibly lethargic. All things being equal, one might expect these proteins to be rather unchoosy as to which reactions they accelerated—in the words of a biochemist, they would be "promiscuous." So too nobody would be surprised if they were relatively inefficient, ticking along and requiring millions of years of successive refinement to be honed into the superb molecular machines that now drive all life. Such examples have been identified, but are they the general rule? In a way analogous to revealing the nature of LECA, by comparing the same enzyme across a variety of groups whose common ancestor existed billions of years ago one can then infer its ancestral structure.

Not only that, but by genetic engineering we can test its effectiveness in a living cell (typical *Escherichia coli*) and so confirm that it was pretty hopeless, gratifyingly primitive, and stupidly simple. However, one such investigation, looking at an enzyme that plays a key role in the synthesis of the amino acid tryptophan, revealed the exact opposite.[26] Far from being all things to all people, the ancestral version was extremely specialized in its operation and little different from its modern-day descendants. Nor is this the only example.[27] An added bonus is that in some circumstances the "resurrected enzyme" has an activity that is consistent with what is inferred to have been substantially higher temperatures and more acidic seawater on the early earth.[28]

To the first approximation, this story of early complexity and subsequent streamlining appears to apply in every major group. Among the animals, for example, the sponges are evidently the most primitive, consistent with an absence of either tissues (including a nervous system) or organs (for example, a gut). Their genomics tell a different story. Once again the byword is "unexpected," in that the sponge's genetic tool kit[29] is much more complex than it seemingly ought to be. So too it has the seeds for a whole range of capacities that will underpin the success of later animals, not least an immune system, cell communication, neuronal activity, gonads, and the associated developmental machinery.[30] As with LECA building on its archaeal foundations, the molecular scaffolding of sponges calls on the machinery of more primitive groups, which in turn reveal their yet deeper inherencies that will help to provide the basic architecture of animals.

Another consequence of Koonin's Big Bang model is that the process of complexification is not only geologically rapid but highly combinatorial. In other words, there is an extensive juggling of different evolutionary solutions. Such a chaotic state of affairs seems difficult to reconcile with the processes of orthodox Darwinism. Even if it transpires that they are, this genomic turmoil would help to explain why recovering a genealogy of descent has proved so difficult, if not actually impossible. Nick Lane has referred to this as a "phylogenetic 'event horizon,'"[31] analogous in a way to a cosmological event horizon, and with the implication that there is certainly an ancestry but one which we will never be able to see. After all, if one set of genes suggests one set of phylogenetic relationships and another set a radically different arrangement, then clearly an appeal needs to be made to a different sort of evolutionary umpire. It also prompts the question of whether the ultimate outcome, which on this planet came to be labeled LECA, was one of chance, or whether, in the final analysis, one or other similar combination is likely to win through. If the rest of evolution is any guide, then the latter is much more likely.

So yes, in Cambridge where I write these words, three billion years ago the same spot would have shown little of material interest other than microbial mats. Since then, the world has become more diverse and from our perspective much more interesting, but life itself seems always to have manifested a deep and often hidden complexity. Self-evidently it was assembled from prebiotic molecules; even at this crucial stage, one suspects that the sum was fast becoming far more than the parts as it propelled itself from the inert to the vital. Whenever and however this process happened, life was set to embark on an astonishing odyssey, of which the oddest end result was a species that could begin to understand itself (chapter 5). Concomitant with this three-billion-year spree, to echo the words of the science-fiction writer Brian Aldiss, has been the universal assumption that this process is effectively without limits. With the critical exception of ourselves, this notion must be abandoned.

THE GREAT WALLS OF BIOLOGY

Thus, a succession of evolutionary Big Bangs transformed the biosphere. For all intents and purposes, however, this set of expansions has been treated as an open-ended process, with the focus of attention revolving around Stephen Jay Gould's concept of "walls."[32] In this metaphor, all life is constrained by a "left wall," effectively reflecting a limitation of size. To be sure, the smallest cells are almost unbelievably tiny, with lengths of less than 0.0005 millimeters (mm),[33] but further shrinkage is unlikely. The real debate is what lies to the right? Is the occupation of this metaphorical space—be it genomic, morphological, or ecological—controlled by the capacity of life to wander off in almost any direction, diffusing more or less at random? This was Gould's view, and as with his emphasis on the contingencies of evolution, it was largely driven by ideological reasons.

Alternatively, is moving away from the left wall inherently directional? Andy Knoll and Dick Bambach,[34] for example, address this issue by identifying six mega-trajectories in the history of life, starting with the origin of life, including eukaryogenesis and concluding with the emergence of intelligence. At each step there is a successive enlargement of ecological possibilities, and the processes do appear to be directional rather than diffusive. It hardly needs to be added that these steps are cumulative, and not ones of dynastic succession. The bacteria are always with us (and are as busy evolving as the rest of the biosphere), and were they not we would be in a very sorry state. A more interesting question, however, has largely gone unaddressed: are there ultimate limits of life, and if so is the biosphere anywhere near such a closure? Curiously, the evidence suggests that with one crucial exception we are indeed near to the boundaries.

Given the immense diversity of life, one might reasonably ask how these limits could ever be coherently defined. To complicate matters, the combinatorial possibilities are so gigantic, with the number of potential alternatives far, far exceeding the small change used by

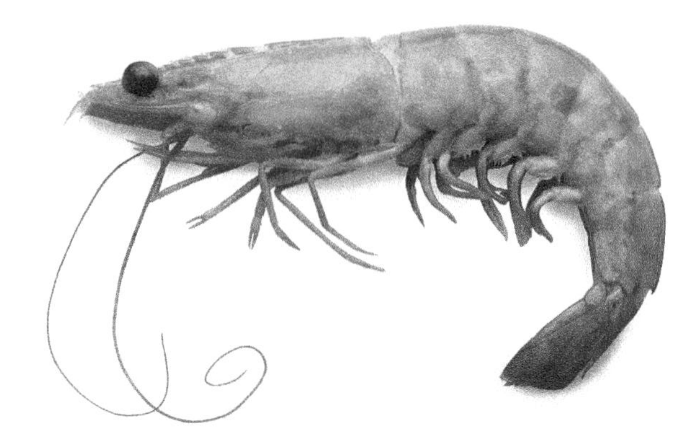

FIGURE 1.6 The familiar shrimp, but also an exemplar of tagmosis with a variety of specialized appendages along the length of the body: from anterior antennae, feeding appendages, walking legs, and finally the pleopods.

cosmologists (such as the number of particles in the visible universe: about 10^{90}). In the latter case, however, convergence comes to our rescue; its ubiquity suggests the regions of biological hyperspace that are actually habitable represent the minutest fraction of what is potentially available. This observation, however, makes no direct claim that all species were or are at the limits of the possible.

These limits, however, appear to be very real. The pioneers in this field, Sarah Adamowicz and colleagues,[35] addressed what is known as tagmosis—how the body can be arranged in specialized zones rather than a serial uniformity. Their specific target was the crustaceans (figure 1.6), familiar from such forms as lobsters and barnacles. The usual notion is that the ancestral crustacean had a largely uniform array of appendages, with little division of labor. Through time, these appendages were allocated to a series of functions, such as those dedicated to feeding. Not only was there clear evidence for increasing complexity that was consistent with directed evolution moving away from Gould's left wall, but not surprisingly it was a recurrent feature in separate lineages. But the important question is: Are there absolute limits to tagmosis? Intriguingly, the evidence suggests "not quite" inasmuch as the values have only shown modest increases for the last

350 million years (Ma) but still fall a little short of the theoretical maximum. Maybe this reflects developmental constraints, but could there be a very slowly receding "Darwin Zone" where evolutionary processes asymptote infinitely?

SHRINKING ANIMALS

Regardless of whether a Darwin Zone really exists, we can now begin to mark the limits of form beyond which no life may travel. Do not, by the way, be gulled by the standard vocabulary of "simplification," "degenerate," and similar terms. Whatever the metric of complexity,[36] the organisms in question are extraordinarily sophisticated. A more ambitious book would be exhaustive in its cataloguing, but what is more important are the general principles. Let us turn to the animals (metazoans).

Animals possess tissues that are linked to digestion, movement, senses, and, of course, sex. The respective organs are constructed with huge numbers of cells. (In humans, the estimated total is about thirty-five trillion.[37]) There is no straightforward metric between number of cells and complexity, as is evident from a particular marine organism (in the form of a dwarf male) that counterintuitively showed "high complexity with low cell numbers."[38] Researchers have found that what can be achieved with about a hundred cells is astonishing. And so, in a different way, is another example of miniaturization, the dicyemids (figure 1.7).[39] They are not a household name, nor do they inhabit a household location: rather, they spend their lives drenched in urine. This is out of choice, mind you: their abode is the kidneys of squids and their relatives (collectively, the cephalopods), where they form a dense felt on the surface.[40] Wormlike and minute, these creatures lack any organs—which is scarcely surprising, given that they are composed of fewer than fifty cells.

The informal group to which dicyemids belong, the mesozoans, were so named on the supposition that they provided a direct link

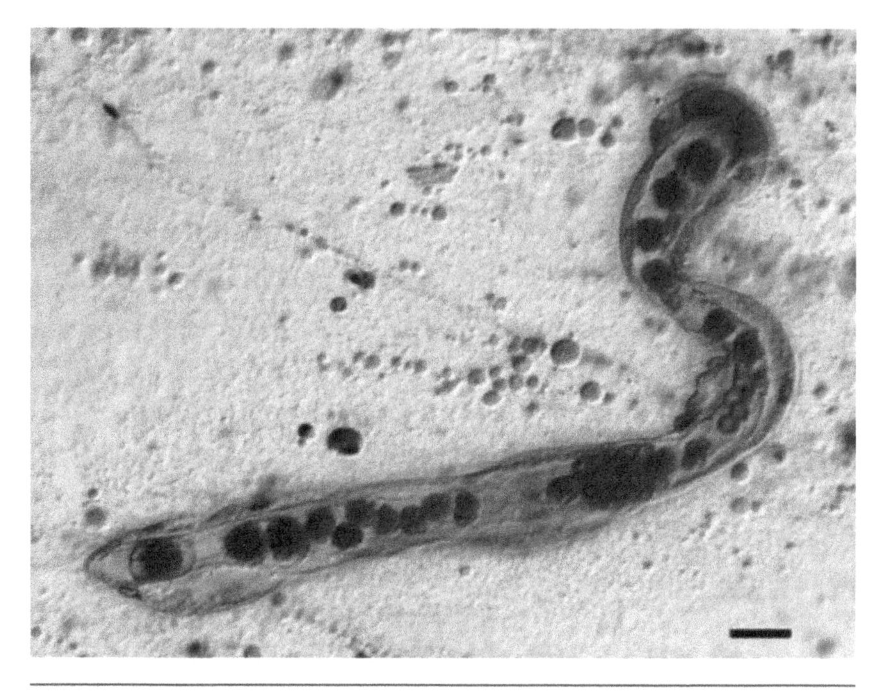

FIGURE 1.7 Almost protistan-like, the remarkable dicyemids epitomize simplification (of a sort) in the animals. Credit: 古屋秀隆 (Hidetaka Furuya), CC BY-SA 4.0 (https://creativecommons.org/licenses/by-sa/4.0), via Wikimedia Commons.

between the protozoans (or protistans) and metazoans. On form alone, which is as effectively an elongated axial cell with a coat of ciliated cells, the wider relationships of dicyemids would be insoluble. Molecular data, however, have suggested that they are derived from more advanced animals and are apparently close to some of the early spiralians (a medley of groups that include the flatworms).[41]

The otherwise similar orthonectids,[42] which are now known to be vastly simplified annelids, are thus most likely convergent with the dicyemids.[43] The crucial point is that in transmuting themselves into honorary protistans, dicyemids have approached the limits of what an advanced animal can turn into. Their status as protistan manqué is reinforced by a striking convergence with the genuine article in the form of chromidinid ciliates,[44] which also coat the kidneys of some cephalopods, have much the same mode of attachment, and perform the same sorts of function. Although they are routinely dubbed as parasitic, both dicyemids and ciliates are much more likely to be engaged in an intimate symbiosis, thus ensuring a high-level kidney function that is one ingredient in propelling the cephalopods to a level of organization that rivals that of the fish.

One might think that stripping an animal to fewer than fifty cells represents an absolute limit. That may be so for more advanced animals, but is the ultimate limit of effectively one cell achievable? Indeed so, although here we need to turn to the cnidarians and specifically the jellyfish (the medusozoans), which show remarkable evolutionary excursions far removed from the more familiar pulsating bells of their free-living relatives. In this case, the group in question are the myxozoans,[45] and again the name hints at a supposed halfway house. These parasites are of more than passing interest to the aquaculture industry, afflicting trout and salmon with so-called whirling disease as they sabotage the nervous system and send the fish into a corkscrew-like motion. Unlike the mesozoans, whose precise origins are rather obscure, the sort of route the myxozoans might have taken to the limits of existence is more obvious.[46] This is because there is a gradation, from more orthodox cnidarians with tentacles (exemplified by *Polypodium*), to curious wormlike creatures that mimic nematodes (*Buddenbrockia*), to the final endgame among the myxozoans in what is essentially a single-celled animal. Welcome to the myxosporeans.[47]

Prior to molecular scrutiny, once again the consensus was that they must belong to the protistans. Skeptical voices pointed out that, although these myxosporeans may have ditched all the multicellular accoutrements and transformed themselves into a bag of cytoplasm (technically a plasmodium), they remained multinucleate. So too they retained the diagnostic stinging cells (nematocysts) of the cnidarians, albeit redeploying them for new functions.[48] Yet protistan-like the myxosporeans certainly are, not least in their amoeboid behavior and radically reduced genome. And just as the dicyemids have their ciliate avatars, so too some myxosporeans have protistan counterparts.[49] One can label myxosporeans as "simple," but this is false perspective, not least because they possess the complex life cycles connected to the dark arts of parasitic infestation.

The intricacy and intimacy of these and other parasitic associations leads us to ask whether at least some are also located on the margins of the possible. Evidence that such might be the case could receive additional support if it transpires that more benign symbiotic associations are also seen to bang up against unscalable ramparts. The handful of test cases discussed here will be at best indicative, but their levels of integration suggest that once again biology has no further room for maneuver.

LIVING ON WATER

All animals feed, but some seek out a diet that anywhere else would spell rapid starvation. For instance, some insects, of which the aphids are the most familiar, tap the sugary fluids conveyed along the phloem in plants. These liquids may be energy rich, but they are severely deficient in many key compounds, not the least of which are vitamins. But this pales in comparison to insects such as the sharpshooters (figure 1.8) that tap the adjacent xylem vessels. It is not pure water that is being transported from the roots to the photosynthetic factories housed in the green leaves, but the concentration of organics is exceptionally dilute, and for any insect it is a dietary nightmare.[50] In addition, because the xylem fluid is under negative pressure,[51] a consequence of it being drawn up by the processes of phototranspiration, extraction by pumping is in itself a real challenge.[52] And not only do prodigious quantities have to be swallowed (roughly equivalent to an adult human consuming twenty-five gallons an hour[53]), but the capture of the organics is achieved with stunning efficiency (sometimes exceeding 99 percent[54]).

A farmer's perspective might be that such capabilities had never been achieved, given many of these bugs spend their time trashing

FIGURE 1.8 Insects such as this sharpshooter live on a seemingly "impossible" diet.

citrus crops and the like.[55] Unassisted these insects would pose no threat, but when the symbiotic bacteria are thrown in, matters change entirely both for the xylem suckers and their sap-feeding counterparts. Deep in the insect abdomen lurks a sort of bacteria hotel, the bacteriome. Hosting bacteria in crypts and other receptacles is by no means unusual, so at first sight the function of these bacteria is quite straightforward: they serve to synthesize the various molecules, notably the so-called essential amino acids and vitamins, to redress the dire deficiencies of the ingested plant sap.[56] But in the bacteriome and its component bacteriocytes we encounter a new level of organization, of unrivaled integration and sophistication that once again suggest we stand on the absolute limits of the biologically possible.

In some insects the symbiosis involves a single strain of bacteria, with animal and prokaryote being completely interdependent.[57] Some of these associations are also very ancient, with molecular data suggesting insect and bacteria have been constant companions from before the time of the dinosaurs.[58] In other cases, however, a second bacterium has been recruited (and in one instance then split into two species that provide "two genomes with the functionality of one"[59]). As Ed Yong has remarked, here is a case of "a bug within a bug within a bug."[60] This, however, is more than a metaphor because the second type (drawn from several different bacteria[61]) is literally housed within the cell cytoplasm of the primary symbiont (and in one case has taken the final step where the two genomes of the bacteria have fused[62]). And this association is far more intertwined than mere cohabitation. First, there is an integration of biosynthetic pathways between the two strains of bacteria, so each is vital in the production of a particular compound—say, the amino acid histidine.[63] Clearly each bacterium requires an array of functioning genes, but in the genomes of the bacteria there has been a shrinkage in its size so drastic that even the pursuit of a cellular life is now impossible.[64] In extreme cases, it is as if a once immense factory has been reduced to a few whirling cogs and gears.[65]

Having painted itself into this genomic corner, the bacteria now face grave risks. Given the symbiotic association is so tightly integrated, it is expedient that each generation of insects effectively infects the succeeding one, usually by ensuring the eggs carry a handful of bacteria. This means, however, that although immense numbers of bacteria flourish in the bacteriome of the adult insect, at each generation these bacteria have to wriggle through a severe bottleneck.[66] At this point, any geneticist will mutter "Müller's ratchet" and with good reason. For the bacteria, this is very bad news because it makes it very difficult to weed out deleterious mutations. But this triple association has persisted; most likely, when one strain finally goes extinct, another is recruited to fill the gap.[67]

With the bacteriocyte providing a factory to supply essential components of its diet, does the insect simply stroke its abdomen and murmur "thank you"? No, the host is also intimately involved with this symbiotic union.[68] It can also acquire a series of bacterial genes from a variety of other sources, and these too play their part in this intricate symbiosis.[69] It is difficult to see how it could get much more complex. If one of the partner bacteria is little more than a ghost, if a flotilla of other bacterial genes have arrived by multiple episodes of horizontal gene transfer, then surely we are very close to the limits of the possible.

What applies to the insects and their bacteriocytes finds parallels elsewhere, in environments far removed from the insect abdomen. One such example can be found in the nitrogen-fixing symbioses,[70] where plant and microbe join forces. Much farther afield are examples from the open oceans: here, two bacteria (the photosynthetic cyanobacteria *Prochlorococcus* (figure 1.9) and a heterotrophic counterpart known as *Pelagibacteria*) have tiny cells and correspondingly reduced genomes,[71] but they account for an astonishingly large biomass (in the case of the alga, estimated as totaling about 10^{27} cells[72]).

How to account for such abundances? In what has been termed the "Black Queen Hypothesis"[73] (a counterpoise to the better known Red

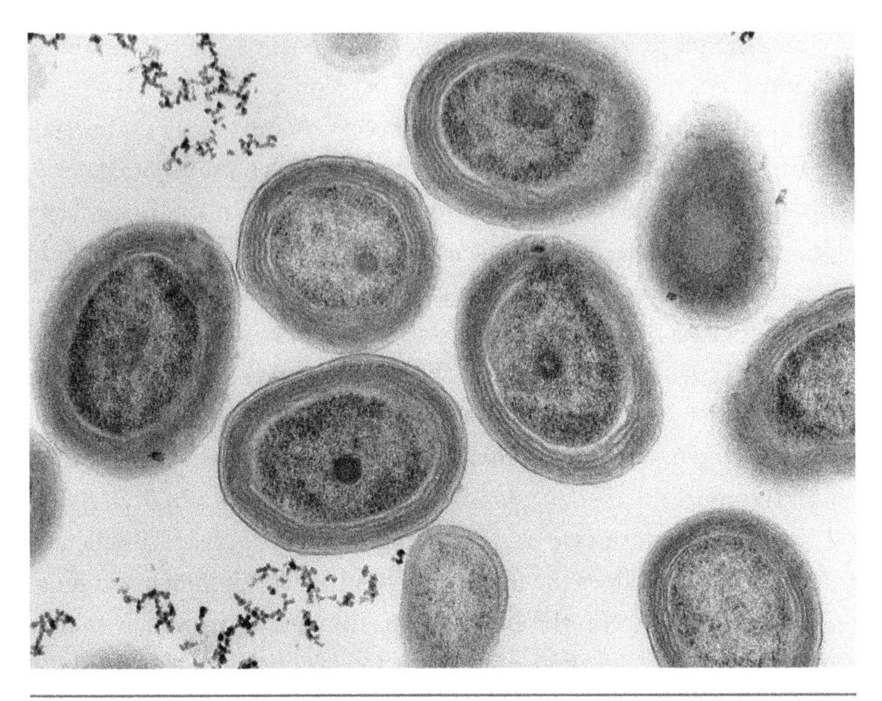

FIGURE 1.9 The cells of *Prochlorococcus* may be minute, but they are masters of the oceans. Credit: Luke Thompson from Chisholm Lab and Nikki Watson from Whitehead, MIT, CC0, via Wikimedia Commons.

Queen Hypothesis, where all species are in a continuous race against each other for survival), these superabundant bacteria compensate for their stripped down genomes by drawing on nutrients "leaked" by other microbes. These metabolically interdependent networks probably extend across the bacterial world,[74] and it is likely that again the limits of complexity have been reached. Beyond these flaming ramparts evolution can never transgress, either on Earth or for that matter anywhere else (chapter 6). We can now begin to sketch a map of life that delineates not only the habitable zones (and as importantly their innumerable convergences, see chapter 2), but from different perspectives other sorts of ultimate boundaries. Once again we can demarcate regions where no life will ever be found.

SWIMMING IN MOLASSES

How do we define lifeless zones? A fertile approach is to look at the constraints of the physical environment. Particularly instructive in this regard are fluids, and most obviously water. Water is essential for photosynthesis, so plants employ a system of conduits to channel it. This is the xylem, whose interest to sap-sucking insects we have just addressed. It is worth emphasizing the xylem is far more than a system of passive plumbing,[75] and in at least some cases its hydraulic design seems to be more or less at an optimum.[76] From our perspective, the fact that in a forest millions of gallons of water are ceaselessly ascending toward the skies may give pause for thought, but this architecture also has its limits. Few of us remain unmoved in the presence of immense trees, but is there any upper limit to their height? Indeed there is, and it revolves around the design of xylem whereby the ability to transport water to ever greater heights needs to be traded off against the increased tension experienced by a column of water more than one hundred meters above the ground.[77] Just as in the bloodstream bubbles of gas can be fatal, comparable embolisms in the xylem could sabotage the system. On this basis it seems unlikely that any tree could ever exceed 150 meters.[78]

A tree this size might also face a testing time when air whistles past it in a gale; although it is much less dense than water, air is just as much a fluid (i.e., it has little resistance to shear). Most intriguing are those counterintuitive worlds that emerge when viscosity becomes the predominant factor and thereby defines another intriguing set of limits. In brief, when an organism is small (typically less than a millimeter) and/or living in a dense fluid (water being about a hundred times more viscous than air), the inertial forces that reflect the Newtonian world of mass and acceleration are much less significant than viscosity—the balance is enshrined in the so-called Reynolds number. The parameters used in the equation for the Reynolds number cancel each other out so that it is technically dimensionless. This has the advantage that a laboratory can use a relatively large object in a high-viscosity fluid such as syrup to investigate how the scaled-down equivalent would behave in a low Reynolds number environment.

Being relatively enormous and moving at speed, we inhabit a world of high Reynolds numbers, as do whales in the ocean and birds in the skies. Entering the world of low Reynolds numbers, on the other hand, involves much more than miniaturization. To us it would be a deeply alien experience. If the stopping distance of a supertanker ploughing its way through its high Reynolds number ocean is on the order of miles, what is the equivalent for a bacterium when it ceases swimming? Less than an inch? Maybe a tenth of an inch? In fact, its stopping distance is about 0.1Å (angstroms; in metric units: ten picometers), the equivalent to one-tenth the diameter of a hydrogen atom.[79] To put it another way, for us to be immersed in a low Reynolds number world would be equivalent to going for a swim in a pool of molasses (figure 1.10).[80] This may sound vaguely uncomfortable, like a super-sticky equivalent of bobbing around in the Dead Sea, but in reality it would be horrifying. In 1919, the unfortunates of Boston's North End found this out to their cost when an enormous vat containing a couple of millions of gallons of molasses collapsed catastrophically and engulfed them.[81] To be sure, most of the victims were killed by the debris caught up in the tsunami of molasses, but those who were simply immersed unwittingly entered a low Reynolds number world: as the viscous forces predominated, orthodox methods of swimming proved entirely useless.

Although the milieu of low Reynolds numbers might be thought to govern only micron-sized bacteria and other minute objects such as pollen and spores, this is by no means the case. Consider the life of larval fish. They commonly suffer massive mortality, a reality that was spelled out by Darwin in the third chapter ("Struggle for Existence") of his work *On the Origin of Species*.[82] He was writing under the long shadow of Thomas Malthus, and although nobody doubts Darwin's

FIGURE 1.10 Honey and similar fluids are familiar enough but actually involve deeply counterintuitive worlds where laminar flow reigns supreme.

formulation of competition as a driving force in evolution, at least among the larvae of fish there are other forces at work. Here, mortality is a result of starvation but not from a shortage of food. At their scale of operation, the suction feeding employed to capture their tiny prey operates within the realm of low Reynolds numbers: the kinetic energy needed to drag their prey through the viscous water toward their mouth is correspondingly greater,[83] making it a terrific effort. As the handful of surviving larvae grow in size, their efficiency of feeding rises sharply.[84] This, however, prompts a Darwinian question: if selection is so ferocious at smaller sizes, would it not be much more sensible to lay larger eggs from which correspondingly bigger larvae could emerge? In an ideal world, this would presumably be the case; but larger eggs involve other risks—such as being more visible or more likely to experience multiple fertilizations.[85] Evolution is full of trade-offs, but here the grim result is that being born into a high viscosity world makes death by starvation a near certainty. The world of low Reynolds numbers represents another absolute limit as to what is possible.

At first sight, the predicament of the larval fish might suggest that any animal that ventures into a world where viscosity rules does so at its peril. All, however, is neither doom nor gloom. Consider flight and what happens when insects such as parasitic wasps explore this low Reynolds number zone. These minuscule creatures include the Hawaiian *Kikiki,* which is less than two hundred micrometers long[86] (and thus smaller than many protistans). In point of fact, some terrestrial arthropods reach half that size,[87] and maybe that too is an absolute limit.[88] But so far as flight is concerned, *Kikiki* may be approaching the boundaries of the possible. Certainly this wasp shows an astonishing transformation of their wings: they abandon the classic blade shape not only to become very narrow but also to bear a series of very elongate spines.[89] They are strikingly unlike an airfoil; in fact, at this scale a normal insect wing would be a positive distraction. In this low Reynolds number setting, the air is too viscous to flow between the marginal spines, so despite appearances it is the tips of the spines that serve to define the margins of the wings.[90] This is, of course, a very economical arrangement (and finds a convergent equivalent in the minute featherwing beetles[91] and perhaps more curiously finds a further parallel in the dandelion seed[92]), but is only one component of a quite remarkable miniaturization.

Despite a drastic reorganization, *Kikiki* still displays full functionality, not least in terms of vision.[93] Nobody is pretending that a minute eye, which in the case of the minuscule *Kikiki* has only about twenty-two lenses (ommatidia), enjoys the same visual acuity as the eyes of larger insects.[94] Even so, the equally tiny featherwing beetles not only manage with as few as ten ommatidia, but the diameter of the individual facets can be as small as six micrometers.[95] It is not disputed that compound eyes not only have inherent limitations,[96] but in a comparison with a camera eye the latter wins hands down.[97]

The more basic question, however, remains. Given there is a wide range of acuities, not only in vision but across the other sensory modalities, can we identify absolute limits beyond which these biological systems can never pass—realms that can never be accessed? Indeed we can.[98]

SEEING TO THE LIMITS

First, let us examine vision. Here, the essence of the operation is the impingement of photons on a transmembrane protein (rhodopsin), the reconfiguration of which leads to an electrical signal and ultimately the capacity to see. Eyes, of course, come in an extraordinary variety of configurations (camera, compound, mirror, etc.), and they can augment their capacities by possessing a fovea, drawing on fiber-optics, showing an exceptionally dense packing of the photoreceptors, and even employing the cell nuclei to focus the light.[99] This is all part and parcel of a well-known Darwinian story that in one way or another concerns the endeavor to capture one or other segment of the electromagnetic spectrum (ultraviolet to infrared). Nevertheless, accessing particular wavelengths is one thing, but more generally what are the limits?

The apotheosis of these capabilities involves those animals that dwell in the darkness, most obviously at night but also in caves and the inky depths of the deep oceans. So far as the latter are concerned, except for bioluminescence light is effectively always absent, whereas with nighttime vision the central problem is, of course, a paucity of photons. The differences compared with full daylight are impressive: illuminated by the moon, the world is about a million times (10^{-6}) dimmer, and when there is only starlight it is a hundred times less bright (i.e., 10^{-8})—and that is before it gets cloudy, when light levels can drop by another order of magnitude. Where we would be blundering around,

FIGURE 1.11 Some species of carpenter bee are not only nocturnal but perfectly capable of flying in near darkness.

there are nocturnal insects that not only fly with precision but use landmarks to ensure a safe return to their domiciles.[100] For example, a species of Indian carpenter bee (*Xylocopa*) (figure 1.11) flies on moonless nights.[101] When we recall that humans become color-blind at an illumination equivalent to the half-moon, it is even more astonishing that both this bee[102] and a species of hawkmoth[103] not only can discriminate colors under starlight but also show what is known as color constancy (i.e., the perception of a given color does not alter as the ambient illumination changes).[104]

That this night (or scotopic) vision might have technological applications has not escaped notice,[105] but how do these insects manage these apparently incredible feats? The problem is even more severe than might at first appear. First, the number of photons arriving at the retina is so small that only a tiny proportion of the photoreceptors are activated. Seeing in what we would justifiably call pitch-dark is bad enough, but if one is flying rapidly through dense jungle on a rainy night searching first for nectar and then for the safety of the home burrow, the challenges are greatly magnified. To begin to understand what is going on, first we need to look at the eyes and their structure.

In insects there are two types of compound eye, respectively referred to as superposition and apposition. In the latter arrangement, effectively each lens serves to channel light to its photoreceptor, whereas the superposition eye is considerably more effective because a number of lenses combine to direct the light via their component rhabdoms to a single photoreceptor. In other words, the apposition eye produces myriad individual images (that the brain then has to interpret), whereas the superposition eye is more sophisticated and more akin to our retinas.[106] Given that this latter sort of eye is substantially more sensitive, unsurprisingly it is found in nocturnal moths (and, of course, numerous diurnal insects). Yet by contrast the carpenter bees rely on apposition eyes during their nighttime excursions. Some compensation is obtained by an increase in their overall size as well as larger facets and much wider rhabdoms. The eyes are also augmented by a sort of additional eye (the so-called ocellus); in the carpenter bee not only is it enlarged but more remarkably the photon capture is enhanced by a reflective surface (analogous to the familiar eyeshine—the tapetum—in animals like cats). Yet despite these upgrades, the overall improvement of sensitivity is only increased roughly thirtyfold.[107] Much more is required to be able to fly beneath the stars.

The fundamental problem is that to be effective the eye has to distinguish real photons from false positives; in other words, it has to enhance the ratio between true signal and spurious noise. In an analogous manner to living in a low Reynolds number world, this desideratum collides with the unyielding realities of physics. In principle, single photons can trigger a signal,[108] but unfortunately not every photon is guaranteed to generate an electrical response (as a so-called bump). The problems of detection are further magnified by various sorts of noise.[109] Thus, biochemical reactions in the photoreceptors can trigger a false response ("transducer noise"). Spontaneous activity that serves to generate "dark noise" further decreases reliability. This potential problem, however, needs to be set against the extraordinary stability of the rhodopsin proteins. At least so far as mammalian rods are concerned there is a false signal about every five milliseconds, but given the huge number of rhodopsin molecules in a rod (about a billion) this means that a given molecule will on average only fire after several centuries.[110] Even so, this nocturnal bee can only obtain its extreme sensitivity if it is willing to risk being fooled by false signals. Therein lies the dilemma. As the bee rapidly flits through its crepuscular world, it needs to discern nearby objects (and obstacles). A reliable assessment of what these objects represent (spatial discrimination) and making a series of speedy decisions (temporal discrimination), however, can only be managed by a series of trade-offs. Broadly, these allow effective navigation but at the expense of acuity. Crudely, you can zip along, but the objects around you will look "vague."

Given all these constraints (eye design, unavoidable "noise," and so on), how then can the insect fly confidently? The solution appears to be to engage in so-called neural summation, which is a higher-order process in the brain whereby information is aggregated or "bundled."[111] In the insect brain, specific regions (such as the first optic ganglion) house specialized neurons with prominent extensions that apparently are instrumental in this summation.[112] In an analogous way among the primates (like the macaque) particular neurons (the so-called parasol ganglion cells) consist of two complementary types (labeled "on" and "off").[113] These respond in different ways: "on" neurons can only react to a flurry of photons (and so behave in a decidedly nonlinear manner). On the other hand, the "off" variety are extremely sensitive to the arrival of a photon. This results in a linear response but pays the penalty that some apparent detections are actually noise and thus false. Combining the on and off mechanisms allows the process of neural summation to hedge its bets and thus confers a reliable response in the deep gloom of the nighttime world.

Even so, it is all very well conjuring up the process of neural summation, but exactly how these neuronal "conversations" translate into an ability to see clearly in starlight is another question. There is a final point, perhaps ultimately the most important one. Given that humans can detect a single photon, might this then allow direct access to the otherwise utterly mysterious quantum world, at least in terms of the famous Copenhagen interpretation whereby the introduction of an observer is believed to be crucial? If so, could we then bring the observer back into experiments involving quantum nonlocality,[114] not least to detect the mysterious process of entanglement?[115] This is a fascinating conjecture, not so much on account of the human eye replacing something like the famous slit experiment, but far more importantly because it might provide the avenue to reunite the "subjective" observer and "objective" science. If so, one might meld the eternal abstractions of quantum mechanics with the imperative "now" of human experience.[116]

Be that as it may, what applies to the sensitivities and limits of vision finds an equal resonance not only across the familiar worlds of sound, touch, taste, and smell, but also among other very different types of modality. To us these may be utterly alien, but they provide the sensory milieu of animals as different as rattlesnakes and sharks. The former, for example, have astonishing sensitivity to heat (of their prey),[117] and when it comes to the infrared detectors in pyrophilous beetles they can register major fires more than fifty miles away—which is at (if not beyond) the limits of thermal noise.[118] Equally impressive are animals that can detect electrical fields. Not that we are immune to electrical shocks, but what we cannot do is "see" an electrical field in the same way we see a field of poppies. But other animals most certainly can—and in doing so they once again appear to be banging up against the limits of the possible.

All animals generate electricity; it is, for example, an unavoidable product of muscular contractions. Not surprisingly a number of animals have evolved mechanisms that can detect what turn out to be very weak electrical fields. Correspondingly a number of groups can detect magnetic fields,[119] with sensitivities extending to the level of nano-teslas (roughly a thousandfold less than the Earth's magnetic field).[120] So far

FIGURE 1.12 The stingray, a master at electroreception.

as sensitivity to electric fields are concerned, the best known are the capacities of various fish, but they are by no means the end of this roster. Among the mammals we count not only the duck-billed platypus[121] but also the Guiana dolphin.[122] But it is in the oceans where this capacity comes into its own, especially among the sharks and their close relatives the rays (figure 1.12). Their methods of detecting electric fields, interesting as they are, need not detain us. What matters in the context of limits is their sensitivity.

The platypus is impressive, given that it can operate in the range of millionths of a volt[123] (μV, microvolts, with the strength of the field measured in microvolts per centimeter, $\mu V\ cm^{-1}$), but the sharks and rays are astonishing. They display sensitivities that are a thousand times greater, operating in the realm of nanovolts (nV, a billionth of a volt), and the claim they can even detect voltages smaller than a single nanovolt[124] has raised some eyebrows. (In laboratory conditions this level of sensitivity requires instruments supercooled with liquid nitrogen.[125])

In a manner reminiscent of the eyes detecting a handful of photons, all things being equal, such tiny voltages would run a comparable risk of being swamped[126] by thermal noise.[127] One can only assume they have some sort of system analogous to the neural summation we have already encountered in the night-flying insects.

If, however, you are racking up memorable experiences then wait until you encounter the stunning effects of the electric eel, which delivers a very short-lived bolt of about five hundred volts.[128] These eels are the nearest thing to a biological car battery, and by no means are they the end of the story. Other fish can also generate their own electric fields, albeit of a much weaker strength. Celebrated in this regard are the freshwater mormyriforms of Africa (figure 1.13) and their convergent counterparts in South America, the gymnotids. These animals live very largely in an electric world: they interrogate the surrounding environment of their aquatic homes by interpreting distortions of the electric field as well as using the same medium to ceaselessly communicate with each other. With respect to the latter activity there is, however, a potential snag. What happens when two individuals simultaneously transmit the same signal, thereby potentially jamming each other? The solution is known as the jamming avoidance response,[129] which also plays a role in the social interactions of some weakly electric fish.[130] Although the precise details are somewhat complex, in

FIGURE 1.13 Mormyriform fish have vision, but for all intents and purposes they live in an electrical world. Credit: Carl Hopkins, extracted from Electric fish and EODs.png, CC BY-SA 3.0, https://commons.wikimedia.org/w/index.php?curid=59289023.

essence (and somewhat counterintuitively) the system does not rely on the fish relying on an internal pacemaker. Rather, the temporal patterns of amplitude and phase modulations are assessed across the fish's body and integrated in such a way that locally unreliable readings do not swamp the overall electrical signal. More importantly the actual response times to shift the signal and so avoid jamming can be fantastically fast—as short as four hundred nanoseconds (a nanosecond being 10^{-9} of a second).[131]

These extreme sensitivities, which include being able to detect outside signals that are a thousand times weaker than the electric field generated by the fish itself,[132] are examples of what is known as hyperacuity.[133] At first sight such temporal sensitivities are impossible because the sensory receptors simply cannot work that rapidly. The solution revolves around a series of neural structures within the brain and the parallel processing of the amplitude and phase information.[134]

Significantly, this hyperacuity of weakly electric fish has striking parallels in other sensory systems—in fact, hyperacuity is ubiquitous.[135] So in vision the same principles apply, inasmuch as the powers of resolution far exceed the limiting size of the retinal receptors, even for those in the fovea.[136] A standard method to assess visual hyperacuity is by using a so-called vernier scale. Imagine two rulers, each engraved with fine scale lines, that are placed exactly opposite. Now slide one ruler until the eye is able to detect the slightest displacement. If the smallest degree of offset was constrained by the size of the photoreceptors, it would be equivalent to about thirty arc-minutes (i.e., about half a degree). This is impressive enough, but the threshold for hyperacuity is far smaller, on the order of five arc-seconds (so equivalent to a thousandth of a degree).[137] As with the weakly electric fish this discrimination is clearly based on neuronal processing,[138] but interestingly in humans by concentrating on the task this degree of acuity can be improved even further.[139]

We see ourselves as predominantly visual creatures, but in reality we are immersed in a riot of sensory perceptions. Take olfaction. The mantra insists that in this regard we are sadly impoverished. In comparison with dogs this may be true, but our being able to recognize "more than 1 trillion olfactory stimuli"[140] hardly suggests an evolutionary anosmia. More significant, however, are the limits of olfactory systems, and they turn out to be extraordinary. Moths, for example, are hypersensitive to airborne odorants. In one case a cardiac response could be triggered by fewer than six molecules arriving on the antennal receptors,[141] and in another analysis a single molecule (of the female pheromone bombykol) was enough to trigger the nerve.[142] Moths may be übersensitive, but mammals are no slouches. By employing the so-called vomeronasal organ, mammals can detect pheromones at extraordinarily low levels (approximately 10^{-11} M),[143] although even this is eclipsed by an aquatic equivalent released by female *Volvox* (an algae) which is an astonishing five orders of magnitude more sensitive.[144] Whether equivalent chemicals trigger sexual interest in humans is still debated, but what is not in doubt is that sperm can be superbly sensitive to the signals emitted by their object of desire.[145] Either way, it transpires that our supposed olfactory deficiencies are as much the result of a series of peculiar metaphysical assumptions stemming from the nineteenth century led by Paul Broca as well as Sigmund Freud, who played his usual baleful role.[146]

So too it is seldom appreciated that we are exceptionally tactile, especially at the fingertips (and somewhat more surprisingly there is an equivalent fovea for pain that again may employ a sort of neural summation[147]). Equivalent somatosensory systems are very widespread, not least in the probing bills of shorebirds. Not only are the star-nosed moles the masters of this realm, but again they appear to be bumping up against the absolute limits of this sensory modality with their so-called Eimer's organs, where the density of nerve endings can exceed seven thousand per square millimeter.[148] This compares favorably with some retinas; but the wider parallels between the star-nose and an eye, which extend to a fovea, have long been appreciated.[149] Remarkably, the degree of congruence extends to the saccades, those jerking movements that allow an eye (and nose) to economically scan the view. In the star-nosed mole these movements come in the form of a series of very rapid and brief touches that have evidently reached the absolute speed limit for such tactile interrogations.[150]

Even if hyperacuity is ubiquitous, it might seem odd to equate the visual capacity with a tactile counterpart—until we realize that the respective receptor surfaces (photoreceptors or mechanoreceptors) are both two-dimensional. Although tactile hyperacuity does not quite rival its visual counterpart, it is still very impressive, and again it underscores the disparity between the size of the receptors and the actual capacity for a much finer degree of discrimination.[151] Not only do the senses of vision and touch interrogate the world around them to a remarkable level of detail, but so does hearing. Acoustic hyperacuity often revolves around using both ears (so-called binaural hearing). This involves the discrimination between two sounds, whether they be separated by time intervals, relative strength, or angular separation. Again the limits are impressive: in the case of time intervals, a separation of ten milliseconds will suffice in discrimination; for volume, a minute difference (equivalent to half a decibel) is also discernible. So too, even if discrimination depends to some extent on the chosen frequency, the finest discrimination between two separate sound sources is equivalent to an angular separation of about a degree.[152] Nor, despite their relatively diminutive size, are insects such as the crickets at a disadvantage; their directional hyperacuity rivals that of both mammals and birds.[153] Acoustic hyperacuity echoes the other such examples whereby the apparent limits of the receptor system are far transcended by the actual sensitivities, pointing in turn to a fundamental identity of principles.

The details of hyperacuity can only be unraveled in the technical literature, but there are some intriguing sidelines of enquiry. One is that although the difference between the auditory capacities of males and females is relatively small, it is undeniably real; for example, females have greater sensitivity to weak sounds.[154] The straightforward observation would be that because the male auditory equipment is on average somewhat larger, this is sufficient to explain their discriminatory powers versus those of females. In reality these differences are much more deeply seated.

In this respect, the most interesting differences revolve around our binaural hearing. Thus, when either ear of a listener is supplied with a pure tone of a slightly different frequency, the psychophysical effect is for these tones to be perceived as a sort of warbling beat. Not only are such binaural beats routinely detectable at higher frequencies by males compared with females, but there is little overlap. Also intriguing is that the capacity of females to detect such frequencies shifts through their menstrual cycle (although whether this is directly under hormonal control is another matter).[155] Given that the binaural capacity shows its most dramatic shift when menstruation commences (and at this point happens to come closest to the male capacity), then, as Jerry Tobias[156] wickedly suggested, it might provide a useful cue if "FM radio stations could perform a public service by broadcasting binaural-beat signals just before they sign off the air each night."[157]

"Deaf as a post," we may say—but even the most acute hearing is circumscribed by unchallengeable limits. No animal can perceive the entire register of frequencies, from the lowest of seismic rumbles to the shrillest of squeaks. In the latter domain bats reign supreme, but even they are outclassed by a moth sensitive to an ear-piercing three hundred kilohertz.[158] By comparison, with an upper range of about twenty kilohertz (which erodes with age) humans fall far short, even though our overall spectrum is still impressive. But in terms of the lower limits, the transductory cutoff operating within the inner ear is determined by thermal noise.[159] Beyond that there is no hearing. The complexities of hearing, not least in the cochlea, require little emphasis, but here the sensitivities are almost unbelievable.[160] In humans, the work (measured in watts) required to initiate hearing is minuscule (about 10^{-12} W/m^2).

When it comes to the actual mechanics, an infinitesimal displacement (equivalent to a tenth of the diameter of a hydrogen atom) in the middle ear bone known as the stapes will transmit a sound to the inner ear, where in turn a deflection in the sensory hairs of about 0.3 nanometers will trigger a nervous signal. Such a displacement is equivalent to shifting the tip of the Eiffel Tower by the width of a thumb.[161] (For whatever reason this somewhat ludicrous edifice has become the gold standard for the metrics of acute hearing.) Thus, when it comes to the male mosquito, the corresponding deflection is equivalent to less than a millimeter.[162] Given their antennae are very closely tuned for the sounds of their female counterparts, such a sensitivity makes very good sense.

Other mechanoreceptors are just as impressive. Many arthropods, for example, can detect airflow, and again the sensitivities are extraordinary. Among spiders the sensory hairs (trichobothria) can be triggered at an energy level smaller than a quantum of green light.[163] However, the equivalent mechanoreceptors in crickets are an order of magnitude more sensitive, hovering on the edge of the thermal motion arising from Brownian motion.[164]

Like any other animal, of course, humans employ a battery of sensory modalities, and our windows of perception only access a small fraction of each sensory universe. Even with neural summations that confer extraordinary hyperacuity, animals have effectively reached the limits of the sensory universe, be it the detection of a photon or the Brownian motion of background thermal agitation. Moths can fly confidently beneath the stars, and so too many animals navigate by using celestial cues such as the Moon and even stars. In this way, dung beetles are even more remarkable because they can employ the Milky Way.[165] But to them it is a convenient band of light, not a potential home for extraterrestrials (chapter 6). Evolutionary convergence (chapter 2) would suggest that what we find on Earth will be a reliable guide to what there will be "out there," including the physical limits. In the context of possible extraterrestrial life, let us meet the so-called extremophiles. Principally drawn from the prokaryotes, especially the Archaea, these bacteria are a major target for astrobiology, not least on the reasonable supposition that most habitable exoplanets are likely to be far less benign than ours. However, when it comes to the physical limits (such as temperature, pH, salinity, and pressure), all the evidence suggests that earthbound terrestrials have once again reached the edges of the possible.[166]

That we humans can address any of these questions is, of course, because we have curiously unique mental capacities. In part, and this is an important proviso (chapter 5), it is because we have an exceptionally large brain. So it is appropriate to conclude this chapter by enquiring into the absolute limits of brain size.[167] Unsurprisingly, protistan-sized insects inhabiting a low Reynolds number world of high viscosities also show an astonishing degree of neural miniaturization.[168] The brain of one such wasp (*Megaphragma*), for example, is built of less than four hundred neurons, but the animal displays full functionality.[169] What about in the opposite direction? Science fiction is replete with mega-brained extraterrestrials (generally with very short tempers and a fondness for laser cannons). In point of fact, given that our brains are about seven times larger than they "ought" to be—were we to follow the normal scaling laws between brain size and body mass—neurologically we are already pretty monstrous.

But are we near any sort of brain size limit (figure 1.14)? Indeed we are. Increasing brain size poses challenges such as energy consumption (remember our brains account for about 2 percent of our body mass but consume a bit under 25 percent of our metabolic energy) and blood supply, not to mention maintaining communication networks. Equally compelling in terms of determining absolute limits is that not only do our brains grow disproportionately fast in early development, but the relative rates of growth (the process known as allometry) of the brain's gray and white matter are markedly different. The net result is that by the time a brain evolved that was about three times larger than ours, it would be almost entirely composed of white matter.[170] Thinking would grind to a halt.

But brain size is only part of the equation. Humans are radically different, and irrespective of the likelihood (or more likely otherwise[171])

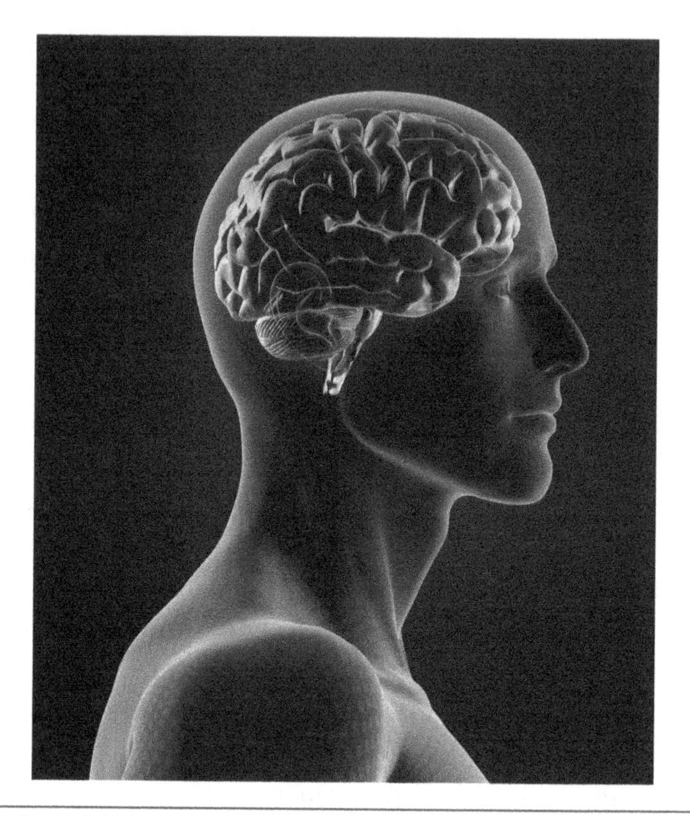

FIGURE 1.14 Relatively speaking, human brains are out of all proportion, but there is limited scope for any further increase in size.

of machine-based consciousness, our entire cognitive architecture is now irreversibly tied to culture and technology. Not only that, but whereas all other animals have reached the limits of perception, humans can unpick the weft of creation, be it with extraordinary imaging techniques or enormous particle accelerators. In contrast, biology will be forever blind to these deeper fabrics. Evolution has delivered us to this threshold, and now we need to understand how we alone peer far deeper into the universe (knowing, for example, that a photon is only one type of boson) and its construction (not least in terms of thermodynamics)—and madly it all seems to make sense.

2

The Myth of Randomness

When it comes to the history of life, the prevailing wisdom echoes the chilling words of Macbeth: "a tale told by an idiot, full of sound and fury, signifying nothing." In evolution, the narratives are delusional, and mayhem rules. Now shift the metaphor to a casino. Watch the croupier shuffle the deck to play the Game of Life. The cards are turned over, and there is a gasp of surprise. The countess reluctantly places her priceless necklace on the table, and she walks into the darkness of extinction. As the board is swept clean, hitherto shadowy figures now join the game. But they are very far from being unexpected arrivistes—rather, they were not only waiting in the wings but turn out to be yet more accomplished players. The rules remain the same but at a yet faster pace.

There is, however, a neglected twist to this narrative. Although through geological time increasing degrees of biological complexity and integration are undeniably the case, superimposed on this is an intriguing cyclicity, revolving around the idea of an eternal return: *plus ça change, plus c'est la même chose.* It transpires that evolutionary history is very far from random.

THE ETERNAL RETURN

The notion of an eternal return is an important component of evolutionary convergence—in other words, the repeated but independent appearance of the same type. In the case of mass extinctions, this has resulted in a rather peculiar sort of melodrama. Zombie-like, the so-called Dead Clades Walking shuffle forward after the mass extinction but never recover.[1] In other cases, a particular form appears to have gone extinct but then "unexpectedly" pops back into the fossil record; as it turns out, they were merely on vacation and forgot to leave a forwarding address. These so-called Lazarus taxa need not detain us, other than to note that taxa that "return from the dead" are reminders of the likelihood that probably many more lineages slip through mass extinctions than is sometimes expected (chapter 3).

There is, however, a more interesting alternative: welcome to the "Elvis taxa." In reality, these are mere impostors. Like Hans van Meegeren or Tom Keating—and the many other gifted fraudsters, recent and ancient, who have turned their hands and artists' palettes to the mass production of Vermeers and Samuel Palmers—in the fossil record we also might be fooled.[2] Elvis taxa really do look like the real thing, but on closer inspection they turn out to be mere imitations. These are convergent forms, which re-evolve into very much the same solutions.[3] But should this be any occasion for surprise? It would only be if convergent evolution was limited to mass extinctions. In that case these Elvis taxa would certainly deserve their moment in the sun; but evolutionary convergence is ubiquitous, and to call them Elvis taxa is hardly informative.

But what about the eternal return on the grand scale, such as those forms that helped to define the Permian world but then went extinct in the greatest of mass extinctions? Hands up if you have heard of the alatoconchids? Most likely you have not, but when it comes to the bivalve molluscs, otherwise familiar to us in the form of scallops and oysters, the alatoconchids are pretty eye-popping. Their name is earned on account of their possessing enormous "wings" (the Latin for wing being *ala* and shell *concha*).[4] Growing up to a meter in length, the extensions would have been analogous to a plant leaf because photosynthetic algae were crowded into the soft tissues beneath the shell.[5] Here, an ingenious system of calcareous optic fibers served to channel the sunlight to the enclosed symbionts.[6] Despite their gigantic size, the alatoconchids were not strictly reef dwellers; their splayed shells allowed them to recline on the soft, limy muds of shallow lagoons scattered across the Permian tropics, both on continental shelves[7] and remote sea mounts[8] scattered across the global-spanning ocean known as Panthalassia.

As it happens, the alatoconchids were not victims of the end-Permian event but met their demise as part of slightly earlier extinctions[9] that most likely coincided with a time of dramatic cooling. Farewell then to the extraordinary molluscs, but no matter—they were neither the first, nor the last. Giant bivalves, the aptly named megalodonts, are a recurrent feature of shallow-water reef environments; they made their debut in the Silurian and flourished until the succeeding Devonian (ca. 430 Ma).[10] After the disappearance of the alatoconchids, there was a 40 Ma hiatus until the megalodont train puffed into sight to deliver the homeomorphic alatoforms,[11] which were once again of gigantic size and now reclining in the tropical lagoons of the Triassic.[12] After their day in the sun, other bivalve avatars would dot the Cretaceous seas in the shape of the coral-like rudists. And after they accompanied the dinosaurs into extinction at the end of the Cretaceous (in the Cretaceous–Tertiary extinction event, or KTE), it was the turn of the giant clams (tridacnids), that flourish to the present day in tropical reefs (figure 2.1). And when they go extinct, after a decent interval we can be sure that another group will provide a substitute.

Not all these evolutionary returns are associated with the really large extinctions. Groups arise, flourish, and then slip into extinction. One such case involves those iconic marine predators of the Mesozoic, the ichthyosaurs. One might have thought these enormous reptiles

FIGURE 2.1 Vividly colored extensions of the body of this giant clam (*Tridacna*) house innumerable photosynthetic symbionts.

would go to the wall along with the dinosaurs and pterosaurs when the Cretaceous came to its sudden and violent termination (chapter 3). In fact their demise was some 30 Ma earlier.[13] This, however, hardly matters because unsurprisingly their place was taken by another group of giant reptiles. This time they were aquatic lizards known as mosasaurs (figure 2.2), closely related to animals such as the Komodo dragon.

Ecological replacement is the order of the day, but what is much more interesting is that some of the mosasaurs then reinvented themselves: some as sharks[14] and others yet more intriguingly as ichthyosauroids. Most striking in this regard was an advanced mosasaur known as *Plotosaurus*.[15] Tracing of this evolutionary excursion is not yet complete, but as Johan Lindgren and colleagues note, "With time, the *Plotosaurus* lineage might even have yielded fully thunniform [tuna-like] animals, had it not been for the aftermath of a bolide impact

FIGURE 2.2 Aquatic lizards known as the mosasaurs pursued many ways of life, some evolving into fast-moving predators not so different from the ichthyosaurs.

that brought an abrupt end to the reign of mosasaurs."[16] This suggestion is by no means unreasonable, given that the ichthyosauroid trend started early in mosasaur history.[17] The end-Cretaceous extinction terminated this particular evolutionary adventure, but certainly not the propensity to reinvent a thunniform shape. In this sense, this move by the mosasaurs toward a tuna/dolphin-like configuration might be regarded as anticipatory.

A much more concrete precipitation of prospective forms comes from another Mesozoic example, but this time it is non-marine. It concerns the mammals (and related mammaliaforms), a name that typically conjures up the two principal groups: placentals and marsupials. Although these were moving into pole position during the Cretaceous (chapter 3), the Jurassic was already very much an age of mammals. Importantly, these earlier cohorts were evolving an extraordinary range

FIGURE 2.3 *Fruitafossor* epitomizes the diversity of Mesozoic mammals. Credit: Nobu Tamura (http://spinops.blogspot.com), own work, CC BY 3.0, https://commons.wikimedia.org/w/index.php?curid=19461591.

of functional solutions (as so-called ecomorphs), even though they were living in a world still dominated by the reptiles.

These forms are important for two reasons. First, they demonstrate that long before their ultimate breakthrough, the mammals were already major players in the time of the dinosaurs. Second, in many cases the various convergences prefigure subsequent mammalian adventures.[18] For example, there were burrowers with hands and feet patently adapted for effective digging.[19] Particularly noteworthy is a late Jurassic form known as *Fruitafossor* (figure 2.3). This creature spent its time digging through late Jurassic soils,[20] and so striking are the convergences with armadillos that the researchers go out of their way to warn that there is no direct phylogenetic connection. The slightly older *Castrocauda* had a beaver-like tail and was evidently an adept swimmer in pursuit of its fishy diet.[21]

More generally, expansions in diet and ingenious reconfigurations of mastication expanded the mammals' repertoire of ecologies.[22] This was paralleled by a growing range of locomotory styles, with hopping being one of the few modes yet to be invented.[23] Even as some of these early mammals were aquatic, others went in the opposite direction—up the trees.[24] And others went yet higher in the skies. In the Mesozoic the latter domain was largely the prerogative of the pterosaurs and, in due course, the first birds, but silently drifting above the heads of the dinosaurs we have found a variety of gliding mammals as well.[25]

It is no more surprising that mammalian senses were becoming more acute, not least in terms of hearing.[26] Even a few years ago nobody would have guessed at such a diversity of mammals, but a more general point is worth making. Just as is the case with "missing links" such as in the sarcopterygian fish and paravian theropods (see chapter 4), the degree of mosaic evolution is striking. Among these early mammals, despite their possessing relatively primitive skeletons, the degrees of functional specialization needed to pursue a wide range of ecologies was not only impressive but, as in the case of *Fruitafossor,* frequently paralleled those seen in modern mammals. Some of these details even extend to their fur—which ominously already showed evidence of ringworm.[27]

CONCENTRATED CONVERGENCE

The examples I have given look at the reinvention of manifestly successful adaptive forms, and they should hardly come as a surprise. Yet however striking these types of recurrence might be, is there not a danger of us being too parochial, too earthbound? This has been a common complaint of convergent evolution. At first sight this is borne out by the almost unimaginable vastness of "biological hyperspace" and the consequent improbability of ending up in any particular region. It is also worth remembering that when it comes to numbers, it is the cosmologists who deal with the small change. How many atoms are in the visible universe? Let us say about 10^{80}, which sounds pretty impressive until the biologists arrive. Their combinatorics can yield staggeringly large numbers of potential biological alternatives, where 10^{200} (or even more) is typical. Given that only something like 10^{17} seconds have elapsed since the Big Bang, then even if a major step was achieved every millisecond (in itself a most unlikely proposition), in principle evolution would still be only able to explore an infinitesimally small fraction of potential biological hyperspace.

The conclusion seems obvious. On each and every planet, assuming, of course, it has any sort of biosphere (chapter 6), the likelihood of it possessing biological forms that converged with their terrestrial counterparts would surely be vanishingly remote. On each of these worlds, hanging jewel-like in the greater void, its biosphere could occupy only one tiny point in this hyperspace—innumerable biospheres, each utterly different. Even taking into account every planet in the visible universe (perhaps 10^{24}), this still would leave nearly all of potential biological hyperspace permanently untenanted.

So how can we find a way forward? We have only one biosphere: *n* equals one, as they say. To be sure, the deep fossil record reveals many additional examples of convergence, one of my favorites being the saber-toothed mammals. Isolated habitats such as New Zealand are further crucibles for convergence. And then there are examples where convergence occurs not just on a handful of occasions, but dozens of times, if not more. Think of C_4 photosynthesis (developed at least sixty-five times, and according to Erika Edwards an "inevitability"[28]), seed dispersal by ants (myrmecochory, maybe 140 times[29]), or aquatic plants (possibly a hundred times[30]). But until we find a planet that allows specific and detailed extraterrestrial comparisons, *n* still equals one.

Another angle is to try to discombobulate what we expect of convergence,[31] especially when it is put in the context of S. J. Gould's now famous metaphor of "re-running the tape of life."[32] Starting, say, from the Cambrian "explosion," would the end result be wildly different or eerily similar? Initiated by the philosopher John Beatty,[33] this discussion has engendered what appears to be an important set of distinctions. In brief, although historical contingencies have taken center stage in the debate as to how likely it is that evolution will rerun the tape of life to arrive at much the same set of solutions, in fact there is an important distinction between random and/or unpredictable events in history (the great what-ifs, such as Halifax becoming the premier rather than Churchill on May 10, 1940, or Pontius Pilate releasing Jesus on April 3, A.D. 33) versus a chain of causality that determines a particular outcome. The distinctions may be valid but how well does either stand up?[34] With respect to the former, so far as evolution is concerned perhaps the most obvious are factors such as mutations or mass extinctions. In the latter case (chapter 3), the evidence is that, rather than their radically redirecting the course of evolution (out go the dinosaurs, in come the mammals), they serve to accelerate what was going to happen anyway. At first sight, mutations might offer more scope for evolutionary mischief—or maybe not. Admittedly it is a complex area, but one striking discovery has been that in at least some cases mutations are very far from random but rather tend to be concentrated in so-called hot spots.[35]

Nor are such evolutionary "hot spots" (or their counterpart "cold spots") restricted to the genome. A recurrent but rather unexplored feature of evolution is that within a given group certain clades show a remarkable propensity to re-evolve much the same solution, while related forms are largely immune to this variant of "Play it again, Sam." From one perspective such hot spots can be filed under the umbrella of convergence, but they still provoke the question as to why only particular clades have a predisposition to evolve toward certain end points despite many other groups being apparently equally favored. This paradox has special force because, despite the phylogenetically restricted adoption of a given novelty, the developmental pathways that are routinely co-opted for a given novelty usually far predate the innovation. If therefore the mechanism is more or less universally available, why do only a handful of clades seize the opportunity? So too, even among the groups that take this high ground, we should ask: Are there multiple pathways or only one? Either way, are the key staging posts effectively signposting the inevitable or is there the possibility of derailment? The background to these discussions is the potential recognition of a deeper architecture that might constrain evolution so that the "biological hyperspace" will be dotted with inevitable destinations.

Perhaps all clades have their hot spots, but some striking examples include microbial associations with the ants (famously the attine fungal farmers, but also such groups as the carpenter ants),[36] and among parasitic plants both the red algae (specifically the florideophytes)[37] and angiosperms (such as the broomrapes and other lamiids)[38] that lose at least partially the capacity for photosynthesis. When, however, it comes to these hot spots it is the plants that engage in nitrogen fixation or employ C_4 photosynthesis that are especially instructive. Nor is it purely theoretical: in their different ways, not only do they both have a profound impact on the global economy (for example, almost a quarter of gross primary production is by C_4 photosynthesizers[39]), but embedding such physiological processes into other crops—and thus dramatically bolstering their productivity—has been a constant lure.

Central to nitrogen fixation in plants is a process of reciprocal signaling that ensures carbon is supplied to the bacteria, and nitrogenous products (initially as ammonia) are dispatched in the opposite direction. The general assumption has been that this capacity must have arisen multiple times. This is because the gram-positive proteobacteria (in so-called rhizobia) and *Frankia* (a gram-positive actinomycete; hence actinorhizal) are respectively housed in different sorts of root nodule.[40] In addition, nitrogen fixation has a relatively scattered distribution among plants.[41] The shining exception to this are the legumes (figure 2.4), but they in turn belong to a so-called nitrogen (N_2) fixing clade of the rosids, where again taxa specifically capable of this process show a generally erratic distribution.[42]

The first difficulty in trying to explain this rosid hot spot is that the signaling pathway for N_2 fixation was recruited[43] from a much older association that dates back to the Devonian.[44] These are the arbuscular mycorrhizae, and this symbiosis is found in the great majority of

FIGURE 2.4 The legumes, a prime example of nitrogen fixation by plants.

plants.[45] The rosids appear to have first acquired this trick, at least with regard to the actinorhizal symbiosis, in the late Cretaceous.[46] But at that time there were plenty of coexisting angiosperms, so did the rosids possess some cryptic predisposition that allowed them to become the N_2-fixation hot spot? Such a precursor could explain the multiple acquisitions and the generally erratic distribution of the nitrogenous nodules.[47] And if ingredient *X* could be identified,[48] that might speed up the genetic engineering of plants that are currently incapable of N_2 fixation. All this is "chasing unicorns," as Jeff Doyle[49] aptly says in his overview of the deep roots of this process and its promise for the future. And the wider question as to why only the rosids took advantage of this highly beneficial symbiosis remains even more elusive.

Maybe a clue will come from the example of C_4 photosynthesis? Here, in contrast to the more widespread C_3 process, the initial organic acid has four carbon atoms. This apparently minor shift from the making of a three-carbon molecule (in the form of phosphoglycerate) is a direct consequence of the vulnerability of plants to a precipitous drop in atmospheric carbon dioxide (CO_2, vital for photosynthesis), as occurred about 30 Ma ago.[50] In photosynthesis the weakest link is the enzyme known as rubisco (ribulose-1,5-bisphosphate carboxylase/oxygenase), which is central to the fixation of CO_2. This molecule has a singular disadvantage: oxygen, generated as a result of photosynthesis, poisons its catalytic activity (in a process known as photorespiration). The solution is to devise a two-compartment system: one where the CO_2 can be captured (a process mediated by carbonic anhydrase) and then transported to a second compartment, where high concentrations of dissolved CO_2 can be delivered to the waiting rubisco. The classic arrangement is called Kranz anatomy, with compartments of mesophyll and bundle-sheath cells, respectively; however, astonishingly this bipartite system can even be constructed in a single cell.[51] As it happens, C_4 photosynthesis is only one of several carbon-concentrating mechanisms, but it has been rampantly convergent, having evolved more than sixty times.[52]

As one of the most fully understood examples of convergence, C_4 photosynthesis provides a litmus test for the predictability of evolution. Although the fundamental shift involves the addition of a carbon atom, the necessary rearrangements for C_4 photosynthesis are complex and operate at all levels, from genetics and biochemistry to anatomy. Although a fall in atmospheric CO_2 was very likely the initial impetus, many other factors linked to climate change (not least aridity, salinity, and elevated temperatures) also have repeatedly favored the C_4 pathways.[53] With sixty-plus transitions, unsurprisingly the exact pathways taken are not identical; in one case where eighteen lineages were investigated, four distinct evolutionary trajectories were identified.[54]

The alternative choices that still lead to the same destination are strikingly apparent in the mutations at key sites in the various enzymes that underpin proper C_4 function.[55] In many groups a key substitution in one enzyme (PEPC) is at site 780 (the amino acid serine replacing alanine),[56] but this is not always the case.[57] Such examples remind us that causal chains are far from inviolate, but even so when Erika Edwards remarks on the "inevitability"[58] of C_4 photosynthesis she is stepping into territory largely unmapped by neo-Darwinians. With all roads contributing to the Via Quatro Carbonio that leads to this photosynthetic Rome, it also transpires that this metaphorical evolutionary road is built as a remarkably smooth incline, with perhaps only the first and last few steps requiring a bit more effort.[59]

And this journey is well worthwhile. Despite the vast majority of angiosperm taxa remaining firmly on the C_3 side of the fence, it is the C_4 plants that account for about a quarter of global productivity on land.[60] But why then are there such obvious hot spots for C_4 plants, not least among the so-called PACMAD[61] grasses and the caryophyllaceans (colloquially called pinks)? To be sure, there is a direct counterpart to C_4 photosynthesis in the form of crassulacean acid metabolism (CAM). In contrast with the spatial system of C_4 photosynthesis (with its two-compartment arrangement), the CAM system is temporal in that the stomata open only at night (also potentially reducing evaporative loss) to allow the CO_2 to be stored (in malic acid); when dawn breaks, CAM photosynthesis roars into action. These parallels are not coincidental; although CAM may be more ancient (and is found in plants beyond the angiosperms), it is possible that it arose at about the same time as C_4 photosynthesis and under similar selection pressures.[62]

The CAM system is also gratifyingly convergent and has its own hot spots, not only among the eponymous crassulacean plants (aka stonecrops) and other succulents such as the cacti, but among plants whose water supply is highly intermittent (especially epiphytic orchids).[63] Overall, CAM plants have a wider distribution across the angiosperms, and the question of phylogenetic focus seems to be especially acute among their C_4 counterparts. And such evolutionary hot spots reveal an interesting tension as to whether the inevitability of form is ultimately unremarkable or points to yet deeper constraints.

In one sense the answer has to be *both*. After all, evolution is never a level playing field, and the fact that C_4 plants regularly crop up in saliferous soils and other challenging environments is no guarantee against C_3 plants winning through in other circumstances. And yet the evolution of the C_4 process involves a complex set of rearrangements that is not only rampantly convergent but also crucially involves all the necessary building blocks (including the suites of enzymes)[64] and networks[65] that are already available in the C_3 plants and as often as not have even more ancient origins.[66] And so far from being a metaphorical Mount Improbable, the path is an easy ascent. Thus, the hunt is on for "enablers": the features that predispose only some groups to step onto the Via Quatro Carbonino. The candidates span all levels, from the genome (including more favorable enzymatic kinetics or mutational paths[67]), to cell biology (such as preadaptations to stress[68]), to leaf anatomy.[69] In many cases, however, the specific enablers are not in themselves unique solutions.[70] The clear implication is that beneath these entirely plausible factors there are deeper organizational principles at work and of which we know very little at present.

Nor is the story of C_4 photosynthesis by any means finished. If classic C_4 crops such as maize (figure 2.5) and sugarcane could have their genetic machinery redeployed in C_3 plants such as rice, then the Malthusian threat of global famine might be forever banished. Might it also assist with forestry? A curiosity of C_4 is that for all intents and purposes there are no treelike forms.[71] Why this is the case is by no means clear, but a combination of C_4 plants generally not having an ability to grow wood-like tissue combined with often hostile habitats may be at least a partial explanation. And perhaps we should be grateful for this evolutionary "failure." Just suppose that with the rise of C_4 vegetation, trees had been part of the mix. As Rowan Sage and Stefanie Sultmanis[72] have pointed out, with the ensuing suppression of savannahs and other grasslands, might the human experiment have been postponed or even derailed?

FIGURE 2.5 Native to the Americas, maize is one of the world's staples; its success in part is because of its C_4 photosynthetic machinery.

SABOTAGING CHANCE

Given that the what-ifs do not get us very far, what about John Beatty's other approach to historical predictability? This is to link the recurrence of evolutionary events to the reliability of a given causal chain. Event E, for example, will not happen unless it is preceded by events A to D. At first sight, this appears to be a much stronger position. Patently there must be historical constraints; even if the pentadactyl limb is a celebrated example of evolutionary virtuosity in the form of hooves, wings, flippers, and the hand that holds my pen, it cannot do everything. So too, as is often asked, if insects are six-legged, why not some vertebrates? Well, we can tip our hat to a gurnard fish that employs hexapodal walking,[73] but Geerat Vermeij's catalogue of "forbidden phenotypes"[74] is a reminder that for one reason or another not all may be possible. But in the grand scheme of things, it seems likely that a deeper

FIGURE 2.6 Nikolai Vavilov, brilliant scientist who fell victim to Stalin's tyranny. Credit: Library of Congress, Prints and Photographs Division, NYWT&S Collection, LC-USZ62-118109. Digital ID: (digital file from b&w film copy neg.), cph 3c18109, http://hdl.loc.gov/loc.pnp/cph.3c18109. Photo taken January 1, 1933.

order of the world may predispose particular chains of causality to be particularly recurrent.

As one example, let us consider the remarkable affinity of phosphorus for oxygen—will this not predetermine all manner of biochemistries? Nor is it surprising that, in the footsteps of the neglected Nikolai Vavilov[75] (figure 2.6) and his prediction of "expected forms,"[76] there is a renewed interest in "periodic table of niches."[77] This is not only in terms of the identification of recurrent ecomorphs, but also an echo of

Vermeij's observation that some niches appear to be potentially available but remain unoccupied. The analogies with the periodic table of chemists are not meant to be precise, but as with convergent evolution in general there is a sense of deeper levels of organization at least partly guiding, if not controlling, evolutionary outcomes.

But might Beatty's point about causal chains rescue the belief in the open-endedness of evolution? As was apparent from the specific case of C_4 photosynthesis, I doubt it. Such chains may be as labile as one desires, but evolution has a further card up its sleeve. In fact, in a sense it *is* evolution, at least so far as innovation and novelty are concerned. Central to any such process is the co-option and deployment of existing systems, parachuting them into new configurations, which can have dramatic consequences, reshaping both the biosphere and the planet. Want to build a forest? Seeds, particular sorts of leaves, water transport along meter-long pipes, all are part of the story, but so too is the employment of a remarkable biopolymer known as lignin (aka wood). Its tight bonding makes it robust in the face of microbial attack,[78] but this and other protective functions (for instance, against ultraviolet radiation) long predated what is now regarded as its canonical use in terms of structural support. Given this, it is much less surprising that the roots for these capacities of lignin go much deeper than the origin of the land-plants. We can see that the evolution of lignin was not instantaneous but followed a marked set of evolutionary milestones.[79] This journey was associated with the recruitment of a set of key enzymes that crucially were already in place among the algae and mosses. Collectively these milestones and recruitment determined most of the pathway that ended up with a final product in the form of lignin.[80] With this machinery assembled long before it was "needed," it is no more surprising that lignin has evolved independently in the red algae[81] nor that different types of lignin are also convergent.[82]

These inherencies, whereby the "instruction manuals" were inscribed long before the novelties themselves made their debut, serve to sabotage Beatty's appeal to causal dependencies that might otherwise forbid various evolutionary options. A related stumbling block concerns so-called deep homology, whereby the near universal employment of a gene in development (such as *Pax6* in eyes) makes the organ or structure in question fundamentally identical, be it octopus, annelid, or human. This is only true at the most superficial level. First, eyes are by no means identical and more significantly not all eyes require *Pax6*.[83] Also, although the original function of *Pax6* may have been visual,[84] it plays many other roles in the anterior nervous system,[85] and its ancestral function was as likely connected to the embryonic head.[86] Recall also that these genes can be redeployed in avian flight muscle[87] and mammalian spermatogenesis.[88] It is a versatile molecule indeed. And should, for whatever reason, co-option fail, then reinvention will save the day. The olfactory proteins of insects, for example, are uncannily like the standard model employed in nearly all sensory systems (these are the G-protein coupled receptors, of which rhodopsin is a familiar example), yet they are of an entirely separate evolutionary origin.[89]

REINVENTING WHEELS

To unpack this area a bit further, let us consider an enzyme vital to all life and accordingly one that must have appeared in the earliest cell and then been transmitted to every descendant, still alive or extinct. I speak of carbonic anhydrase, an astonishingly efficient enzyme. Carbonic anhydrase is utterly essential: if it had not evolved, it would have certainly have had to have been invented. We see it playing roles as disparate as assisting in the concentration of CO_2 for photosynthesis[90] or carefully controlling the transport of the same gas as a respiratory waste product in our blood and lungs. Hence, the natural assumption would be that given its fundamental function—specifically the hydration of CO_2 (or the reverse reaction)—the same enzyme would be employed across all life. But that is not the case at all. Carbonic anhydrase is splendidly, exuberantly convergent: as Anders Liljas and Martin Laurberg put it, it is "a wheel invented three times."[91] (Or actually, seven or eight times.) Of these, α-carbonic anhydrase is the most familiar, and in humans its malfunction can lead to severe medical consequences. There is a rough correlation with organismal groups, but the pattern is far from clear-cut.[92] Nevertheless, β-carbonic anhydrase can be claimed to be characteristic of plants, algae,[93] and archaeal bacteria,[94] while γ-carbonic anhydrase[95] is typically but not exclusively bacterial. It is perhaps not so strange to find γ-carbonic anhydrase also associated with the mitochondria of plants, given the bacterial origins of these organelles.[96]

In addition to the canonical α, β, and γ carbonic anhydrases (which are self-evidently unrelated[97]), there are at least five more: the δ, ε, ζ, and θ varieties[98] as well as the more recently recognized ι variety. The ε variant is found in bacteria, although it has been argued that, although distinctive, it is really a member of the β-carbonic anhydrases.[99] This may be true, but the ε variety has no sequence similarities and also has a number of distinctive features not found elsewhere. So why is it allied to the β class? The identity crucially depends on the striking similarity of the catalytic site. But how sure can we be that this is not convergent? After all, the α and β varieties have effectively identical catalytic sites, with the crucial distinction that they are mirror images.[100] This sort of question haunts discussions of evolutionary convergence—are they similar because they descend from a common ancestor or because the solution is a jolly good way of doing things?

With at least three unequivocally independent inventions of carbonic anhydrase on the block, a few more would not in themselves make a great deal of difference. What then of the δ, ζ, and θ[101] carbonic anhydrases? All occur in the diatoms. Although the δ variant[102] is like nearly all other carbonic anhydrases in being zinc-based (but otherwise is quite distinct), the ζ variety is particularly interesting because it employs the chemically similar element cadmium.[103] In point of fact, when zinc is available it is readily substituted, and the enzyme then operates at astonishingly high efficiencies. In addition, this carbonic anhydrase shows not only the expected functional convergence but also a striking structural similarity to β-carbonic anhydrase.[104] Given that cadmium is associated with toxicity, why then use this element if zinc is more effective (and otherwise near ubiquitous in the carbonic anhydrases)? Evolutionary chutzpah wins over chemistry: with the relentless demand for zinc not only to use in carbonic anhydrases but as an element embedded in many types of protein (including the so-called zinc fingers), might not the competitive edge go to any organism that can manage to scavenge cadmium? By learning this trick, the diatoms gained a decisive advantage, so explaining their immense abundance in the sea.[105]

Is this just another technical fact, far removed from our everyday world and its pressing concerns? Perhaps not. It is estimated that of the gigantic quantity of inorganic carbon that is fixed each year (more than 40×10^{15} grams), at least a quarter of this total is captured by the diatoms.[106] Given that an increasing amount of that "inorganic carbon" is from the burning of fossil fuels, both diatoms and their cadmium carbonic anhydrase may be worth a little more attention. The story of carbonic anhydrase has yet another twist when we come to the ι variety. First detected in diatoms but now known to have a wide microbial distribution, this variety employs neither zinc nor even cadmium, but manganese.[107]

So this is a wheel invented three times—or perhaps as many as eight? (That is if we recognize the ε variety and regard the cadmium-based and manganese-based carbonic anhydrases as also distinct.) Where, then, does the convergence lie? It is, of course, in the crucial catalytic site. This is effectively defined by a zinc ion associated with three amino-acid residues—histidine (and sometimes cysteine) are particularly prevalent. What seems extraordinary is, given that carbonic anhydrase is an essential prerequisite of life, why should nature keep reinventing it? Conceivably, specific molecular configurations offer particular advantages in different contexts, but if so why is the distribution of the different classes so promiscuous? Whatever the explanation, there is a much more important point to be made. So specific is the catalysis of the reversible hydration of CO_2 employing an active site with zinc and three amino acids that we can make a specific prediction: "Wherever in the universe there is carbon-based life, it *must* employ carbonic anhydrase." It is not only the molecule of choice, but it will turn out to be literally universal.

In circumstances where oxygen is lacking (anoxia), carbonic anhydrase will also employ iron[108] (because of its availability in the soluble form of Fe^{2+}) rather than the canonical zinc. More generally, iron is otherwise essential to life. This is not withstanding its propensity to generate dangerous radicals such as rocket fuel (aka hydrogen peroxide), and in the presence of oxygen iron also becomes promptly insoluble (as Fe^{3+}).[109] Despite all this, nature is adept at parcellating iron, not least with respect to respiratory proteins. Here, hemoglobin bestrides the molecular stage, but there is a less well-known alternative. This is a protein known as hemerythrin (figure 2.7), and paradoxically its very obscurity makes it a convincing test case in the question of evolutionary choice.

To be sure, there are crucial differences between hemoglobin and hemerythrin, inasmuch as the molecular structure of the latter[110] consists of a striking array of helical bundles that surround its active site with two ions of iron. On the other hand, the shifts in valency that either bind or release the oxygen are the same, and more notably there are a series of remarkable parallels between hemerythrin and hemoglobin. Thus, the former protein is also involved with oxygen transport in the blood; one day it may serve as the basis for artificial blood, given it would be less susceptible to unwanted side-effects involving oxidants.[111] And just as myoglobin is the muscle-based equivalent of hemoglobin, it is shadowed by myohemerythrin.[112] Hemoglobin has yet more counterparts, located in the nervous tissue and hence termed the neuroglobins.[113] They evidently play a critical role in brain function and are apparently involved with molecular stress (including oxygen radicals)[114] and protection against strokes. So neuroglobin finds its avatar

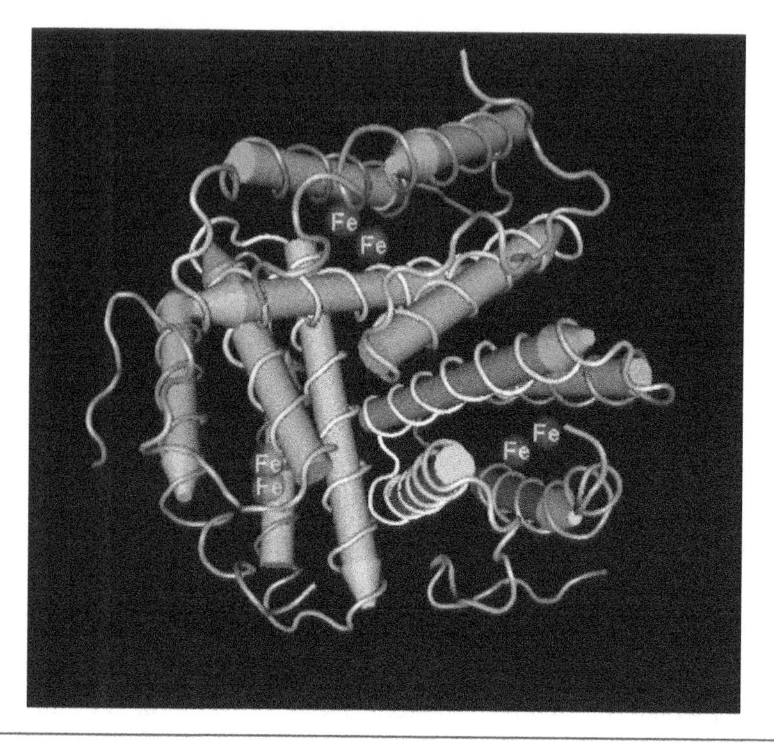

FIGURE 2.7 The respiratory protein hemerythrin has a very different structure to the more familiar hemoglobin, but operates in an almost identical fashion. Credit: Ben Berserker at English Wikipedia, https://commons.wikimedia.org/w/index.php?curid=3663562.

in neurohemerythrin,[115] and here too a role against stress (and bacterial infection) can be identified. One cannot help but wonder if the parallels between hemoglobin and hemerythrin will produce the final trump card. Along with neuroglobin, other identified proteins are cytoglobin[116] and, lurking in the testes, androglobin.[117] Will we one day find a cytohemerythrin or androhemerythrin? Correspondingly, will the yolk protein ovohemerythrin[118] find its counterpart among the globins?

To the untutored eye, hemoglobin and hemerythrin might look equally acceptable: after all, both employ iron and both store and transport oxygen. Is not the choice simply a matter of chance? Think of that compelling evolutionary metaphor of the "frozen accident." A "decision" made perhaps billions of years earlier embeds evolution not only in an imperfect world, but—after having inherited some other frozen accident—now in a quite different place. How then might we ever know? The solution is to study the distribution of these molecular twins. Not only does hemerythrin form a shadowy understudy to hemoglobin, but turning to the distribution of hemerythrin among the animals we encounter what at first sight seems a bizarre distribution. It occurs in many annelids (including the sipunculans[119]). The other groups where hemerythrin occurs include the relatively familiar brachiopods,[120] as well as the priapulids (or penis worms),[121] sea anemones,[122] and quite unexpectedly mammals.[123]

Annelids and brachiopods are quite closely related, but all in all the overall evolutionary distribution of this protein makes little evolutionary sense. So what is going on? The most probable solution lies in the repeated recruitment by different groups of animals, at different times, of hemerythrin. This seems a likely scenario because like many proteins hemerythrin is evidently very ancient and occurs in a wide variety of bacteria.[124] Conceivably hemerythrin was ancestral but just happened to be lost in nearly all animals. Repeated recruitment, however, seems the more likely explanation. Irrespective of whether animal hemerythrin was derived from a single[125] or multiple sources, metaphorically this respiratory protein is on the shelf and is hardly ever put into the evolutionary basket. It is there, always available, yet it simply is not the molecule of choice. As Xavier Bailly and colleagues have noted, hemerythrins "exhibit a considerable lack of evolutionary success in metazoans," and they remark also that in comparison to the remarkably versatile globins their counterpart has "barely maintained a foothold in living organisms."[126] Why so? They speculate that hemerythrin may have been more vulnerable to deleterious mutations, especially at the all-important oxygen-binding site.

Arguably, carbonic anhydrase comes in perhaps eight different flavors, but otherwise has no counterparts. On the other hand, hemoglobin will be by far the most frequent choice, but it is not quite ubiquitous.

More generally, when it comes to respiratory proteins, hemoglobin (along with its understudy hemerythrin) and hemocyanin, in the words of Donald Kurtz, represent "three solutions to a common problem."[127] That is, each protein has a metal site (copper in hemocyanin) that must handle the chemistry of dioxygen (that is, O_2). Kurtz points out that other solutions may yet be found, but if they exist we can already predict a key feature because, as he graphically writes, any of the three known proteins "apparently represents Nature's tiptoeing along the edges of the energetic barrier separating reversible O_2 binding from O-O cleavage without crossing it."[128]

This powerful metaphor encapsulates the knife-edge of biological existence, the narrowest of tracks that thread their way across an otherwise desolate landscape of nonviability. And the immense wastes of nonviability, where nothing biological will ever work, have two grim faces. On the one side it is gas-like, ceaseless chaos, bereft of order. On the other side, there is too much order, a crystalline-like world where everything is locked into immobility. And if we look more carefully, we will see that almost without exception these tracks have been repeatedly traversed by different evolutionary travelers, sometimes as chattering throngs and other times by the solitary journeyer. For throngs, read hemoglobin, for the solitary wayfarer, look to hemerythrin.

This in turn hints at the real nature of the unimaginable vastness of biological hyperspace, with its potentially almost infinite number of alternative possibilities. They may be mathematically imaginable, but practically all are biologically uninhabitable. Thus, whether we talk of alatoconchids, Jurassic gliding mammals, carbonic anhydrases, or respiratory proteins, in their different ways each suggests that only tiny areas of biological hyperspace are repeatedly populated. That the infinitesimal fraction of vitality in the hyperspace is so specific might in turn point to a deeper order of the world. To physicists and chemists, such an order in particle physics or the periodic table is hardly novel, but for biologists to abandon the idea of historical indeterminacy seems to be a bit more of a struggle.

IMMUNE BRAINS

If the examples we have discussed of "eternal returns" and "molecules of choice" remain insufficient to persuade the skeptics of an order that underpins evolution, then let me have one more go. This concerns the brain and thereby touches on the knotty problems of cognition (chapter 5). The brain is remarkable enough in itself, but a hint of a deeper order is evident from an unexpected but striking link between the immune system and the brain. The connection arises because each communicates using a similar syntax. In the words of Edwin Blalock, they "speak a common biochemical language."[129] Some have gone so far as to suggest that this terminology may be more than analogous, especially if either viral or microbial genes were to become involved not only with cognitive processes (and disorders), but even language itself.[130] Presupposing a materialist view of mind, the correspondences might include comparisons between the "memory" of the adaptive immune system and the recall of words. Or, to be even more adventurous, we could draw a correspondence between the B and T lymphocyte cells with nouns and verbs. This is all highly speculative, of course, but as we will see when it comes to language there are much more severe obstacles to this approach (chapter 5).

What is not in doubt, however, is that at a more biological level the various proteins classically associated with the immune system—notably those known as the major histocompatibility complex (MHC)—also play a central role in how the brain develops. Even if one accepts this as another example of evolutionary versatility, at first sight this dual role is unexpected. Yet these similarities go much deeper, pointing once again to some fundamental organizational principles in life.[131] This is because, in essence, the unfolding of the brain is an engineering nightmare: how do something like 10^{12} neurons make the thousands of necessary connections (via the junctions known as synapses) so the most complex structure in the universe can remain specific in function but show flexibility in the face of change? No genome could possibly code for such complexity; it can only provide the instructions for the basic structure of the brain. If one considers, however, the manner by which the adaptive immune system operates, then we have a vital clue. Its business, of course, is to test a vast number of alternatives in the form of the constant blizzard of pathogenic antigens, constantly probing our systems of defense. So too the synaptic web is constructed in an analogous fashion—it too is one based on endless trial and error. Accordingly it is no surprise that the MHC (and other immune)[132] molecules play a central role in not only the immune system but in the construction of the brain.

The connections between how to build a brain or run an effective immune system go even farther. Central to the functioning of the synapses are molecules known as protocadherins. As their name suggests, these are involved in cell adhesion, and they ensure the synapses are effective in neural transmission. What is remarkable, however, is both the molecular structure and genomic construction of the protocadherins are strikingly similar to the immunoglobulins.[133] Again, effectively the same rules of engagement are central to the versatility of the immune system and the construction of the immensely complex network that decides how, when, and why the neurons will connect by their synapses and ultimately make a brain. As they do in the vertebrates, the analogous molecules in insects (the immunoglobulin *dscam*) play a central role in the neuronal wiring of the brain.[134] Moreover, even if the development of a brain and the immune systems go hand in hand, it is more likely that the molecules first evolved in the context of the evolution of the nervous system. Their shared use, however, reinforces the sense that evolution is seeded with inevitabilities.

So much will become apparent from a very different angle as we move to explain why mass extinctions are paradoxically creative, bringing to fruition what was sooner or later going to happen but now is well ahead of schedule.

3

The Myth of Mass Extinctions

In humans an eschatological tremor runs deep, with its twisted fascination in the "end times." If the world does not end in fire, then rest assured one or other cataclysm will do the trick. Small wonder then that mass extinctions exert such a lurid grip, representing as they do disasters beyond human imagination. A vast battery of scientific data reveal not only devastated worlds but a series of grim scenarios revolving around titanic volcanism, asteroid impacts, colossal tsunamis, global superwarming, acid rain, wildfires, mega-pollution, ozone destruction, and asphyxiated "Strangelove oceans." Mass extinctions may all be safely in the past—and accordingly great fun to teach—but they also encourage the pointing of accusatory fingers at our reckless folly unless we rapidly mend our ways and sign yet another Greenpeace petition. But consider the contrarian position. Not the reality of mass extinctions—no dispute here, they most certainly happen. And they are indeed the apocalyptic agents of both scientific and popular imagination: extremely nasty and most definitely to be avoided. That's all true. Nevertheless, the role of mass extinctions has been vastly exaggerated—yet paradoxically they still remain central to the history of life.

First let us look into some background. When we talk of mass extinctions it is customary to identify the "Big Five."[1] Although in the last 500 Ma there have been many other major perturbations (depending on your yardstick), I will concentrate on the catastrophes that terminated the Permian (ca. 252 Ma) and Cretaceous (ca. 66 Ma)[2] periods, here identified respectively as the PTE (Permian–Triassic extinction event) and KTE (Cretaceous–Tertiary extinction event). Why only these two? To begin with, they are the best known and arguably by far the most devastating. Second, respectively they provide two tipping points: between the Paleozoic and Mesozoic and then between the Mesozoic and Cainozoic (or Tertiary), and thus each is a fulcrum in the history of life. So the post-Permian world (beginning with the Triassic) ushers in what is effectively the beginnings of the modern world. After the Cretaceous debacle, the dinosaurs perish, and the mammals seize the opportunity. If that had not been the case, I would not be here writing about it nor you reading about it.

First, let us look at the end-Permian event (PTE), an occasion you simply cannot miss. At much the same time as the Cambrian "explosion" was perturbing Darwin,[3] so the Oxford geologist John Phillips broke new ground by depicting a curve of how biological diversity had changed through geological time.[4] His sketch was necessarily approximate, but it distinguished the Paleozoic and Mesozoic worlds. Presciently he also wondered how much of the dip in diversity that separated these two eras was not original but actually an artifact that reflected our lack of knowledge of ancient life. Even so, any paleontologist will sense a profound difference across the boundary:[5] in the oceans, away with the trilobites and the brachiopods, and welcome to the molluscs (especially bivalves and snails) along with the crustaceans, more advanced fish, and more strikingly huge predatory reptiles. On land, perhaps the shift was somewhat less dramatic, but now see how the Earth begins to shake with the footsteps of giant reptiles, while stirring in the undergrowth the premammals are becoming ever more alert.

It is not yet the modern world, but clearly it is well on its way. Not only is the connection with the Paleozoic world permanently sundered, but without this mass extinction this new world could never have been ushered in, or so the argument goes. This is the mantra of mass extinctions: unless the evolutionary and ecological slates are wiped more or less clean then, in the case of the PTE, the Mesozoic world would never have materialized. More famously if there had been no end-Cretaceous mass extinction, the dinosaurs would have continued to reign supreme, and those pesky primates would have never got the green light. In the history of life there would seem to be no more obvious cause and effect. To be sure, there will be bravura repeat performances, but to the first approximation only by clearing the block can it be repopulated by the new crowd.

How then are mass extinctions yet another evolutionary myth? Have I lost the plot? Self-evidently they are catastrophic, episodes when the President of the Immortals raises a hand to radically redirect the history of life into new, unforeseen, and unpredictable directions. Here, I argue the exact opposite. In the short term, mass extinctions are indeed destructive. Paradoxically, however, they are creative because they serve to accelerate the inevitable. What was going to happen sooner or later, bringing biospheres to new levels of complexity, now happens but crucially far ahead of schedule. Where most perceive only disaster and mayhem, I invite you to look beyond the turmoil and contemplate the bigger picture.

SO WHAT REALLY HAPPENED?

To begin with, however, prepare to steel yourself: shut your eyes during the worst bits, and let us go for a tour of the post-Permian battlefield. No doubting it is a time of unredeemed catastrophe for some, but not for all.[6] Often cited is David Raup's calculation[7] that perhaps 96 percent of all marine species went to the wall. The scenarios are appropriately grisly; as for explanations of to "How to Kill (Almost) All Life,"[8] the foremost culprit is the supereruption, especially the vast magma outpourings in what is now Siberia. In ways that are not completely understood, this volcanism then engendered a series of bloodcurdling consequences[9] that encompassed oceanic anoxia and acidification, extreme physiological stress (hypercapnia), super heat waves, and faunal dwarfing (Lilliput taxa). These views have not passed unchallenged, and it would be a mistake to believe that there exists a single consensus. Nevertheless, the central trope expressed by Raup—"with such severity of extinction, chance elimination of certain biologic groups would have been probable"[10]—has become now a canonical truth.

The decimation of species and tumbling ecologies are the hallmarks of mass extinctions, but there is also repeated emphasis on the thread by which survival so often seems to hang. Here we are invited to watch those handful of species, perhaps represented by only tiny populations, just managing to slip past the closing gates of extinction. "Imagine," thunders our erudite professor as she bangs the table, "if this one genus of echinoid, those handfuls of ammonoids, had not survived the PTE—and despite what you might read in that deeply misguided book by that dangerous maverick Conway Morris, I can assure you it was by the merest fluke that they did pull through—*Class!* Pay attention and consider! No sea urchins? End of story! Not one to impale that careless tourist on vacation. No ammonoids? How very different would have been the Mesozoic world!" The class sits appalled—could it be that they too are merely flukes of evolutionary circumstance? Things are not quite so simple.

What of the echinoids and the group that encompasses them, the echinoderms? First, groups such as the brittle stars (ophiuroids) were already modernizing in the Paleozoic, and there is little evidence for their being squeezed through a bottleneck as the Permian gave way to the Triassic.[11] Next, at least two groups that helped define the Paleozoic but had received end-Permian obituaries in fact turn up in the Triassic.[12] The particular focus, however, has been on the sea urchins (echinoids) and sea lilies (crinoids), where the standard mantra is that either group pulled through only by a whisker.[13] In the former case, it is now clear that multiple lineages survived the debacle.[14] The crinoid story (figure 3.1) is more open, but their "unexpected" diversity in the earli-

FIGURE 3.1 The sea lilies (or crinoids) are among the most iconic of fossils, but the notion they came within a whisker of total destruction at the end of the Permian is overstated.

FIGURE 3.2 The ceratitids typify the Triassic, but their evolutionary roots extend into the Permian. Credit: H. Bernd, CC BY-SA 3.0, via Wikimedia Commons.

est Triassic points either to an astonishingly prompt recovery or as likely several lineages pulling through the maelstrom.[15]

What about the ammonoids? Again the received wisdom is that they barely managed to pull through the PTE.[16] The real story is considerably more interesting. It revolves around a particular group known as the ceratitids (the name comes from a Greek word for horn). Although these ammonoids were abundant in the Triassic (figure 3.2), they actually made their debut well *before* the PTE. There are many lineages of ceratitids, and working out their evolutionary connections relies on a method of analysis known as cladistics. This approach is far from faultless, but it has the advantage that the resultant evolutionary tree (the cladogram) means that in principle you can infer the entire history even when (as usual) the fossil record is seriously incomplete. In the case of the ceratitids,[17] such analyses suggest that perhaps as many as thirteen lineages slipped across the PTE, and it is clear that this event simply accelerated a prior trend of diversification.[18] The odds of ammonoids surviving the PTE do not look too bad, and subsequently they continued their boom-and-bust history.[19] Some two hundred million years later, they finally went over the KTE cliff in the company of the dinosaurs.

The journey of the ammonoids and echinoderms is telling both in terms of the frightening inadequacy of the fossil record and the exaggeration of the squeeze points. To anticipate one of the central themes of this chapter, we see group after group are already moving into pole position ahead of the catastrophe.[20] This is not to deny that a clearing of ecological space as a result of mass extirpation provides a real advantage, but this is an acceleration, not a radical redirection. So too it is axiomatic that the rapid radiations that ensue are contingent on this opportunity. This seems very likely, but in cases such as the first ichthyosaurs[21] the scope and rapidity of the diversifications cannot but make one wonder whether some of the players were already poised to take advantage of the new world before the Day of Doom.[22]

ZONES OF SAFETY

There is another neglected aspect of mass extinctions. However extreme the disaster, there will always be regions on the planet where things were not nearly so bad, even zones where the disasters passed almost unnoticed. In the case of the PTE these places of greater safety were located in the higher latitudes of the Southern Hemisphere, and not only in the oceans[23] but also on land.[24] More striking is what was happening at equivalent latitudes on the northwestern margins of the supercontinent known as Pangea. There, a diverse record of trace fossils shows that even if the deeper waters remained in a perilous state, nearer the shorelines one would hardly know there was a mass extinction.[25] Such refugia are the latent crucibles that provide the vital springboards to recovery and the restocking of the planet.

Not for a moment am I suggesting that the early Triassic is a preferred destination for the time traveler. Your tour group would be greeted by shocking scenes, but the cognoscenti would see which way the wind was blowing. Yes, life is in a pickle, but things are not quite as bad as they seem. If refugia are a token of resilience, so too is the evidence that the rates of recovery were more rapid than once thought. Again, this does not negate the idea of a mass extinction but rather invites a somewhat different perspective. Overall the older ideas of a cripplingly slow post-PTE recovery phase have taken some hard knocks.[26] First, revisions of the geological time scale show that the key intervals in the Lower Triassic were shorter than once thought.[27] Just as important, groups such as the ammonoids[28] and conodonts[29] (which are important for the biostratigraphic resolution of this interval and in general were little affected by the PTE[30]) promptly reradiated. So too the actinopterygian fish seem to have been pretty unfazed by this perturbation.[31]

In ways not well understood, this rapid recovery may be linked to these groups pursuing a pelagic existence. There also are telling exceptions elsewhere. For example, evidence from the Middle East suggests that even in the tropics matters were not invariably an unmitigated disaster,[32] while in south China (the source of some of our most important insights into the PTE) the earliest Triassic has yielded surprisingly diverse mollusc faunas.[33]

A TICKET ON THE *TITANIC*

It is time to move on to more controversial ground and defend my paradoxical thesis that mass extinctions are fundamentally creative. There are two strands to explore. First, are there long-term trends in

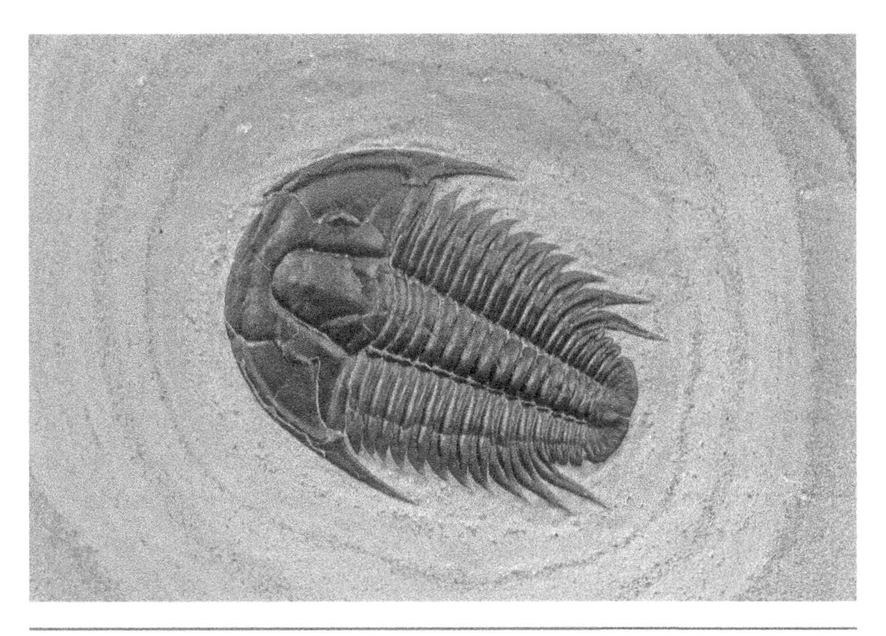

FIGURE 3.3 Trilobites almost define the Paleozoic, and although they were victims of the Permian mass extinction, the writing was on the wall long beforehand. Credit: Copyright 2017-R. Weller.

the history of life that will ultimately guarantee the demise of particular groups and ecologies and correspondingly the emergence of others? The rub here is "ultimately," and as it turns out mass extinctions are of huge assistance. The complementary strand is based on the observation that although explosive evolutionary radiations routinely followed a mass extinction and the cleansing of the ecological slate, almost invariably the major players that would inherit the new world were not patiently waiting to be invented. To the contrary, they were already waiting in the wings of the evolutionary theater,[34] and as often as not they were already beginning to diversify before the mass extinction.

What then of the first strand, those groups whose time was up? Let us start with that iconic group, the trilobites (figure 3.3). This group was a staple of Paleozoic marine life, but ostensibly also was one of the principal victims of the end-Permian extinctions. Technically this is correct, but irrespective of their final fate, their days were always going to be numbered. Why this was the case is made clear by the deceptively simple query of Dave Raup: Was the extinction of the trilobites a matter of "bad genes or bad luck"?[35] Should we commiserate with them, that as a group their history was like some unfortunate families, dogged by recurrent mischance and accidents? Alternatively should we narrow our eyes and concede that in the long term they simply were not officer class? Trilobites clearly fall into the latter category. Their story is one of early glory (especially in the Cambrian and Ordovician), followed by steady diminishment.

Raup's model certainly depends on various assumptions, but his analysis suggests that as time wore on the trilobites were facing ever mounting odds. By the Carboniferous they had become uncommon; by the mid-Permian, you would have to look hard to find one.[36] Trilobites did enjoy a modest resurgence in the Upper Permian,[37] but the fact remains that their doom was unavoidable. As the catastrophe loomed, maybe their tendency to dwell in reefs[38] was akin to holding a ticket on the *Titanic,* given the concomitant spectacular collapse in reef ecosystems that was an invariable in mass extinctions. This, however, only served as an immediate cause. It was the long-term factors that meant the death knell for the trilobites had already been sounding a hundred million years before their final extirpation.

To explain why it was "bad genes" that sealed their fate we first need to remember that like other arthropods trilobites periodically had to shed their rigid skeleton to encompass continued growth. This process (of ecdysis) represents both a time of stress and, until the new carapace can harden, one of vulnerability. Trilobites, however, seem to have been at a peculiar disadvantage because in contrast to other arthropods, somewhat oddly they never learned how to resorb their skeleton.[39] As a consequence, they discarded valuable biomass (even though this was an advantage for paleontologists: during its life an individual trilobite would conveniently generate multiple fossils). This extravagance of exoskeleton production may have been one nail in the coffin of the trilobites, but their main problem was finding themselves in seas that were becoming increasingly dangerous as predators diversified and became more effective. Countermeasures could be taken, including enrollment[40] and molting in protective burrows,[41] but evidently these did not suffice.

No doubt the Permian extinctions were the last straw. But even if that time of trauma had been conjured away, one is still entitled to ask: How much longer would the trilobites have survived? We can only speculate, of course, but given that the odds mounting against them rose ever more steeply, I believe we can be reasonably confident that the trilobites would have been exceedingly lucky to greet a Jurassic dawn.

SHIPS THAT PASS IN THE NIGHT?

If some groups were doomed to fail, for others their stars were in the ascendant. Again it was not by chance. If the contrast between the Paleozoic and Mesozoic oceans is to be epitomized, it is how, to the first approximation, the former was the era of brachiopods (figure 3.4) and the latter the day of the molluscs (especially bivalves, but also snails).

FIGURE 3.4 The heyday of brachiopods was in the Paleozoic, but to imagine the Permian mass extinction as the fulcrum point in their evolution is simplistic.

That there was such a transition is evident from even a casual inspection of the fossil record. But how and when was it achieved? Did the PTE catastrophe decide the winner, or was the respective decline and rise the result of eons of unrelenting competition?

So far as Brad Calloway and Stephen Jay Gould were concerned it was emphatically the former explanation. As exemplified in the lambent subtitle of their article ("Ships That Pass in the Night"),[42] they point out how these two major groups of invertebrates had much in common. Both possessed a bivalve shell and extracted suspended food from the seawater by using a filtering device known as the lophophore (in brachiopods) or ctenidia (in bivalves). So they were similar and yet so different because if we follow the thesis offered by Calloway and Gould each group effectively inhabited an unrelated world. They never interacted, so their respective histories of decline and success can only be understood by the pivotal role of mass extinctions. Rather than each group being involved in an epic boxing match, slogging it out for tens if not hundreds of millions of years, for these workers the Permian debacle becomes the defining moment. The bivalves simply did better, and the day of the brachiopod was over. This is a seductive argument—but it is gloriously wrong.

To explain why it is wrong, we need to look at the wider picture. That the brachiopods suffered during the PTE is self-evident. They took a double hit,[43] and in some areas they were well and truly clobbered.[44] More generally, however, matters are far from black and white. Some major groups, notably the orthids (so named because of their strikingly straight hinge line), did indeed hit the wall. This, however, is far from being some unexpected catastrophe because, in a manner reminiscent of trilobites, the orthids had declined exponentially during the Paleozoic, and interestingly perhaps on account of competition with other brachiopod groups.[45] If we ever find a Triassic orthid (or trilobite), this will be no occasion for surprise for their doom was pronounced much earlier.

FIGURE 3.5 The diversity of the bivalve molluscs far overshadows that of the otherwise similar brachiopods. Credit: CC BY 2.5, https://commons.wikimedia.org/w/index.php?curid=164753.

Not that extinctions cannot be ignored, but when it comes to the brachiopods and bivalves, their respective histories are consistent with competitive interactions.[46] Exploring the ways in which bivalves might be superior to brachiopods has provoked a minor industry,[47] but it is self-evident that the ecological repertoire of bivalves far outstrips that of the brachiopods (figure 3.5). Brachiopods do one thing well: anchor themselves to the seafloor and suck in seawater. So do many bivalves (e.g., mussels), but they do a host of other things as well; some can swim and glide for short distances (handy when a predatory starfish approaches), while others are themselves carnivorous. Teasing apart what bivalves do is of great interest, but when we wish to consider their overall history, it is legitimate to lump them together.

But what of the more general point: that brachiopods and bivalves metaphorically steam past each other, oblivious to each other's evolutionary destiny? Drawing on an improved data set, Jack Sepkoski suggested that their respective histories may be better described as "Things That Go Bump in the Night."[48] In essence, Sepkoski argued that the brachiopods had already begun to decline in the later Paleozoic, and although they were hard-hit at the PTE they actually rebounded in the Mesozoic,[49] including a reoccupation of ecospaces.[50] Indeed, in some Triassic sediments brachiopods retained their dominance over bivalves, and the likelihood is that the whole story of their respective evolution was more protracted and complex.[51]

What happens if we take an even wider perspective? Far from their diversification being triggered by the PTE, the inevitable rise of bivalves began in the depths of the Paleozoic. This becomes clear when we look at a transect from the tidal zone into increasingly deep waters at various intervals during the Paleozoic (and beyond). Key to this analysis is to plot the distributions across this seafloor transect of mega-assemblages that comprise three distinct evolutionary faunas, respectively those of the "Cambrian," "Paleozoic," and "Modern." These faunas reflect the overall "character" of an ancient seabed and so are identified statistically. This means that they are not restricted to a given time interval, such as the Cambrian. Crucially, each of these evolutionary faunas originated in shallow waters, and by expanding its range into successively deeper waters displaced the preceding fauna.[52]

In the context of this discussion, the Cambrian evolutionary fauna is of less significance, and not surprisingly for much of the Paleozoic the shelf seas were dominated by the Paleozoic evolutionary fauna (broadly, shelly faunas rich in brachiopods and trilobites). Lurking, however, on the shallow margins of these oceans was the Modern evolutionary fauna, consisting of assemblages rich in bivalves (and gastropods) that persist to the present day. As the Paleozoic wore on, it was the latter fauna that started to shift offshore and occupy more and more of the available space. Yes, the rate of migration was slow, but as Sepkoski presciently remarked, "I favor a hypothesis that brachiopod diversity dwindled as the Paleozoic way of life in the oceans was replaced by the Modern way . . . a process *accelerated* by, but not contingent upon, the end-Permian mass extinction."[53] In other words, the bivalves were well on their way to domination long before the PTE. All this event did was provide a helping hand.

Another line of evidence that very much complements this view arises if one compares the respective body sizes of the brachiopods and bivalves. Body size matters because it serves as a proxy for metabolic activity and thus the ability of a group to corner the energy market. From this perspective we see that by the end of the Devonian the bivalves had already outstripped the brachiopods; by Lower Triassic times they were gobbling up 95 percent of the shared total.[54] In a parallel way, key adaptations in the bivalves (such as anchoring the shell with tough threads [the byssus] or employing siphons to direct incoming and exiting water) not only evolved long before the Triassic (respectively in the Ordovician and Devonian), but in the latter case proved to be a hallmark of bivalve ecologies—that is, an existence as a free-burrowing suspension feeder. Such forms were not only predominant before the PTE,[55] but by being concealed within the sediment they became much less vulnerable to predation.[56] When one widens this scope of enquiry and considers the overall range of functional types among the invertebrates during the Permian and Triassic, one is hard pushed to see much change.[57] In fact, only one of the twenty-nine modes of life disappeared. This loss becomes almost ludicrous when we learn that it involves a suitably obscure group of molluscs (the rostroconchs) that were already rare in the Permian and probably had tottered off the stage well before the PTE.[58]

LET'S MAKE A MAMMAL

So much for the oceans. What was happening at the same time on land?[59] Volumes could be written on archosaurs and plants,[60] but there is one assemblage in which we can claim a special interest. The umbrella group is called the synapsids. Our specific interest, however, is with the therapsids and especially the eucynodonts (figure 3.6) because it is from this group (and specifically the probainognathans) that the mammals arose. Our focus may be chauvinistic (or not: see chapter 5), but it is only by tracing a story, which begins with our sprawling predecessor emerging out of the coal swamps,[61] that we can set the scene. This is to explain how in due course the mammals had reached a stage of evolution that ensured that they were already beginning to seize the high ground before they and the dinosaurs walked into the inferno of the end-Cretaceous (KTE) catastrophe.

Given, however, that the emergence of mammals centers on events in the late Triassic and early Jurassic, at first glance looking much

FIGURE 3.6 Cynodonts such as the Triassic *Brasilitherium* were already assembling the features that will define the mammals. Credit: Smokeybjb, own work, CC BY-SA 3.0, https://commons.wikimedia.org/w/index.php?curid=27486820.

deeper into geological time might seem to be of marginal relevance. But this is not the case and for two reasons. First, a good part of the mammalian body plan had already been assembled *before* the PTE. Second, although this mass extinction cannot be ignored, it is equally difficult to see what long-term difference it would have made. With or without the PTE, mammals would have eventually risen to dominance.

Just as important, their evolution corresponds to what Tom Kemp has identified as a "correlated progression."[62] In other words, the evolution of such key features as locomotion, feeding, as well as enhanced senses and physiology, set the groundwork for subsequent success as these features began to intermesh, and as such provide the so-called missing links (chapter 4). To put this progression in its visual context, Kemp has invoked not only an adaptive landscape but specifically one rather like a mountain ridge. In using this metaphor we can further envisage the synapsids on a journey, strolling past the signposts "mammal-like" and "more mammal-like," ever onward toward the destination marked "true mammal."[63] This is not, as Kemp would insist, a linear excursion, but neither is it a random walk.

What then does this history show? Early synapsids were sprawling beasts; they are often referred to as pelycosaurs and are exemplified by *Dimetrodon*. Survivors, in the form of the carnivorous varanopids,[64] appeared to have held their own against the newly emergent therapsids,[65] but in the Middle Permian the latter group underwent a dramatic modernization. This involved a transformation from a rather archaic model to a sleek organism with enormous potential.[66] Naturally, a great deal of evolutionary footwork separated the early therapsids from the more advanced, mammal-like forms, but there also has been a widespread sense that this story relied on a rather special set of historical circumstances. Two interpenetrating strands of evidence suggest the opposite view.

First, if there is a leitmotif to synapsid evolution in general, it is the striking degree of convergence (chapter 2). New tricks do not appear once but typically several times independently. Most tellingly, perhaps, is the "convergent evolution of a neocortex-like structure" in the enlarged brain of one such therapsid.[67] The second point is the degree of evolutionary mosaicism (chapter 4), meaning that advanced features can sit cheek-by-jowl (sometimes literally) with more primitive ones.[68] This in turn allows a remarkable freedom for the group to experiment with different configurations and combinations of anatomy. In this way the boundaries of evolution are being constantly tested. At a fine level of analysis, every trajectory is unique; but viewed more broadly, again and again we see how certain combinations inevitably win out (chapter 2).

Nobody supposes that as a group the therapsids define a simple *scala naturae,* but in their recurrent ways the principal lineages tended toward an increasing degree of "mammalness." Not only is the near-simultaneity of these steps striking, but in terms of key milestones evolution managed to achieve a great deal between the Middle Permian and Lower Triassic. Not only were the demands of heavy-duty herbivory soon cracked,[69] but some of the Late Permian anomodonts displayed impressive dentitions with molariform teeth, nipping incisors, and belligerent-looking saber canines.[70] In addition, there was a transition from a lumbering gait to an upright posture,[71] which was reflected in a skeletal reorganization of the hands and feet.[72] Increasing agility was reflected also in the first, but far from last, arboreal excursions.[73]

So too we see a capacity for binocular vision,[74] while an overall simplification (chapter 2) of the skull not only conferred strength but led to an increasing effectiveness in feeding.[75] Of equal moment was the independent development of the palate,[76] which allowed a separation of breathing from feeding. In terms of skeletal evolution, a dramatic hallmark was the progressive loss of nearly all the bones in the lower jaw, so in the mammals finally only a single bone (the dentary) remained. This story of simplification was not monotonic, but its culmination saw the redeployment of two of the jaw bones as the minute incus and malleus of the middle ear. Here is a story that might well be connected to more acute hearing[77] (chapter 1).

More intangible, but an all-important move, was the move to endothermy. Given that warm-bloodedness is fundamentally a cellular process, at first glance looking to the fossil record for guidance might seem a forlorn hope, but oxygen isotopes have pointed to multiple acquisitions.[78] The clinching evidence would be the discovery of fur. To find fur in superbly preserved Jurassic mammals[79] is not so surprising, but the tantalizingly hairlike structures found in Late Permian coprolites (fossil feces) hint that even then the therapsids had some sort of pelt.[80]

Skeletons also can be informative, both in terms of histology[81] and more specifically in the form of the so-called turbinal bones. Located in the nostrils, some of these thin bones are involved with smell, but the maxilloturbinals assist with warming the incoming air and recovering moisture as the breath is exhaled.[82] Endothermy is a complex physiological state, and it must have emerged by stages; yet it seems that by the Late Permian therapsids not only had fur but were warm to the touch.[83] What of that other mammalian hallmark, female lactation? The evidence is circumstantial, but paleoneurological evidence suggests that milk was available to greedy young synapsids no later than the Lower Triassic.[84]

All this evidence unequivocally shows that the thrilling project of "Let's Make a Mammal" was already well underway by the Late Permian.[85] Then the therapsids ran smack into the PTE. What happened next? In a nutshell, not a lot. Certainly there is no convincing case to be made that this particular evolutionary train was diverted, let alone derailed. Yes, there was a crisis, and "disaster taxa"[86] stalked the tortured landscape; however, even the most bloodcurdling details need to be qualified with some important provisos. Detailed information is available from only three regions,[87] and the ghost lineages that have been revealed by cladistic analyses remind us that some groups must have slipped unobserved beneath the wire.[88] To be sure, the dicynodonts were generally hard hit, but the cynodonts were much less affected.[89] And it was this latter group that carried the banner toward the mammals.

There is another aspect of this story that also demands attention. The Early Triassic was a world of cynodonts, whereas toward the end of this interval, as the dinosaurs strode to dominance, the roles were reversed. One explanation is that the process was analogous to those ships that pass in the night inasmuch as the reptilian behemoths may have possessed some innate advantage such as a physiology that promoted gigantism. The story remains complex,[90] but the fact remains that archosaurs in the form of dinosaurs *did* rise to dominance, even if in the latest Triassic an early theropod could find itself in the company of an elephant-sized dicynodont.[91]

There is then a pervasive sense that, as the mammals retreated into the shadows to avoid being squashed by a passing leviathan, their evolutionary future could hardly be described as rosy. A. W. "Fuzz" Crompton approached this question from one angle: reviewing the declining size of the protomammals and the rise of reptilian groups. Thus he wrote, "It is clear that if the mammals had become the dominant land animals shortly after they had arisen, the history of life during the last hundred million years would have been profoundly different. Who knows what mammals, including man, would have done with an additional one hundred million years of evolutionary development?"[92] That for a good part of the Mesozoic the mammals did not even play second fiddle but were relegated to the timpani section of the evolutionary orchestra has not been in dispute. But paradoxically was their small size[93] actually an escape clause and their ultimate key to future success?

What, in other words, is inherent in the ultimate success of any group? As already noted, a good part of the mammalian tool kit was already in place before the PTE. Subsequently, an increasingly energetic lifestyle and voracious appetite, a small size (thus, a low surface area to volume ratio), and an insulating pelt would help to drive fully fledged (so to speak) endothermy, which in turn is the *sine qua non* for a larger

FIGURE 3.7 Mammaliaforms such as *Megazostrodon* are the heralds of another evolutionary revolution. Credit: Nordelch-Megazostrodon Natural History Museum, CC BY-SA 2.0, https://commons.wikimedia.org/w/index.php?curid=4295242.

brain (figure 3.7).[94] If only the Mesozoic reptiles had known, the writing was already on the wall for them. No doubt a dinosaur would have begged to differ: "Mammals? Shy, charming little chaps, except their bite, a painful nip, but simply dealt with . . ." Unbeknownst to this correspondent, and as we shall shortly see, both the actual history and as importantly any counterfactual tell a rather different story.

Not so long ago the claim was that the entire Mesozoic record of mammals (I include the preceding mammaliaforms, because many of their evolutionary innovations prefigure mammalian success) could be housed in a single matchbox. This was on account of their rarity and the fact that the story almost entirely depended on tiny teeth. Such remains are reflected in a welter of toothy names such as the docodonts, morganucodonts, symmetrodonts, and eutriconodonts. There is still much that we still do not know, but even if the hymn sheet is still a bit crumpled, the evolutionary framework is increasingly robust. Spectacular fossil material has begun to reveal a far richer history than was ever imagined, one of not only remarkable diversity but of repeated evolutionary experimentation (chapter 2). By any definition of innovation, the Mesozoic was as much the Age of Mammals as it was of the reptiles.

Despite the cliché that the Mesozoic record consisted of shrew-like creatures shivering beneath the cycads, in terms of ecological diversification, locomotion, feeding, and sensory systems the mammals were forging ahead long before the KTE. Of course, some were tiny. The early Jurassic *Hadrocodium* weighs in at a mere two grams,[95] but by the time we get to the Lower Cretaceous *Repenomamus* had reached the size of a dog (about 14 kilograms). Nor was it a mammal to meddle with, given it had baby dinosaurs for breakfast.[96] Not unrelated was the invention of the so-called tribosphenic molar, which had an arrangement of cusps and basin-like hollows that led to significant advances in feeding, allowing both shearing and grinding.[97] As dentitions evolved thus the mouths of early mammals were open to an increasing range of diets, including the herbivores with the invention of high-crowned teeth (hypsodonty).[98] When it comes to eating, teeth may be on the frontline, but coordination with a mobile tongue and muscular throat are also key factors. Importantly a reconfiguration of the relevant bones (the hyoid) shows that this innovation again appeared early in mammal evolution[99]

Hearing also became more acute, with the sound space allowing some mammals to access higher frequencies,[100] while detecting low-frequency sounds was mandatory for those that had opted for a life underground.[101] All these developments involved those mammals that were evolving into the so-called therians—that is, the marsupials (metatherians) and placentals (eutherians). The latter in particular would inherit the world after the dinosaurs and sundry other reptiles got the chop during the KTE. But again much of the evolutionary legwork had been achieved long before the Cretaceous world ended.

THE HIDDEN MAMMALS

What is clear is that the torch leading to the advanced mammals had been ignited deep in the Mesozoic. Thus, a putative eutherian (*Juramaia*)[102] from the mid-Jurassic indicates that the divergence that separated the marsupial from placental may have started almost 100 Ma before the KTE. These and similar discoveries[103] suggest that such early forms already showed a significant diversity of form, not least in terms of locomotion (including arboreal taxa[104]). A health warning, however: the fossil record of Mesozoic mammals may have greatly improved, but the example of how a key insight into the dispersal routes of some early marsupials revolved around the discovery of a *single tooth*[105] is a timely reminder as to how crucial evidence can hang by a thread. One's admiration for such paleontological detective work (which necessarily involved sieving vast quantities of sediment) is counterbalanced by the existentialist fear that we are mostly clutching at straws. In any event, this particular marsupial tooth is a little under two millimeters in size, reinforcing the almost universal perception that Cretaceous mammals were shy little chaps that spent their short lives scurrying through the undergrowth.

The more fundamental question remains: were these Mesozoic mammals doing anything more than waiting in the wings, or did at least some of the modern groups make their debut before the KTE? It all depends who you choose to talk with. Those who put their money on the fossil record have been hard pressed to see anything interesting before the KTE. David Archibald, for example, did not mince words: although he acknowledged the fossil record is far from perfect, he continued that the "chances that paleontologists had missed any possible trace of this earlier radiation seemed to border on the absurd."[106] The molecular biologists beg to differ.

Not that this divergence of opinions should be seen as either willful or (despite the occasional rhetoric) some silly academic spat. Both sides need each other, but they cannot both be right. The orthodox view[107] is clear enough: the rise of the mammals is inextricably linked to the KTE, and without this catastrophe the road to humans (and aardvarks) was blocked. For example, as Felisa Smith and colleagues noted, "The extinction of dinosaurs . . . was the seminal event that opened the door for the subsequent diversification of terrestrial mammals."[108] Did, however, the KTE define a *de novo* playing field, or was it more like a field invasion where the spectators turn into players? To protest, as some might, that there were no Cretaceous bats or whales is to miss the point. What we need to know is how far mammalian evolution had progressed *before* the KTE. If a good part of the legwork had already been achieved (as is clearly the case with the preceding mammaliaforms), then it is more plausible that, rather than setting the agenda, this mass extinction simply lent a helping hand. Like travelers on a jet with a strong tailwind, the passengers may find themselves disembarking well ahead of the scheduled arrival (in this case, perhaps 40 Ma?) but still at the same terminal.

This is where molecular evolution points to a conclusion at odds with the fossil record. Not that there is an exact consensus, and the methods employed that arrive at the sometimes-conflicting conclusions vary. In any event, these matters are not decided by a vote. The discovery of a single fossil tooth could, in principle, change the entire picture. Why then should we trust an approach that looks at data from living animals? First, the phylogeny of mammals is very largely dependent on molecular data. When this is woven into the geological time scale, not only is it evident that the initial diversification of the placentals was underway a good 10 Ma before the KTE,[109] but this process may have continued without particular interruption across the extinction event.[110] Sending the dinosaurs packing obviously facilitated the subsequent set of diversifications, but (as we will see) if the KTE had never happened, it still would not have made any long-term difference.

The mammalian genome also is informative in other ways. In brief, certain aspects of the genome[111] turn out to be correlated with population size and life span and thus to body size—so elephants are rarer than mice, and they live approximately fifty times longer. By using the agreed mammalian phylogeny one can then reconstruct the nature of the ancestral genome and thereby infer the longevity and body size of the early mammals. The conclusion is at odds with the fossil record, strongly suggesting that these animals were decidedly larger than your average mouse.[112]

So where on earth are they? Is this genomic story somehow flawed, or has the fossil record let us down? Although many species of Cretaceous mammal clearly were small, the fossil record makes it clear that there were exceptions. In this respect, teeth again are informative; molar size correlates reasonably well with body size, so even isolated teeth can be informative. Thus, graphical equations derived from living "primitive" marsupials can be employed to suggest that a Cretaceous form (*Didelphodon vorax*) might have weighed in at perhaps 1.5 kilograms.[113] A more recent discovery of the entire skull suggests a substantially larger animal with an impressive bite,[114] which would have been comparable to the dinosaur-gobbling triconodont *Repenomamus.* Another indicator that Cretaceous mammals were very much on the move can be found in a diamond mine in Angola, where a series of fossil trackways correspond to an otherwise unknown beast much the size of a raccoon.[115] So too a stroll along the sidewalks of Araraquara in Brazil and other neighboring towns can be highly instructive.[116]

Can we yet say exactly what happened as the Cretaceous mammals took their first steps into the limelight? Not precisely—one problem is that at this early stage the diverging groups still look pretty similar to each other, so distinguishing what will ultimately turn into a bat, hippopotamus, or monkey is not at all straightforward.[117] The balance of probabilities suggests, however, that not only had at least three of the major mammalian groups evolved before the KTE[118] but specific orders such as the eulipotyphlans (broadly insectivores such as shrews, hedgehogs, and moles),[119] glires (rabbits and so on),[120] and the primates had as well.[121]

Rodents are sometimes included in this roster, but other evidence has pointed to an origin a few million years after the KTE.[122] In this particular case, a Cretaceous absence is somewhat academic, given that another group of mammals (known as the multituberculates) are in many respects rodent avatars.[123] It does not really matter so much who is doing the chewing as long as somebody is,[124] although the post-KTE demise of multituberculates might be linked to the rise of the rodents.[125] But the success of the former in the Cretaceous,[126] including the appropriately named *Bubodens magnus*,[127] is another indication that the evolution of Mesozoic mammals was not simply controlled by the predominant dinosaurs. Intriguingly, it has also been suggested that this evolutionary burst of the multituberculates was linked to the rise of the flowering plants, technically the angiosperms. This group serves as another reminder that in many respects the Cretaceous world was one that was fast modernizing.

THE ROAD TO TULIPS

Without flowering plants our world would be inconceivable. They accompany us through our lives, from the daisy chains of children, to the cautiously proffered violets. Then follows the posy of the bride and the carnation of the groom. Married life settles down but the naturalist is awestruck by the intricacy of the orchid. Not forever; at the end our biers are piled high with wreaths, and ultimately the apple tree beside the grave takes our mortal remains back to wood and soil and air. The flowers have been our constant companions, but they were also there for the dinosaurs and other denizens of the later Mesozoic, if not before.

Angiosperms definitely appeared not long after the beginning of the Cretaceous,[128] although a Jurassic origin cannot be ruled out.[129] There are even a few indications that angiospermy may have evolved in the preceding Triassic.[130] This is by no means an academic point and for two reasons. First, while a Cretaceous debut means of course that they were in place long before the KTE, a leitmotif of this chapter is that in many cases groups were under starting orders long before their advertised appearance. There is, moreover, a deeper point. We can think of angiospermy not so much as the history of a specific clade but as a general biological property. This is much more momentous because it suggests biological systems have an inherent order, guaranteeing certain evolutionary outcomes (chapters 2 and 4). So it is we see another group of plants, the gnetaleans, that are in many convergent ways approximate to the angiosperms but are closely related to pine trees.[131]

In the end it was the angiosperms that won out, but in the broad scheme of things angiospermy was going to emerge come what may. Either way, their arrival was instrumental in transforming the world. "Dark and disturbed" is how Taylor Feild and colleagues[132] described the modest beginnings of the flowering plants, arguing they were shrub-like and preferred shaded, wet, and unstable habitats such as stream banks. Thereafter the angiosperms did not waste much time in spreading out from their tropical cradle.[133] In doing so they drove some competitors to extinction but also provided new opportunities to other plant groups.[134] Long before the dinosaurs tottered to extinction, many of the major angiosperm groups had emerged. These included the asterids (so daisy and milkweeds are now on the cards, not to mention a nice cup of tea)[135] and rosids (including all-important nitrogen-fixing groups; see chapter 2).[136] A time traveler to the Cretaceous would hardly feel disorientated: there are water lilies[137] and palms[138] to enjoy, grasses[139] to walk on, and grapes[140] to eat (and anticipate a glass of Château Petrus), orchids to admire[141] (and so begin the trail to Darwin's home, Down House), and the inferred presence of parasitic plants to confirm,[142] not to mention those handy lianas[143] to ascend when avoiding that charging dinosaur.

What underpinned this remarkable success story? There need not have been a single explanation—or rather, once step A was achieved, the next step B may have been far less of a challenge. Maybe their first masterstroke was to move out of their crepuscular cradle to more sunlit pastures.[144] Another striking feature of angiosperms are the improvements in water conductance along the xylem with the development of the so-called vessels.[145] Perhaps the evolution of the xylem facilitated new types of wood, providing in turn greater strength and so an advantage in height. Water conduction, however, may have provided an even greater impetus in terms of the evolution of leaves.

An important innovation (albeit one that evolved several times) was to move to a so-called diffuse style of growth because this permits the evolution of much more complex leaf venations. In Cretaceous angiosperms, however, a further crucial development in the leaves was a significantly higher density of veins.[146] Again, angiosperms were not alone in this regard, but the angiosperms not only went the extra mile but traveled it several times independently.[147] The consequences were dramatic in two rather different ways. First, with this hydraulic breakthrough the water essential for photosynthesis was brought much closer to the photosynthetic reaction centers, with a corresponding surge in productivity.[148] Second, a further consequence of increased leaf vein density was that rates of transpiration were correspondingly elevated.

Today, as is well known, there is an intimate connection between local climate, intense rainfall, and locations where rainforests flourish (figure 3.8). As Kevin Boyce and colleagues have stressed, "Angiosperms

FIGURE 3.8 Rainforests epitomize biological diversity, but their origins go back to the time of the dinosaurs.

helped put the rain in the rainforests."[149] Yet the idea of Cretaceous rainforests has seemed to founder on lack of fossil evidence. Were they present or absent? Once again it seems to be a question of the reliability of the fossil record. Phylogenetic patterns and inferred ancestral habits in a wide range of angiosperm groups all point to Cretaceous rainforests being a reality.[150]

THE NEW WORLD

Nobody is pretending that these Cretaceous floras were completely modern. But the donkeywork had been achieved, and with three consequences. The first involves what might be taken as the apparently destructive role of fire. Fossil charcoal shows that plants began to catch fire almost as soon as they left the water; by the Carboniferous period, wildfires were commonplace.[151] The arrival of angiosperms, however, changed things significantly.[152] While the broadleaved forests remained relatively immune, other areas burned repeatedly, and in doing so helped to provide a mosaic of habitats suitable for rapid recolonization.

Nor was this propensity to flammability accidental. This is because the second consequence was that angiosperms are not only more productive but produce leaf litter that is both more nutritious and compostable.[153] Fire and humus promoted positive feedbacks for the angiosperms; among the gymnosperms, the rates of decomposition are far less, and the soils on which they grow are correspondingly impoverished. Gymnosperms had dominated the plant world since at least the Permian, but from the Cretaceous onward the story became one of displacement[154] with the result that today only the boreal realms host coniferous forests.

The third point stems from the first two, and it is that with the arrival of the angiosperms the world simply became a much more interesting place. To be sure, various gymnospermous groups declined, but others continued to flourish either in association with angiosperms or where angiosperms never managed to get a real root-hold. But the gymnosperms were not stuck in a rut; rather, they were busy modernizing as well, as is apparent from the debut of pine trees in the Lower Cretaceous.[155] And a similar story applied to the ferns. These plants, which are regarded as agreeably archaic, were once festooned across the coal forests. Now ferns lurk in damp and crepuscular recesses with toads as companions—but not all do. As the angiosperms cast increasingly long shadows, the so-called polypod ferns (which date back to the Triassic[156]) seized the initiative,[157] radiating across the forest floor and climbing the trees to become epiphytes.[158]

If, then, the angiosperms (accompanied by the ferns) and the mammals were grasping new opportunities in their different ways, what then might be the link between them? That the Cretaceous radiations of mammals and angiosperms were interlocked is reflected by the major changes in dentition among the former.[159] Nor is it likely to be accidental that seed size increased significantly toward the end of the Cretaceous,[160] even though near the beginning of the angiosperm radiations many seeds already had fleshy fruits.[161]

Just as the angiosperms and vertebrates were often interreliant, so too were the myriad insects. Their role as pollinators, of course, long predated the appearance of the angiosperms,[162] with the Lower Cretaceous thrips dusted with gymnosperm pollen (most likely from ginkgos) providing direct evidence.[163] But with the arrival of the angiosperms nearly everything changed. Probosces still probed, and along with the beetles[164] some insects managed the transition from gymnosperms. So too it was no accident that butterflies had already appeared in the Jurassic; nearly all the major groups were in place well before the KTE.[165] Others were newcomers, such as the bees. Their diversification at much the same time as the eudicots[166] makes it easy to imagine a series of escalating and reciprocal interactions.[167] Evidence for more sophisticated modes of pollination is also apparent from the floral anatomy of various angiosperms[168] with features such as clumped pollen upping their chances of successful fertilization.[169] In both these cases, their coevolutionary partners were probably bees, but the beetles[170] and flies[171] also played important roles. Nor was this show entirely stolen by the angiosperms. As already noted, some ferns had a renewed lease on life but then had to withstand attacks by chewing sawfly larvae.[172]

Not only was the Cretaceous air alive with insects, it was also swarming with birds.[173] Showing a wide range of flight capacities,[174] they sometimes achieved an impressive stature.[175] Much of this diversity, however, revolved around relatively primitive groups, notably the enantiornithines, although gratifyingly they showed convergences with modern birds consistent with advanced flight.[176] A more central question, however, is when did the modern birds emerge? The debate has striking parallels to the contested history of the pre-KTE mammals.

To unpick this story, it is necessary to outline the classification of the living birds (collectively the neornithines).[177] Here, broadly, one distinguishes the paleognathans from the neognathans. The former, including such fascinating examples as the kiwi, diverged from the neognathans well before the KTE.[178] As for the neognathans they included the primitive galloanserines—so named because they comprise farmyard representatives such as ducks (anseriforms) and chickens (galliforms)—and more advanced neoavians, which included a mass of passeriforms (including the songbirds and crows) and other notables such as parrots. Putting our trust in molecular clocks[179] gives us confidence that numerous lineages of modern birds had already emerged in the Cretaceous.[180] The diversity of birds in the early Tertiary—such as the giant penguins wandering around New Zealand only a few million years after the KTE[181]—makes better phylogenetic sense if their neognathan antecedents had appeared substantially earlier.

All this can be inferred notwithstanding the shockingly bad fossil record of Cretaceous neognathans,[182] for which the interpretations have been divisive and contentious.[183] Not only is much of the material fragmentary, but the recurrent features of evolutionary mosaicism stand ready to gum up phylogenetic analyses.[184] But despite some of the candidate galloanserines,[185] notably *Vegavis*,[186] still being subject to debate, definitive evidence now seems to be available in the guise of *Asteriornis*.[187] Even the fossilized syrinx of *Vegavis* suggests the bird made a honking sound much like today's geese and ducks.[188] These discoveries, along with considerations of neoavian phylogeny, make the discovery of Cretaceous loons[189] less surprising. Like their modern-day descendants, these Southern Hemisphere birds were adept divers, but in due course they collided competitively with the penguins. The loons are still with us, but only in Northern Hemisphere lakes.[190]

Loons and the like are relatively primitive, but molecular clocks suggest that more advanced groups in the form of passeriforms[191] and parrots[192] also dotted the landscape. Yet after what was announced as a lower jaw of a Cretaceous parrot,[193] the skeptical reception followed a familiar pattern:[194] "too old," "unexpected features," and so on. Well, the jury remains out, but reasons for optimism are not difficult to find. As with the mammals, Cretaceous fossil trackways hint at a hidden diversity, with some prints suspiciously similar to those of shorebirds.[195] Particularly intriguing are some early Cretaceous footprints that in principle look as though they could have been made by a roadrunner (a largely ground-dwelling cuckoo).[196] The type of bird foot, with two toes facing forward and two back (zygodactylus), responsible for such a distinctive trace fossil would raise no eyebrows in the Tertiary; however, these tracks appeared about 50 Ma ahead of schedule. This is not to suggest, of course, that there were actual Cretaceous roadrunners, nor that any of these trackways were necessarily made by neoavians, but if they were we should not be too surprised.

CLOSING THE CIRCLE

Here then is a world that in many ways is already strikingly modern. No doubt the Cretaceous had its ups and downs, and despite the evidence for millions of years of holiday weather, rude interruptions such as cold snaps also occurred.[197] Life ticked on, but unbeknownst to its diverse denizens disaster loomed. Far out in the solar system an asteroid was slowly tumbling through space even while deep within the Earth itself thermal instabilities were starting to generate an ascending plume of hot rock. Together they spelled the doom of the dinosaurs.

On an early June day[198] an asteroid roared across South America and plowed into the Yucatan.[199] It could not have come at a worse time because almost antipodally vast volcanic outpourings (the Deccan flood basalts) were already poisoning the world. The seismic shock waves generated by the asteroid's impact triggered a renewed frenzy of volcanic eruption.[200] Although not quite the same scenario as the PTE, once again the world was spiraling into a catastrophe. At this point any other book would dwell in lingering, if not loving, detail on the catastrophic consequences: "Strangelove" oceans devoid of life, immense tsunamis, global wildfires, acid rain, a shattered landscape bestrewn with ferns, blooms of "disaster taxa," disorientated dinosaurs, quivering mammals, and so on. There can be no disputing it was all deeply unpleasant, but were all bets now off?

The simple answer is "yes." But also "no." And even more provokingly, "not at all." Some groups seemed to have looked up and said, "Oh, another mass extinction?" and carried on. Spiders, for example, continued to spin[201] and lice to infest.[202] Meanwhile out in the oceans, the pelagic fish swam on unconcerned.[203] Nor should we assume a uniform explanation for this apparent immunity. Given, for example, that many spiders are generalist predators, this might well explain their relative invulnerability. This is not to deny that they too might have a pretty torrid time as ecosystems crashed around them, but the point remains that, as with the birds, in no way can the KTE be regarded as a fulcrum point. Such is also apparent with the insects, where (with the caveat of a rather meager fossil record) all in all the KTE seems to have passed them by.[204]

Although it is nevertheless tempting to conjure up lurid visions of devastation, driven by a morbid fascination with end-Cretaceous gore, and then consign the entire globe to biological perdition, that was not quite what happened. As with the PTE, there were zones of relative safety. In this respect—and notwithstanding the location of the Deccan eruptions at about 25°S—some parts of the Southern Hemisphere would have been good choices for hunkering down. In these oceans, recovery was remarkably speedy.[205] Sections of Patagonia have revealed a time of stress, but not one that led to severe extinctions.[206] Indeed, if one focuses on plant forms (such as leaf architecture[207]) rather than busily compiling those endless lists of casualties, then it looks more like a time of mass death rather than mass extinction. In many cases the rates of recovery were also pretty prompt. In Colorado by about 1.4 Ma after the impact the rainforests had bounced back,[208] and at much the same time the Patagonian floras were flourishing.[209]

Similarly, however hard hit they had been, groups as diverse as butterflies and bees[210] as well as vertebrate mammals and birds[211] also promptly recovered. "Vertebrates did you say? Ha, ha! What about all those dinosaurs, or for that matter marine mosasaurs, or flying pterosaurs? Don't see many of them in the Tertiary, do you?" There can be no doubt that the KTE was very bad news for the dinosaurs, even though they may have staggered on for a few hundred thousand years[212] (as perhaps did their iconic counterparts in the oceans, the ammonites[213]). So are the dinosaurs the reptilian elephant in the room? No, not necessarily: over the years, there has been an inconclusive, if not tiresome, discussion as to whether the dinosaurs were already in decline before the KTE.

During the Mesozoic period, the dinosaurs had their ups and downs, but even if a decline were to be identified,[214] that does not indicate a hopeless spiral to extinction. Beyond North America there are few parts of the world with anything like an adequate record,[215] and the evidence is that the mass extinction certainly did them few favors. However, this does not mean that each dinosaur group was listening to the same drummer,[216] nor—to extend the musical analogy—can we assume that different areas of the world were singing from the same score.[217] But it also is possible that the late Cretaceous world was already changing in ways disadvantageous to the dinosaurs.[218] Either way, 66 Ma ago the dinosaurs went to their doom, and so did their reptilian companions in the form of the marine mosasaurs[219] and winged pterosaurs.[220] As far as we can tell, no mosasaur (see figure 2.2) surfaced into the post-catastrophe world, but for all we know their survival might have been on a knife-edge. Although these gigantic lizards were principally marine, a freshwater example from Hungary has been found.[221] To be sure, this animal predates the KTE by about 20 Ma, but the estuarine mosasaur lived substantially closer to the KTE.[222] Given that freshwater communities were little affected by the mass extinction,[223] Tertiary mosasaurs would not seem to be an impossibility.

If the mosasaurs had held their own in the seas, what was to be seen in the skies above them? There would certainly be birds, but they were feathered dinosaurs (chapter 4) in that these reptiles never relinquished their aerial supremacy. Nor were they alone: their co-fliers were the pterosaurs. But do they and the birds represent just another example of ships that passed in the night, coexisting in harmony?[224] The evidence suggests not. When the birds flew into the theater of evolution (chapter 4), the course of pterosaur evolution was profoundly redirected. It seems to be no coincidence that the Cretaceous pterosaurs (figure 3.9) were often enormous, some having wingspans comparable to a Cessna aircraft.[225] This achievement, however, was driven by competitive replacement; the nascent birds led to the corresponding demise of the smaller-bodied pterosaurs.[226] Not only was this pterosaurian gigantism convergent[227] (chapter 2), but as a general rule large-bodied species have smaller populations and are more exposed to extinction. So having been driven into a corner, these huge animals were probably at much greater risk when the times got bad.[228]

If broadly the Mesozoic was the realm of reptiles and the succeeding Tertiary a mammalian world, it was far from a simple game of musical chairs. After the KTE snakes remained spectacularly successful, as is evident from constricting boas, venomous mambas, and sea

FIGURE 3.9 Dominating the Mesozoic skies toward the end of their history, the pterosaurs became increasingly gigantic as the birds took off.

kraits, not to mention gliding chrysopelids. Their squamates in the form of lizards also present a credible roster, ranging from wall-hanging geckos to speedy monitors and even viviparous skinks (complete with placenta). That the squamates suffered grievously during the KTE seems clear,[229] but to regard this episode as some sort of fulcrum point is hardly compelling. Stirring in the Jurassic,[230] both snakes and lizards showed a diverse Cretaceous history, already modernizing,[231] capable of bipedal running,[232] and also showing spectacular oceanic excursions in the form of the mosasaurs.

The snakes underwent adaptive radiations that heralded the ecological versatility of the serpentine body plan.[233] Correspondingly, the same was the case for major groups of lizards such as the iguanians and scinomorphs,[234] not to mention the iconic geckos[235] (already in the pole position for their gripping history[236]) who also had their Cretaceous debut. Even within the Tertiary the supremacy of the mammals did not go unchallenged. In the stiflingly hot Eocene of southeast Asia, a giant lizard held its own against herbivorous mammalian competitors.[237] To the south and almost within living memory,[238] the Australian *Megalania*, a carnivorous giant monitor lizard, was more than a match for its marsupial counterparts.[239]

THE ASTEROID MISSES, AND . . .

Despite all the evidence that mass extinctions accelerate the inevitable, the consensus remains with Gould's insistence that for the mammals (and so ultimately us) the rogue asteroid was our lucky star (albeit with more than a helping hand from the Deccan eruptions). With the ecological decks cleared, the mammals could seize the high ground—as they did. But to put an astrological slant on their luck is a massive misreading of the evidence. Can we be serious? Does everyone not agree that until the day of the disaster the mammals were eclipsed by the dinosaurs? Had the mammals attempted to take to the air they could never have evolved into bats. They would have simply been preempted by the birds, and also greeted by ferocious and agile pterosaurs. A similar fate would have awaited them in the oceans, which were swarming with huge crocodiles and giant lizards in the form of mosasaurs. Any such mammalian excursion as a fledgling whale would have been swiftly snapped up. The Mesozoic period was emphatically a reptilian world, where the mammals remained on sufferance. The facts stare us in the face: the doom of the Mesozoic reptiles allowed the mammalian star to rise. If no asteroid impact, then no whales. If no Deccan eruptions, a world without monkeys.

But most importantly, so the argument goes, if no mass extinction, then no smart dolphins or hyperintelligent humans. In terms of history, what actually happened was profoundly important. But if one considers how the grand evolutionary narrative will play out, then we come to a very different conclusion. To understand why, let us consider a counterfactual world, an alternative history where the asteroid sails harmlessly by and the Deccan eruptions are conveniently plugged. We will not have to wait long before the storyline is back on track: come what may, the dinosaurs and their reptilian cousins ultimately were doomed.

As actually happened, oblivion came in the form of massive volcanism with a *coup de grâce* in the form of an errant asteroid—but the KTE only brought ahead of schedule their inexorable Ragnarök. If the asteroid had skipped across the top of the atmosphere before vanishing forever, or if the volcanic episodes of the Deccan had been less violent, the dinosaurs would have only enjoyed a brief stay of execution. On a human time scale this was an immense interval, but nevertheless it was finite. There are three ingredients to this argument. The first applies not only to the KTE but also to the PTE and most likely to all mass extinctions. As repeatedly stressed, during extinction events most of the actors that will inherit the new world were already gearing up in the old one. Are mass extinctions of no importance then? On the contrary! When it comes to clearing space for rapid radiations *within* major groups the results are impressive, but still they only accelerate what was going to happen anyway. Thus, mass extinctions in themselves are not the fulcrum point of evolution.

The second point is to remind ourselves that periodically the Earth shifts from its usual greenhouse state to the opposite icehouse. As a result, periodically major glaciations occur. Some have been so severe as perhaps to enshroud the entire planet, hence a "Snowball Earth." Others have been centered on a pole, such as the ice sheets that covered the Southern Hemisphere in the Late Ordovician and again in the Permo-Carboniferous. Our current ice age is bipolar, and it culminated in ice sheets that stopped a bit north of Cambridge in England and around New York on the other side of an iceberg-strewn Atlantic Ocean. This path to refrigeration began, however, much earlier in the Eocene; so some 20 Ma after the KTE, glaciations in Greenland and Antarctica (along with sea ice) were already underway.[240] Just as today the centers of reptilian diversity are in the tropics and subtropics, in our counterfactual, asteroid-free Tertiary world the rainforests and grasslands would have swarmed with dinosaurs.

But what about the higher latitudes? In the Mesozoic it is clear that dinosaurs were not only polar but must have overwintered;[241] but this of course was a greenhouse world, and at lower altitudes it was more or less free of snow and ice. But in our counterfactual world how would the dinosaurs have fared in a place equivalent to northern Siberia or Patagonia? It really depends how similar they were to mammals, especially in terms of warm-bloodedness. Bone structure[242] and oxygen isotopes[243] (which fractionate according to temperature) are both consistent with dinosaurs having some sort of warm-bloodedness. The calculations of their energetics may have some imponderables, but overall they still suggest that dinosaurs depended on a scaled-up reptilian physiology that fell short of the mammals.[244] And in one sense this whole question is otiose because one group of dinosaurs, the theropods (chapter 4), did become fully fledged endotherms: we call them the birds, and significantly their warm-bloodedness evolved independently of that of the mammals.

If the question hinged only on warm- or tepid-blooded dinosaurs then that might be the end of the matter. After all, oxygen isotopes extracted from the bones of mosasaurs indicate that these reptiles also had some sort of endothermy.[245] As the oceans chilled, the mosasaurs may not have been particularly inconvenienced; conceivably they might have represented a serious obstacle to the reinvasion of the oceans by the mammals we call whales. Similarly, in our counterfactual world where the good times carried on unabated, the "Tertiary" skies would have been populated by soaring pterosaurs, some of them vulture-like,[246] while the trees would be a world crowded with modern birds.

So in this counterfactual world, with at most a muted KTE extinction (we will allow the Deccan to do some damage), the subsequent history as the world moved to icehouse conditions would have differed, but only around the margins. Early on, the warm tropical zones would have remained broadly Cretaceous in aspect, but this world was ultimately doomed. Indeed, the agents that ultimately will lead the reptilian victims to the block were already stirring in the forests of the early

Cretaceous. They were not just any old group, but one that was warm-blooded, had a brain steadily increasing in size, exhibited extended parental care and formed complex social units, and had the capacity to manipulate objects. The fate of the dinosaurs would have lain not with asteroid impacts or massive volcanism but literally in the hands of the mammals.

As it happened, the KTE did provide judge, jury, and executioner, but to think that without this event the mammals would never have seized the high ground is a serious misreading of the evidence. Certainly it would have taken longer, maybe another 50 Ma. It also may have required a major ice age. But the seeds of mammalian success were sown during the Mesozoic era, if not earlier. Cognitive capacity, advanced vocalizations, and ultimately toolmaking were all beginning to come together. Far from the KTE radically redetermining the trajectories of evolution, it achieved something much more creative: accelerating by perhaps 50 Ma the inevitable march toward a cognitively complex animal, equipped with sophisticated tools. In our counterfactual world, these animals would, in due course, remark, "More gravy? More and more difficult to get decent dinosaur these days. Please help yourself, this burgundy is drinking exceptionally well."

4

The Myth of Missing Links

"Missing Links Sensation!!!" screams the banner headline of the newspaper made famous by the novelist Evelyn Waugh, the *Daily Beast*. Charles Darwin saw it coming, and now in the glare of the studio the rumpled paleontologist struggles to explain to the bemused audience, "Yes, the fossil skull is fragmentary. No, it has nothing to do with Adam and Eve." Now he's trying to keep his temper: "Yes, yes, a bit more of the skeleton would be useful." To the professional, fragmentary evidence is all work in progress, and such research is vital when it comes to documenting what are, on the face of it, astonishing transformations in deep time. From swimming fish to walking amphibian, from running dinosaur to soaring bird, and indeed from inarticulate ape to deeply irritated paleontologist. None supposes that the fossils in museum drawers are the actual intermediates that link ancestor and descendant, which in most cases are separated by eons. That is not the point. Rather, the genealogy of life demands "missing links," and despite the distortions and willful misreadings of the so-called creationists, the fossil record shows in considerable detail how the land was invaded and the skies populated. How then can they be a myth? Are you disputing Darwin's account of descent by modification? Of course not.

Missing links can, of course, be found at any level, from those that connect closely related groups such as earwigs,[1] to ones that deservedly have earned an iconic status as they serve to link utterly disparate groups. Not, of course, that there is any reason to think the evolutionary processes operating at either end of this spectrum are fundamentally different. Each is step by step, and definitely there are no hopeless monsters. But in the cases of profound transformation, the consequences are enormous. So where better to start than with the origin of tetrapods—four-legged creatures potentially capable of terrestrial excursions, the precursors of animals as different as dinosaurs, frogs, bats, and apes.

What the Devonian fossil record has revealed is now standard fare for textbooks, but without it even the most inspired biologist would have been hard pressed to sketch out more than a vague outline of what actually happened. As with nearly all missing links, the problem is that the living forms that ostensibly lie at the starting and end points are themselves highly derived. This is true of the amphibians, for example. Think of the frog, whose nearest aquatic relatives enjoy the label of "living fossils" (in themselves another evolutionary myth); but neither the coelacanth (figure 4.1) nor lungfish can give more than a few hints as to how the move to land was achieved. To be sure the coelacanth's moniker of "Old Fourlegs"[2] captures the limblike appearance of its pectoral and pelvic fins, yet it transpires that the lungfish are closer still to the tetrapods.[3] Apart, however, from their eponymous lungs, these fish give few clues as to how their distant ancestors hauled themselves ashore.

LEAVING THE WATER

Now let us look to the fossil record, where the picture changes dramatically (figure 4.2). Key fossils with names like *Acanthostega*[4] (figure 4.3) and *Tiktaalik*[5] take star roles, demonstrating not only how fins morphed into limbs, but the manner in which the rest of the body traded the buoyancy of water for the tyranny of gravity. So too were there dramatic changes in the sensory systems (including olfaction) and

FIGURE 4.1 The limblike fins of the coelacanth hint at the connection to the tetrapod limb, but in reality its sister group, the lungfish, are the more closely related.

physiology (not least controlling water loss and shifting kidney functions). Echoes of these profound changes can be found in the genomes of their living descendants.[6] Correspondingly, fish typically employ the lateral-line system (which is sensitive to pressure changes in the surrounding water), and this system is retained in some amphibians. So

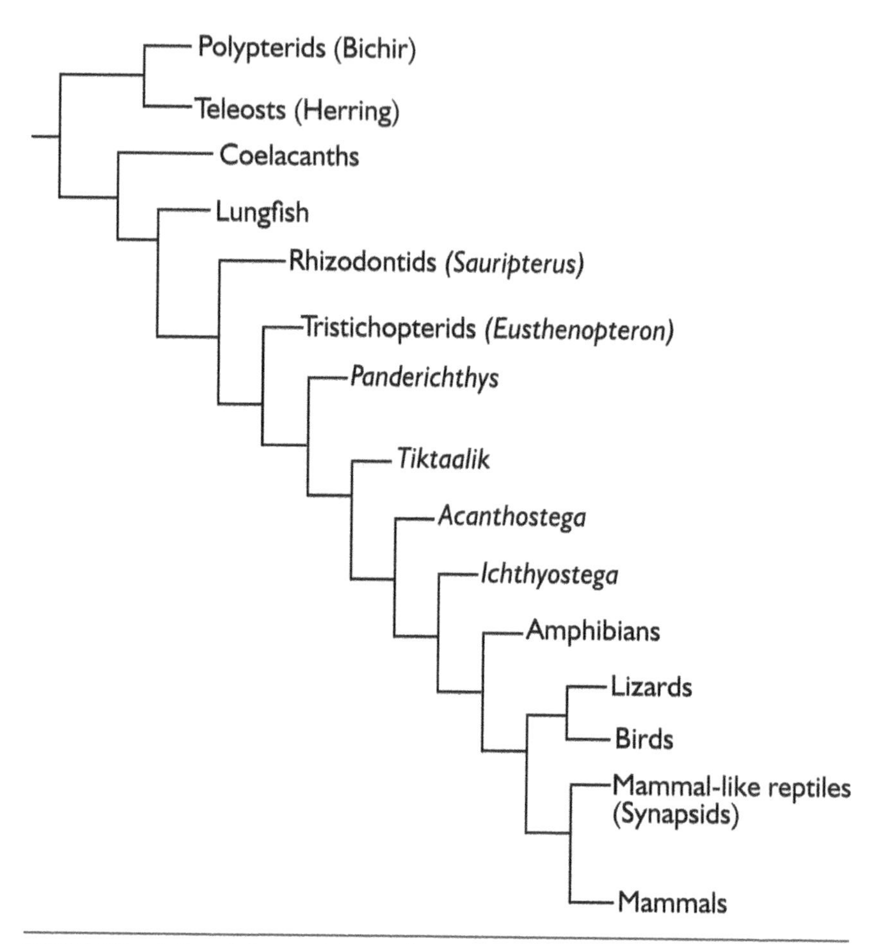

FIGURE 4.2 Outline phylogeny of the vertebrates, with an emphasis on the sarcopterygian-tetrapod transition.

too some fish complement this arrangement with a capacity for electroreception, and this ability may also have been present in some of the proto-tetrapods.[7]

Among the most surprising of the early discoveries was that even by the stage that fins had become limbs, the animals remained aquatic.[8] This is evident not only from the retention of a lateral-line system but even more tellingly the gills.[9] Why then evolve limbs that were never

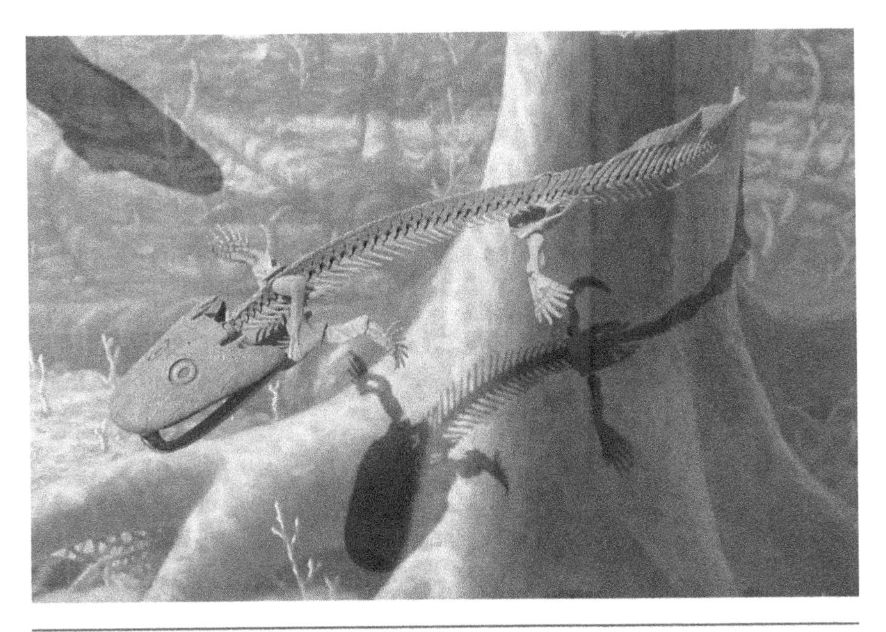

FIGURE 4.3 With its four limbs this skeletal version of *Acanthostega* sculls through the water, but it is all set for the invasion of land. Credit: By Ryan Somma, Acanthostega gunnari, CC BY-SA 2.0, https://commons.wikimedia.org/w/index.php?curid=6738126.

used for walking? In point of fact, a number of fish can stroll across the seabed,[10] although the styles of progression differ widely. In the case of cavefish, their walking up the waterfalls in Thailand is assisted by a tetrapod-like pelvic girdle.[11] Although this does not preclude a similar behavior in the proto-tetrapods, the general consensus is that these animals were ambush-predators, using their limbs for precise maneuvering in weedy habitats. Reorganization of the muscles to execute these sculling movements was, of course, an important precondition for the more precisely controlled requirements of terrestrial locomotion. That such limbs evolved before they were "needed" was presciently suggested many years ago by the American biologist Theodore Eaton,[12] but in the wider scheme of things too often it is forgotten that missing links and co-option of existing structures go hand in hand (so to speak).

So is that the end of the story? A series of dramatic discoveries, pioneered by investigators such as Jenny Clack[13] and Neil Shubin,[14] may in hindsight seem unsurprising, but these fossils unequivocally delineated the story of not only where terrestrial vertebrates came from but how they also lit the fuse that led ultimately to one very odd species (chapter 5) wishing to understand its own history. This grand narrative will, of course, continue to be refined as new fossils emerge, but more importantly it is the concept of missing links that is long overdue for reexamination. This is not with respect to the fossil record per se, which naturally is incomplete but not disastrously so. Nor will everything necessarily turn out to be quite what has been expected. For example, the dramatic evidence of what has been interpreted to be the earliest evidence of tetrapod footprints may transpire to be fish nests or feeding traces.[15] Overall, however, the fossilized trackways serve to complement the bones, and collectively they help to tie down the time scale of terrestrialization.[16]

Do we have it all under control then? Not exactly: the fossil record has revealed something far more interesting. Although the overall story of tetrapodization is clear enough, in detail the story is very far from some sort of monotonic narrative. This is because the fossils in question are almost always a puzzling muddle of characters, revealing a striking degree of mosaic evolution. In other words, in any given species we see relatively archaic features co-occurring with more advanced ones. For example, the pectoral fin may be closer in structure to the tetrapod limb, but the remainder of the body remains far more fish-like.[17] This does not please everyone—or at least not those who seek neat and tidy answers.

A telling example involved the discovery of a new form from the classic locality of east Greenland. Surely this was a cause for rejoicing? Indeed it was—until in the cladistic analysis a *single* character state had to be changed. As a result, the previously well-supported phylogeny effectively collapsed[18] (no fault, of course, of the investigators). More generally this mélange of advanced versus primitive features in any given species means that deciding exactly where a given proto-tetrapod falls in the evolutionary tree is often finely balanced: metaphorically the animal will be pulled both crownward (i.e., toward its living descendants) and downward in the direction of its phylogenetic roots. This has considerable implications not only for reconstructing the original order of events, but much more importantly for the unavoidable conclusion that the story of tetrapodization is one riddled with parallelisms.[19]

So in this regard new fossil discoveries will make no material difference. As soon as one apparent parallelism is shown not to be a convergence but in fact a character derived from a common ancestor, then another apparently monophyletic character will promptly transmute into a convergence. The more crucial point is as follows: there is not one missing link but *several* because independently a number of groups of fish are showing the same trend toward tetrapodization. The exact order by which they achieve this development differs—after all, this is evolution not in the textbook but in the real world. Nevertheless, the overall trajectories are closely parallel.

In this context perhaps the most telling example comes from a relatively primitive group known as the rhizodontids. These were powerful-looking customers, ferocious predators that grew up to twenty feet in length. Even though this group lived near the dawn of tetrapodization and were also totally aquatic, their pectoral fins possessed fishy fin rays[20] that were already decidedly limb-like. So what is in a limb? I write with my right arm (using a fountain pen) and so rely in part on my humerus (technically the stylopod) and ulna/radius (the zeugopod). Here, the transformation from fish fin to protolimb is not so problematic inasmuch as the former already possesses an array of bones that in principle can be stripped down to the single stylopod and double zeugopod (respectively, these words are derived from the Greek for a pillar and paired).[21] In the former case, for example, it is the suppression

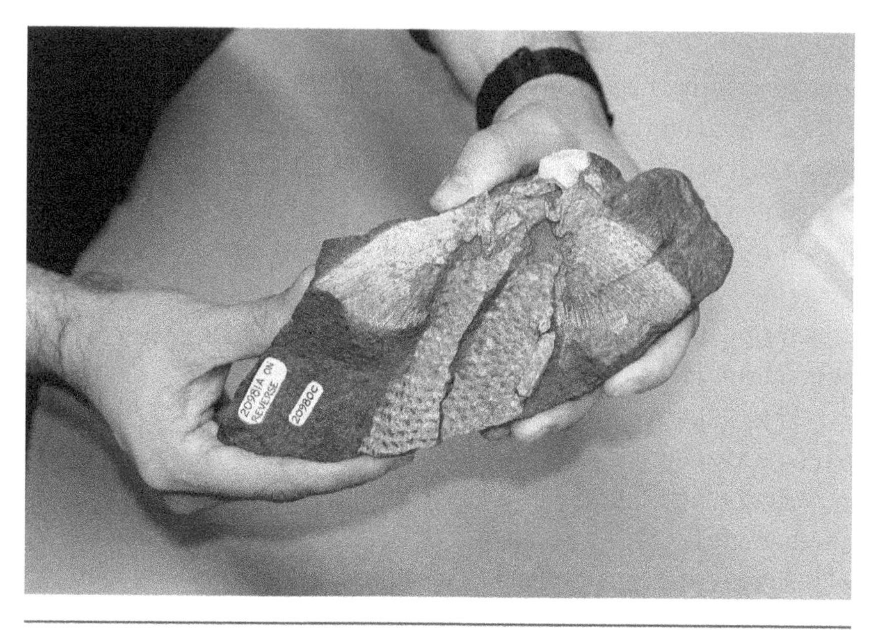

FIGURE 4.4 The fins of the rhizodontid *Sauripterus* herald the development of a fully fledged tetrapod limb. Credit: Matt Wood. March 25, 2020. NSF Multimedia Gallery.

of the anteriorly positioned bones that ultimately ends up with only the humerus.

However, the origin of the unit that holds my pen—the so-called autopod, comprising the wrist bones and fingers—is much less obvious. These are integral to a terrestrial life-form, be they found in paw, hoof, hand, or (by aquatic reversion) flipper. One would not expect an autopod to have any role in a fin, until we look at a rhizodontid from Pennsylvania known as *Sauripterus* (figure 4.4). Is this not, in the words of Edward Daeschler and Neil Shubin,[22] "a fish with fingers"? Enclosed as they are in a stiff, scaly fin, this set of digits nevertheless articulates with a proto-wrist. So what was their function? Maybe like the other bones[23] they served for muscle attachment, imparting a maneuverability to the fin that eventually would be co-opted for terrestrial locomotion. But the crucial point is that the relative phylogenetic position of the rhizodontids means that these digits must have evolved independently of the other proto-tetrapods.[24] Nor is *Sauripterus* alone in this respect—other rhizodontids also have evolved digits,[25] while the convergences with other proto-tetrapods extend to other features such as the lower jaw.[26]

Looking more generally across the proto-tetrapods, we can see how their evolutionary leitmotif is a persistent, unmistakable tendency toward tetrapodization in a number of separate aquatic groups. Aside from fins morphing toward limbs, a series of correlated changes resulted in a body form that, from our privileged perspective, will prefigure the amphibian, but so far as these animals were concerned defined a life as an aquatic ambush predator.[27] There are two consequences. First, repeatedly we find features that ultimately will prove central to terrestrial life such as a neck,[28] or in the case of *Eusthenopteron* a sturdy humerus with functional marrow.[29] The latter animal was for a long while considered the poster fossil for pretetrapods, but the ubiquitous parallelism among the proto-tetrapods has demoted *Eusthenopteron* to just one of several essays toward a tetrapod-like animal.[30] There is a yet more important point as well: by the time we reached even the precursors of the early rhizodontids, one way or another, sooner or later the evolution of a limb that could be deployed on land was an inevitability. This also was true of the rest of the anatomical panoply required to pursue a terrestrial existence. This is the myth of a missing link: long before the final breakthrough, the seeds of success were not only being sown, but multiple times. What will happen, must happen.

Why stop at the rhizodontids? How much deeper in time can we pursue this thread of inevitability? "Not much farther" would probably be the conclusion of most investigators. For them, the story of tetrapodization begins not long before the advent of the rhizodontids. But in fact we need to look into the depths of the Paleozoic to see the first stirrings of the tetrapods, and even then these fish in turn are drawing upon yet more ancient templates.[31] In this enterprise of chasing the evolutionary regress there are in fact two somewhat different approaches: look at extant taxa or turn to the fossil record.

FIGURE 4.5 The polypterid fish looks like another piece in the jigsaw puzzle of tetrapod origins, but in reality the similarities are the result of evolutionary convergence. Credit: By ぱぷじ〜, CC BY-SA 4.0, via Wikimedia Commons.

In the former case, one can infer what the common ancestor most likely possessed by assessing what characters are shared in common by its living descendants. There are, however, pitfalls for the unwary. Consider, for example, a group of rather strange-looking fish that live only in parts of Africa, referred to as the bichirs (the name might be of Arabic origin) or more technically the polypterids (Figure 4.5). They look convincingly archaic, and by general consensus they are survivors of some of the most ancient bony fish (the actinopterygians). By being quite close to the sarcopterygians, they are potentially informative about the fish–tetrapod transition. They are equipped with lungs,[32] so they even breathe through a modified gill opening known as the spiracle in much the same way the proto-tetrapods are inferred to have gulped air.[33] Nor do the similarities with the tetrapods end there: the musculature of the fins is more like a sarcopterygian[34] than a typical fish. The degree of analogy to tetrapods was reinforced when individual fishes were forced by investigators to pursue a terrestrial existence;

the fins they usually employ for swimming adopted a distinctive style of locomotion.[35] Yet strangely, for such seemingly archaic animals, the fossil record fails to tally—and with good reason. When a group of Triassic fish (the scanilepiforms) are added to the phylogenetic equation, it transpires that far from being "living fossils" the polypterids are actually quite derived.[36] Ironically their tetrapod-like features are effectively convergent.

Polypterids reinforce our sense that if one group of fish had not turned into tetrapods, then most likely another would have done so. For our immediate purposes, however, of tracing the ever-deeper missing links, it is more sensible to stick first to the sarcopterygians and then their immediate relatives. Turning to the living representatives of the sarcopterygians in the form of the coelacanth and lungfish, let us start by tabulating the muscles of their respective fins. At this point one needs to explain that the relative closeness of the tetrapods to either coelacanths or lungfishes was not easy to establish because the latter are highly specialized whereas the coelacanths have strikingly limb-like fins. It is, however, clear that the lungfishes are the more closely related to the tetrapods. What this means, of course, is that when it comes to the complex musculature of the fins, this arrangement must be very ancient and so predate any proto-tetrapod.[37] In other words, the remote common ancestor of the coelacanth and lungfish already possessed a complex musculature that strongly prefigured the arrangement that will be required by a limb.[38]

Living descendants offer other clues that limbs were on the cards from a very early stage of fish evolution. Given limbs derive from fins, it would hardly be surprising if at least some of the developmental machinery in the genome was carried over. This turns out to be the case. It transpires, however, that to a marked degree it is the overall developmental template that is conserved. This is the territory of so-called deep homology, but here one needs to exercise some care. Recall that genes do not actually make things, they only instruct (and often very circuitously) cells and tissues to engage in their construction. It can be entirely misleading to say that because the same developmental genes are being used to "make" structure A it is the *same* as structure B, not least because the same genes can pop up in all sorts of different circumstances that are unrelated to either structure A or B. On the other hand, if a developmental template is conserved, it gives us confidence that the assembly of a given novelty, say a limb, is not only likely but might easily happen independently several times.

Thus it is that in both fin and limb the genes that swing into action show an early stage of activity, followed by a later wave.[39] Respectively, these two waves promote first the formation of the proximal region (in tetrapods, the humerus plus ulna/radius) and then a distal zone (the wrist and digits[40]). The activity, moreover, is not only from shoulder to fingertip (so to speak), but it moves also from anterior to posterior. Effectively this presages not just the formation of a series of digits but also, with such a deep-seated asymmetry, can, among other things, prefigure a thumb. In due course this could come in very handy![41] Seen from this perspective, the formation of fingers in an animal like the rhizodontid *Sauripterus* is less dramatic than it might seem, even if the consequences were ultimately monumental.

None of this tells us, of course, why the developmental pathway was tweaked so as to deliver "fingers," but in effect the transition from a fin to a limb is all about moving from a bodily projection that has a largely external casing of so-called dermal bones to one where the bones form an internal array.[42] In these changing configurations, it may well be that in a sense the genetics hung onto the coattails of an anatomy that was evolving to explore new possibilities. Nor would it be surprising if the quantitative genetic changes that allowed a limb to form at the expense of a fin were in themselves trivial, even though the anatomical rearrangements are profound. And the wider point cannot be too strongly reemphasized: by the time an early sarcopterygian fish was sculling through the Devonian lakes and seas, a terrestrial tetrapod was only a matter of time. No wonder the missing links were queuing up. Nor, of course, does the story end there.

ASCENDING INTO THE SKIES

The challenges of hauling oneself onto land, molding fins into limbs, and rethinking reproduction—not to mention shifting aquatic sensory systems to ones that can function beyond the strand line—are not for the faint-hearted. Yet establishing a comparative metric for separate major evolutionary transitions is problematic. How does one judge the scale of relative difficulty? Is fish to amphibian any more impressive than descendant vertebrates transforming themselves from earthbound creatures to denizens of the air, equipped with airfoils that allow a non-stop flight from Alaska to New Zealand,[43] reach altitudes more familiar to long-haul flights,[44] or see the aerial plunge on an unsuspecting prey at 200 miles per hour?[45] These trophies all belong to the birds, and in these respective cases refer to the bar-tailed godwit, bar-headed goose, and peregrine falcon.

Not that birds are alone because in their different ways pterosaurs and bats also show that once a tetrapod stood on land then sooner or later the skies would be populated. Although the fossil records of both bats and pterosaurs are more than adequate, as it happens neither has much to say as to their precise origins from, respectively, a lizard-like diapsid reptile and perhaps a shrew-like mammal. The origins of the birds, however, are well-documented. Tens of millions of years before they launched themselves aloft, their template of future success was not only being assembled (figure 4.6), but if one avenue had closed, just as with the proto-tetrapods, another pathway most certainly would have been found.

At one level the project "Let's Make a Bird" was relatively straightforward, although the details need careful unpicking. Among the desiderata are becoming bipedal (so freeing up the forelimbs), fusing the ankle region (the so-called arctometatarsus, to reinforce the legs), shrinking in size (thus miniaturization), developing airfoils for lift and maneuverability (invent feathers), and taking a crash course in high-energy metabolism and hence an effective system of respiratory

FIGURE 4.6 A diversity of feathered theropod dinosaurs, such as the ornithomimosaur *Caudipteryx*, reveal not only the steady acquisition of birdlike characters but also that in many cases these were convergent. Credit: Emily Willoughby, CC BY-SA 4.0, via Wikimedia Commons.

ventilation (hollow bones and air sacs). That was just for starters. Later developments, such as shrinking the tail (and fusing caudal vertebrae into a bony nubbin, the so-called pygostyle), reconfiguring both hind- and forelimb musculature, reducing the digits, and it is usually argued stripping out the teeth (and so shifting to a keratinous beak, the rhamphotheca) would help bring the project to completion.

Any attempt to delineate this project will involve drawing some somewhat artificial boundaries. It is perhaps worth distinguishing, however, elements of deep ancestry versus the more proximate histories that collectively ensured that birds were written into the fabric of life. As with our examination of the sarcopterygians, the search for ever-deeper evolutionary roots could involve a more or less indefinite regress. Here it will be convenient to start with what paradoxically would at first be taken as a hallmark of the birds themselves: feathers. To begin this journey it would seem logical to turn first to the reptilian scales, familiar on a snake or lizard. After all it is a commonplace that feathers are highly modified scales. This is not quite true. In tracing evolutionary likelihoods, if not inevitabilities, it is sensible to probe deeper in geological time.

Despite their obvious differences, feathers and hair share a fundamental identity. Nor is this so surprising, given that ultimately both birds and mammals have reptilian origins. They are, however, otherwise distinct. The furry mammals emerged from the synapsids, and as with feathers the hair is an outgrowth of the integument. More significantly both feathers and fur employ the same developmental pathway,[46] with the fundamental difference revolving around hair retaining a glandular capacity. Both, of course, employ proteins known as keratins, whose resilience and toughness are essential to their proper function. It is no less surprising, therefore, that the molecular linkages that confer this strength to feathers and hair are convergent.[47]

In discussing the origin of feathers, two other health warnings need to be issued. First, quite how deep their ancestry goes is an open question. This is because a group of dinosaurs known as the ornithischians are also known to possess something akin to feathers,[48] as do the pterosaurs.[49] Is this convergence or common ancestry? In the case of the dinosaurs the arguments have gone both ways, but a reanalysis of their relationships suggests that the ornithischians are more closely related to the theropod dinosaurs than had been generally thought.[50] If so, the first feathers may have sprouted deep in the Triassic (and perhaps this is also consistent with their presence in the more phylogenetically remote pterosaurs). Either way, among the various groups of dinosaurs it was the theropods that ultimately went aloft. In this respect, feathers are almost the dinosaurian *sine qua non* for leaving the ground, but we have a glimpse of an alternative route.

What is even more extraordinary is that among more advanced theropods, belonging to the somewhat enigmatic group known as the scansoriopterygids, it appears that even feathers are not essential. The animal in question (figure 4.7) has one of the shortest names in science: *Yi* (Chinese for wing). It possessed wings, but remarkably they

FIGURE 4.7 Almost bat-like *Yi* demonstrates the diversity of flying theropods.

were membranous, in a way reminiscent of those of bats. These wings were supported by its finger bones and more curiously by a prominent additional bone that finds no counterpart among other theropods.[51] To build an airfoil with a membrane is common (think of the pterosaurs and numerous gliding mammals); however, in its theropod context *Yi* (and the related *Ambopteryx*[52]) is unique, although tantalizingly a related form (*Epidendrosaurus*) also had a remarkably elongate digit.[53] Although powered flight cannot be ruled out, in the case of these membranous theropods gliding across the tree canopy may have been more likely.[54] Not that *Yi* (nor another relative known as *Epidexipteryx*[55]) lacked feathers, but they were formed of elongate, stiff filaments and most likely had no aerodynamic function. In fact, among the theropods there is a striking diversity of feathers, both in shape and location on the body, and this is consistent with feathers originally evolving for display—be it the seductive "come hither" or hostile "go thither" variety.

Yi (and quite possibly *Epidendrosaurus*) reminds us of the alternative routes potentially available to aspiring theropods, but feathers would prove to be the airfoil of choice. Not that a fully fledged bird emerged overnight; even when they were fully volant, further evolutionary steps were made.[56] Not only was the assembly of the features necessary for flight relatively gradual, but in the wider context the actual order of events was hardly of relevance. We know this because, just as with the sarcopterygians, the theropod adventure was rife with mosaic evolution, so the component taxa typically are seen to display a "puzzling" combination of characters. Convergent evolution is equally rampant. In their overview of the roots of theropodan flight, Corwin Sullivan and colleagues go so far as to remark on how the "surprising anatomical novelties . . . exemplify the complex and even *chaotic* nature of [this] early history."[57]

By demonstrating that many of the requisite features required for life in the skies were either available at an early stage of theropod evolution and/or evolved multiple times, the thesis that the evolution of a birdlike form was effectively inevitable is reinforced (figure 4.8). Thus, in early theropods unequivocal evidence for feathers is evident from both the so-called quill knobs (on the arms of *Concavenator*[58]) and actual plumage, which in *Sinosauropteryx* provided a "bandit mask" face (perhaps to protect the eyes from glare) and an elongate striped tail (possibly to confer camouflage).[59] These feathers are typically filamentous and far from being aerodynamic, but even at this early stage of theropod evolution there is evidence for the associated follicles.[60] There are other interesting sidelines as well. Not surprisingly in at least some taxa the plumage was accompanied by regions still bearing scales.[61] But the theropods in question were typically rather small, and the cliché has been that the leviathans were appropriately scaly. Certainly the thought of the iconic *Tyrannosaurus* possessing feathers verges on the comic; indeed, the fossilized impressions of the skin of these immense beasts confirm their scaly appearance.[62] But the smaller tyrannosaurids (such as *Dilong*) were indeed feathered,[63] and most likely the plumage was discarded as they became increasingly colossal, in much the same way as giant mammals such as the elephant have become almost hairless.

Feathers (and ultimately elongate forearms, aka wings) were only one of a series of anticipatory steps necessary for the theropods to become airborne. In the skeleton there were significant developments. One was so-called pneumaticity, whereby some of the postcranial bones hollowed out and eventually contributed to a highly effective respiratory system.[64] This is associated with highly energetic flight, but pneumatic bones have other advantages. Not only are they lighter, but the metabolic burden of maintaining bone is alleviated, especially with respect to the marrow.[65] In a related vein was the development of medullary bone. This has been assumed to be unique to birds, in which it serves to buffer the demands for egg formation in the females; yet a counterpart has been identified in the tyrannosaurids as well.[66] This has a link to the likelihood of brooding eggs, and it is also evident that

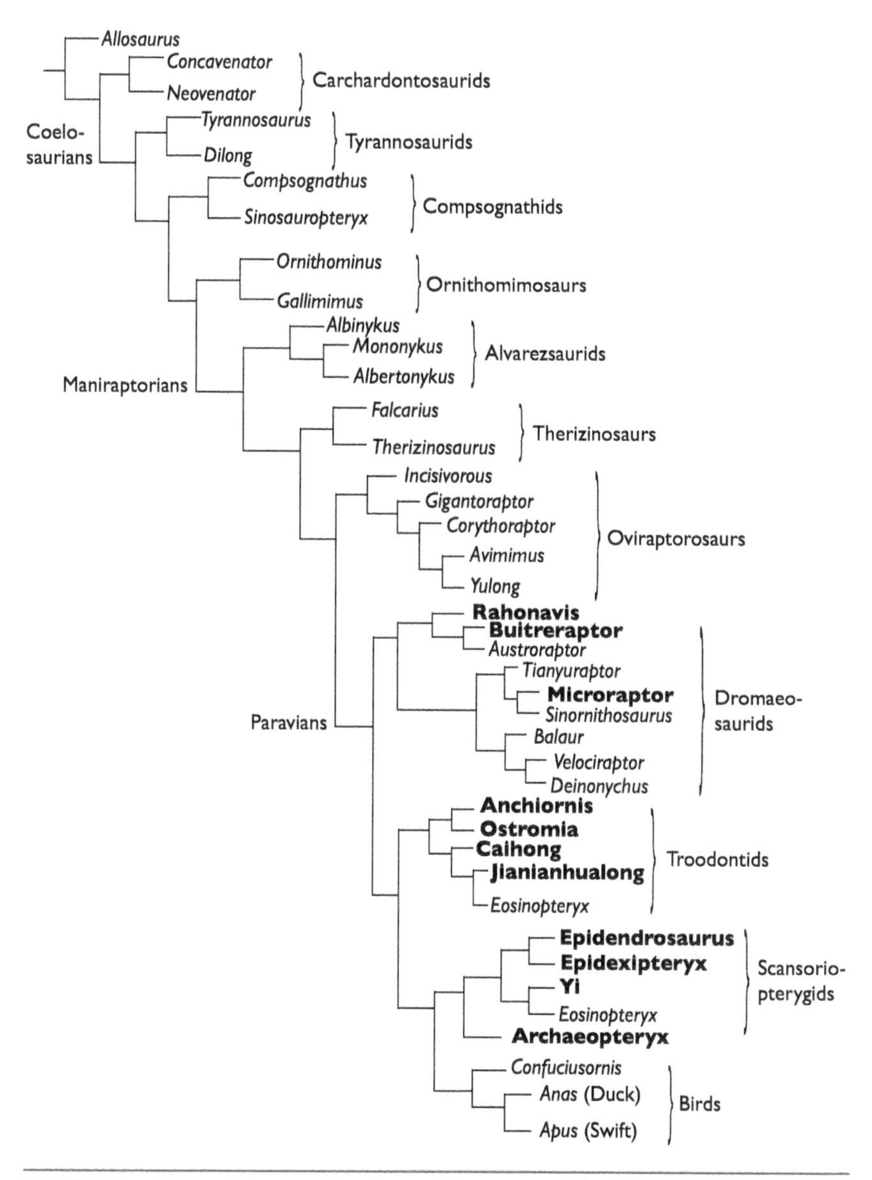

FIGURE 4.8 Outline phylogeny of the theropod dinosaurs and their immediate descendants, the birds. Theropod taxa excluding the birds that were or possibly were capable of aerial locomotion are denoted in bold. Although this phylogeny represents a broad consensus, the exact details remain somewhat fluid. Figure based on various sources.

an avian-like resting posture goes back at least to the dawn of the theropods.[67]

On account of the osteocyte cells being readily preserved, theropod bones have provided other insights. It has long been appreciated that bird genomes are small, and a corresponding genomic convergence is seen in both the bats and pterosaurs.[68] Because cell size is proportional to genome size (the so-called C-value paradox), the latter can be estimated in theropods; it transpires that they too have a greatly reduced genome.[69] A further link has been made between smaller genomes and warm-bloodedness,[70] and this suggests another crucial element of bird organization was in place at an early stage.

Enhanced levels of activity are also evident from not just a bipedal stance, but in terms of locomotion increasing both speed and agility among the theropods. Ultimately this would play an important role in leaving and returning to the ground, but key to this process were the rearrangements of the tarsal bones of the foot. Ultimately this would lead to the so-called arctometatarsus, which evolved in conjunction with a ligament that contributed to the elastic rebound of the leg. The fossil record of the theropods not only shines a light on the transitional forms that saw both longer hind-limbs and increasing speed, but crucially this shift toward the arctometatarsus happened five times independently.[71] As we saw (chapter 3) in the case of large but otherwise cryptic mammals leaving their fossilized trackways across the Cretaceous landscape, the counterpart footprints made by the theropods are also informative. Thus, even at a very early stage of their evolution the locomotor posture we associate with the birds was already present in nascent form.[72]

As the postcranium was morphing toward new possibilities, so too was the skull. In birds the remarkably thin bones and the bulbous configuration that accommodates a disproportionately large brain and large eye sockets (with associated acute binocular vision) may all have emerged later in theropod history, but the antecedents were already evident in the relatively primitive coelosaurians.[73] So too the diagnostic beak—which effectively involved a progressive loss of teeth and the extension of a horny covering (the rhamphotheca)—arose among the theropods several times.[74] Such adjustments heralded significant changes in diet, not least in herbivory and ultimately the pecking of seeds. Integral to their feeding process was also the tongue. For a long time it was thought that the associated internal support structures, known as the paraglossalia, were a defining character of advanced birds. But this is not the case: they now have been identified in a Cretaceous dinosaur, with the implication that their apparent absence is more likely because the paraglossalia generally have a cartilaginous composition[75] that is unlikely to fossilize.

Frequent reference has been made to key theropod features evolving multiple times. There is also a broader trend inasmuch as this recurrence involves more than the "eternal return." Not only are some representatives in each of the major theropodan subgroups strikingly avian, but more significantly as one progresses up this evolutionary tree these excursions become increasingly birdlike. As one motors through theropod history there is a compelling sense that "the bird" moves from "possible" to "likely" to "very probable" and ultimately "inevitable." It would, however, be entirely incorrect to force these steps into some sort of orthogenetic straitjacket, where every species of theropod is marching toward a destination labeled "Birds." The reality is very much the reverse: notwithstanding a rather conservative body plan, theropods show an exuberant range of ecological strategies.

For example, the alvarezsaurids (a group originally described from South America and named after an Argentine doctor) include the bizarre-looking *Mononykus*. With its stumpy forelimbs,[76] this beast was never going to fly. Nor among the more advanced deinonychosaurs would the convergently short-armed troodontids[77] and dromaeosaurids. In the latter case, the taxa in question (*Austroraptor* and *Tianyuraptor*[78]) are, however, near evolutionary neighbors of volant dromaeosaurids, respectively, *Rahonavis* and *Microraptor* (figure 4.9). In terms of ecological chutzpah, shrinking forelimbs to almost

FIGURE 4.9 At least capable of gliding, if not active flight, the four-winged *Microraptor* is just one of several ascents into the sky by the theropods.

stump-like proportions is one strategy, but another would be to become amphibious, complete with a swanlike neck and "wings" that are now employed for swimming.[79] Each and every one of these examples is not only a fascinating evolutionary excursion but as often as not provides further and striking instances of convergence. The alvarezsaur *Albertonykus*[80] (and *Mononykus*), for example, has been compared to myrmecophages such as the mammalian anteaters, with their stout forelimbs well adapted to rip open wood that is infested with nutritious termites.

Becoming airborne may be just one more ecological adventure, but the invasion of the air literally opened new realms to the hitherto earthbound theropods. To unpack what the multitude of missing links show (and noting that even among the archaic neovenatorids some Cretaceous taxa are becoming curiously birdlike[81]), it is simplest to take the narrative from relatively primitive groups of theropods to ultimately full-fledged birds. But such a *scala naturae* is misleading because, just as with the sarcopterygians, there are rampant parallelisms and mosaic evolution. As Evgeny Kurochkin remarked, what we are seeing is an "ornithization of theropod dinosaurs."[82] From this perspective, flying dinosaurs are an inevitability. To begin this journey, let us turn first to the appropriately named ornithomimosaurs. These, the "ostrich dinosaurs," showed various convergences to the birds, not least a pneumaticity of the bones which could become highly developed.[83] Just as striking was the ostrich-like plumage of *Ornithomimus*, which had an array of complex feathers on its "wings."[84] There is no suggestion that either this animal or any other ornithomimosaur could fly, although the "wings" of *Caudipteryx* (figure 4.6) would have generated some uplift when it was running full pelt.[85] In any event, should a volant ornithomimosaur turn up, we should not be too surprised.

Next in phylogenetic line are the alvarezsaurids. Tellingly, when they were first identified they were thought to be primitive and flightless birds.[86] In fact, they were considerably more basal, but the extreme morphological convergences[87] that led to their erroneous placement among the birds are a common enough source of confusion in biology. Ironically, when they were first described the skull was confidently regarded as avian, and the various similarities to theropods were seen as no more than convergences.[88] Along with the more or less obligatory feathers,[89] the catalogue of avian-like features is impressive. These include a delicate skull with strikingly large eyes, which in *Shuvuuia* provided highly effective nocturnal vision and was linked to an acute hearing to rival the owls.[90] The bone articulations in the skull also prefigured the kinetic arrangement characteristic of birds.[91] The postcranial region is equally instructive. In the derived *Albinykus* we see a keeled sternum (in birds this is employed for attachment of the flight muscles), ossification of the ankle region, and vascularization of the bone. All this is also reflected in a dramatic reduction of body size, a trend that is most likely independent of that seen in the main evolutionary branch leading toward the birds.[92] To be sure, the alvarezsaurids have their own peculiarities, and like the ornithomimosaurs so far

FIGURE 4.10 A staging post toward the birds in the form of the therizinosaurs.

as is known none took to the air. But neither does it seem completely out of the question.

Similar remarks apply to the next group along the evolutionary line: welcome to the therizinosaurs (figure 4.10), a name that refers to the dagger-like appearance of some of their claws. Again there are feathers, usually in the form of broad filaments.[93] A significant development, however, was the fusion of some of the tail vertebrae to form a pygostyle.[94] Subsequently this structure would be important for the attachment of tail feathers (the rectrices), ultimately to be employed in flight maneuvers—although this seemed to be a development that was beyond the therizinosaurs. They did, however, take the same convergent road to an edentulous mouth and a corresponding development of a beak, again suggesting a significant shift in food processing.[95] As with the birds, the skeleton was permeated with air sacs,[96] although within the skull things were also stirring with the brain and converging in a number of respects with that of the birds.[97] What is perhaps most striking, however, is not only the rapidity of evolutionary change from the basal types but also some striking convergences between the advanced therizinosaurs and the equivalent taxa in the next group to consider, the remarkable oviraptorosaurs.[98]

These convergences reinforce our sense that the evolutionary footsteps of the oviraptorosaurs (a slight misnomer as few, if any, were "egg-stealers") were closely superimposed on the therizinosaurs. It is hardly surprising, therefore, that they would pass many of the same evolutionary signposts pointing in the same avian direction. To label one oviraptorosaur *Avimimus* tells you quite a lot,[99] although true to form it[100] and other oviraptorosaurs[101] showed the customary mixture of primitive and advanced characters.[102] That, however, the development of avian-like characters was genuinely independent in the oviraptorosaurs is self-evident from looking at their most primitive representative. Known as *Incisivorous*, this beast was more like a general, all-purpose theropod, although as the name suggests its dentition is gratifyingly convergent on the incisors of various mammals, notably the rodents and multituberculates.[103] But in the more derived forms we see a trend to loss of the teeth so that ultimately the jaw is edentulous.[104] In some forms, it has a parrot-like beak.[105] And once again the tails transform into fused pygostyles.[106]

There are, however, a series of further and crucial signposts. For the first time we see feathers that are really plumes[107]—that is, pennaceous, with a shape that prefigures an aerodynamic capacity. Equally important, these feathers arose from both the "wings" and the tail. And in the latter we see another significant development: the tail began to shrink, and in doing so led to some remarkable consequences.[108] In birds, which have reduced their tail to a stump (such as the "Parson's nose" of a chicken), the muscles for walking and hopping are necessarily concentrated at the top end of the leg. In reptiles, however, a significant proportion of this musculature remains accommodated in the tail. As the tail reduces (and also becomes more rigid as a pygostyle), a rather odd configuration emerges with the knee playing the principal role in the bending of the leg. (So much will be obvious when a bird next hops past you.) In the oviraptorosaurs, the forelimbs were not only equipped with feathers but already showed the distinctive style of folding about the wrist that birds show.[109] Not only were elements of the skeleton again pneumatic, but the ribcage could possess distinctive projections (the uncini) that in birds are an important adjunct to respiration by giving a mechanical boost to the ribs.[110] The oviraptorosaurian neck was typically elongate, and the heads of some species had spectacular head crests, which in *Corythoraptor* was reminiscent of the cassowary.[111] Not only was the skull decidedly birdlike, but the brain (as revealed by endocasts) also had avian characteristics,[112] including a convergence in the reduction of the olfactory area[113] (perhaps oddly another convergence, this time with the waterbirds[114]).

What about the chicken and egg question? Skeletons preserved in an avian-like brooding position over oviraptorid nests[115] are consistent with birdlike incubation temperatures calculated using oxygen isotopes.[116] The arrangement of the nests is well understood, and nobody imagines they were entirely avian.[117] An important question, however, is whether in the usual reptilian mode the eggs were entirely buried as compared with being at least partially exposed. Here the fossil record can be unexpectedly informative. Eggs require gas exchange and so are porous. Unsurprisingly eggs in covered nests have a higher porosity than those that are open to the elements. Using this line of evidence it suggests that in the oviraptorosaurs the eggs were at least partially exposed.[118] This has been neatly complemented by the discovery of pigments in the shell. Some eggs were originally blue-green in color, which provided convenient camouflage when the parent was absent and confirms the nest being open to the skies.[119]

To repeat the question already asked of the alvarezsaurids and therizinosaurs: could any of the oviraptorosaurs use their many avian characters to full advantage and actually fly?[120] Given that this group were once regarded as birds[121] this is by no means an unreasonable inquiry. Despite some strikingly birdlike features, the aptly named *Gigantoraptor*,[122] weighing in at two tons and dwarfing the local tyrannosaurids,[123] would have been useless at the end of a runway. But major changes in body size, including miniaturization, were common; for example, the oviraptorosaurian *Yulong* was chicken sized.[124] Like their evolutionary predecessors, the oviraptorosaurs probably remained earthbound, but as we finally move to the deinonychosaurs (that is, the dromaeosaurids plus troodontids) and birds, we find that not only did volant forms evolve, but they evolved at least three times independently.

Or did they? Here, caution is warranted. The broad outlines of theropod evolution are robust enough (figure 4.8), but the devil is in the details. In particular, the all-too-familiar mixture of advanced and primitive features seen in most (if not all) taxa[125] throws continuous wrenches into the phylogenetic machine. A strange-looking, island-dwelling beast from Romania exemplifies these difficulties. This animal has been labeled *Balaur* (and unless your archaic Romanian is getting rusty, you will recall this means "dragon"). The question is whether this theropod is relatively primitive (and some sort of dromaeosaurid[126]) or, as is now argued, is a bird more advanced than *Archaeopteryx* but flightless on account of its island habitat.[127] And speaking of *Archaeopteryx* (figure 4.11), even this poster child of early bird evolution is not exactly secure. As a basal bird, this iconic fossil has become a lynchpin in the popular imagination as to the reality of evolution. Some, however, have seen *Archaeopteryx* as a basal deinonychosaur[128] and thus relatively far removed from the avian line. They both cannot be right. Much depends on precisely which of the characters in *Archaeopteryx* are deemed to be convergent, so other investigators have shooed this animal back into the phylogenetic coop labeled "Birds."[129]

From a wider perspective, all this chopping and changing is not that material for two reasons. First, as we have repeatedly seen, a great deal of what it takes to make a bird was assembled long, long before they took off.[130] So too, in this the last lap (so to speak) we see a reiteration of avian-like features such as a characteristic sleeping posture,[131] significantly larger brains,[132] and further hints of a kinetic skull.[133]

FIGURE 4.11 *Archaeopteryx*, the most iconic of the flying theropods.

Second, and more importantly, the dromaeosaurids, troodontids, and birds are all pretty similar. This, of course, is reflected in their decidedly wobbly phylogeny combined with the observation that the earliest birds do not form some sort of morphological outlier with respect to their theropod cousins.[134] To reemphasize, not everybody is going in the same direction, and not all species are going to turn into birds. The troodontids were more nimble than the dromaeosaurids and sought smaller prey, whereas the latter were more robust and probably more like ambush predators. It was during the handling of prey with clawed feet and the process of savage dismemberment that the necessity arose of maintaining balance by flapping the "wings." And this may have been a crucial prerequisite to actual flight.[135]

Plumage of one sort or another was by now de rigueur, and, sad to say, ticks had already struck up their blood-sucking association.[136] But feathers alone do not make flight itself an inevitability. The iconic *Velociraptor,* for example, possessed feathers (as is evident from the quill knobs, in the same manner as *Concavenator*) but was itself much too large to fly.[137] So too *Epidexipteryx* had feathers, but their shape meant that this animal was not in any condition to fly either. The greatly reduced plumage of *Eosinopteryx*[138] meant this theropod was also permanently grounded. But with so many birdlike forms waiting at the end of the metaphorical runway, this aerial adventure was rapidly becoming a certainty. One important ingredient in this process was not only a remarkable persistence in miniaturization of body, along with the skull becoming increasingly birdlike,[139] but one that happened on multiple occasions. Birds were in the air (so to speak), but exactly how many times? It is difficult to be sure, but at least three times.[140] Depending on which phylogeny you adopt and whether you are persuaded the anatomy is consistent with flight, then it was perhaps six times or maybe more. Each and every one is a missing link.

But in joining the dots it is sensible to have a packet of aspirins on hand. Excess ink has been spilled on debating the route(s) to flight: broadly, between running ever faster or falling out of trees. As already remarked, however, flapping to maintain balance may have been a vital step toward liftoff. This is consistent with the second digit of such theropods being equipped with a dramatic sickle-like claw, which has long been assigned to the melodramatic role of slashing and disemboweling although a more plausible function was to help grip the victim.[141] More broadly, the question of the origin of flight falls into the evolutionary conundrum of what is the point of half of something—in this case, a wing? Feathers for display and incubation before their co-option for flight are also surely part of the answer. Here, the activities of modern chicks provide a useful parallel. Although they lack, or have at best rudimentary development of, the key anatomical features necessary for flight, their proto-wings make an important contribution to running up slopes, controlled descent, and even fluttering episodes aloft.[142]

Deciding which theropods looked like proto-birds but never flew versus which actually ascended is not straightforward. Elongate forelimbs bristling with long pennaceous feathers may be important clues, but more particularly much has been made as to whether the latter had an asymmetrical outline and in principle could serve as an airfoil. The mantra, however, that if the feather was asymmetrical then the theropod it was attached to could fly may be too simplistic.[143] This is because the degree of asymmetry (simply calculated as the ratio between the respective widths of the leading and trailing edges) is generally much less pronounced among the theropods than it is in extant birds. Nor is this the end of the story because even among extant birds the feather form (including the morphology of the central rachis) may not be a reliable guide to their mechanical efficiency.[144] Other factors also need to be remembered.

In *Archaeopteryx* (figure 4.11) and at least some other theropods the arrangement of the feathers on the wings suggests a different sort of airfoil,[145] yet the skeleton was highly pneumatic, pointing to a bird-like system of respiration.[146] Nor, we need to remind ourselves, was there necessarily a single style of flight.[147] Nor need the prominent feathered tail have been an impediment to these activities, given its probable role in maintaining stability and aerodynamic control in these first steps to the mastery of the skies.[148] The tails of some dromaeosaurids also possess a series of bony rods running parallel to the vertebrae. These are strikingly convergent[149] with the arrangement found in rhamphorhynchid pterosaurs which in this group was integral to their flight. Their presence in *Velociraptor* could point to a secondary flightlessness, and equally these rods could be involved with other functions. But their presence in the microraptorines (figure 4.9) makes it an appropriate moment to turn to these remarkable theropods.

If *Archaeopteryx* is a cynosure of a feathered dinosaur, the microraptorines[150] come a close second. Principally this is because of a remarkable four-wing configuration complete with elongate feathers,[151] but also on account of their inferred iridescence[152] and gut contents pointing to the capture of an impressive range of agile prey.[153] As dromaeosaurids, the microraptorines are relatively remote from the birds and almost certainly evolved their capacity for flight independently. But were they any more than adept gliders?[154] Intriguingly, the feathers show what is known as sequential molting[155]—that is, they are shed when worn and damaged, as is characteristic of birds today. The forewings also possessed what appears to be a precursor of the so-called bastard wing (alula),[156] which in birds plays an important role in low-speed maneuverability.[157] Judging from the rib cage, microraptorines also had an avian-like respiration.[158] Nor is this the only example of a volant dromaeosaurid. Among unenlagiids we find another convincing example of flight evolving independently. This animal, known as *Rahonavis,*[159] was about the size of a raven but came with a sickle-shaped claw. Although only the skeleton has been found, the quill knobs on the wings point to the attachment points of former feathers.

So here we see three separate aerial adventures: the lineage that gave rise to the true birds, the dromaeosaurid microraptorines (figure 4.9), and the rahonavids. Did other theropods find the ground falling away from beneath their feet? It may never be possible to confirm unequivocally that these candidate fliers actually patrolled the skies. At any rate, if theropods took off three times, do a few more make that much difference? Remember Goldfinger's warning to Agent 007: "Once is happenstance. Twice is coincidence. The third time, it's enemy action." Three fliers will certainly suffice, but every additional example helps to reinforce the myth of missing links.

So let us meet the candidates. First in line is *Yi* (and very likely its relative *Epidendrosaurus*), as previously introduced, whose membranous wings more than hint at a separate aerial adventure. Frustratingly their bizarre appearance makes the phylogenetic position of the group to which they belong (the scansoriopterygids) difficult to pinpoint, but they may be close to the birds. Then there is the crow-sized *Anchiornis,* which was perhaps a primitive troodontid.[160] Both its body proportions and feathered arms, complete with a prominent fold of skin (the propatagium) on the leading edge, are indicative of at least a gliding ability.[161] To be sure *Anchiornis* lacked a sternum (upon which the flight muscles are usually inserted), but then so did *Archaeopteryx*.[162] In the latter context, the so-called Haarlem specimen is now reinterpreted as an anchiornithid (*Ostromia*) and was possibly volant.[163] And what about *Jianianhualong*? Again displaying striking mosaic evolution, this troodontid was unlike *Anchiornis* in that it possessed asymmetrical feathers, making it another candidate glider.[164] There are also two more possibilities. First was *Caihong*,[165] which was suitably small with feathers, some asymmetrical and longer than in *Anchiornis*, as well as a possible bastard wing; collectively, these characters are suggestive of some sort of flight. And second there was the unenlagiine *Buitreraptor*[166]—phylogenetically close to *Rahonavis*; conceivably was this the eighth time theropods took off?

THE REAL MISSING LINK

The notion of missing links, be they sarcopterygians clambering onto land or theropods ascending into the air, conceals more than it reveals. It was, however, the quest to determine what sort of beast might serve to link humans to their antecedent apes (figure 4.12) that gave this concept wide currency, even though the idea of missing links not only predated Darwin but found a wider employment beyond questions of evolution.[167] Even so, in the backwash of Darwin's *Origin of Species*, these were the convoluted discussions that accompanied, for example, the discovery of the Neanderthals, which provided the test cases.[168] This made things much clearer—or did it?

To be sure, the hominin evolutionary "bush" is being continuously decorated with new finds. The number of species could be in excess of twenty,[169] and in this and many other respects the hominin story is no different from any other evolutionary transition.[170] At least three obser-

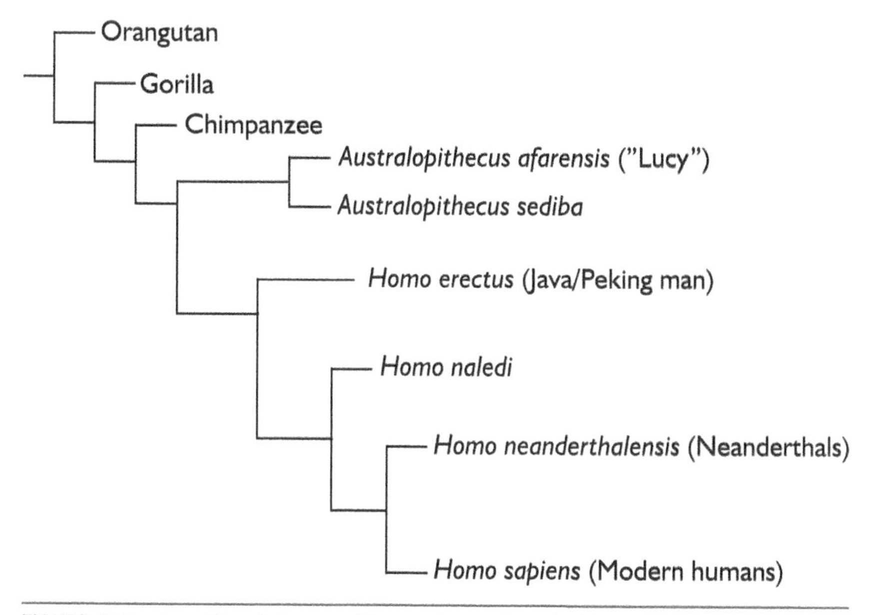

FIGURE 4.12 An outline phylogeny of the great apes, *Australopithecus and Homo.* (Numerous other species of hominin are known, but they are less material to the immediate discussion.)

vations bear this out. First, the journey to ourselves is far from being a simple monotonic trend, not least in terms of brain size.[171] Directly related to this is the extensive mosaicism—that is, in a given species the "puzzling" combination of advanced and primitive characters.[172] These apparent conundrums only serve to ensnare those who remain trapped in a cladistics hall of mirrors. In this strange world evolutionary convergences are at best a tiresome distraction rather than a clear pointer to foregone evolutionary destinations.[173]

With respect to evolutionary mosaicisms, both the australopithecine *A. sediba*[174] and much younger hominin *Homo naledi* provide particularly telling examples. In the former species the arms and shoulders point to arboreal excursions reminiscent of an orangutan.[175] On the other hand, the pelvis structure indicates bipedality,[176] although the manipulative ability of the hand would in principle allow tool manufacture.[177] Even so the hand shows mosaicism, as does the thorax

FIGURE 4.13 An unexpected addition to hominin evolution in the form of *Homo naledi*, not least in terms of its puzzling mixture of anatomical features and place of burial. Credit: Cicero Moraes (Arc-Team) et alii, http://arc-team-open-research.blogspot.co.at/2016/11/voi-chascoltate-in-rime-sparse-il-suono.html, CC BY 4.0, https://commons.wikimedia.org/w/index.php?curid=53285147.

where the apelike upper region is juxtaposed to a lower region more like that of a human.[178] This species is evidently just one of a series of australopithecine adaptive "experiments," some of which foreshadow the transition to *Homo*.[179] Mosaicism does not stop at this hominin boundary, but of the others perhaps the most remarkable was *H. naledi* (figure 4.13).[180]

With a likely date of between about 250 and 300 kyr,[181] *H. naledi* are coeval with ourselves and Neanderthals, but in contrast they show a disconcerting degree of mosaicism that both points back to australopithecines but also to advanced species of *Homo*.[182] Echoing the case of *Australopithecus sediba*, their arms suggest an arboreal proficiency.[183] So too do their elongate and curved fingers, but in principle the wrist could allow movements consistent with tool making.[184] But if we look toward the ground, both legs and feet suggest a striding gait.[185] So far, so odd, but oddest of all is that in their main locality—the Dinaledi Chamber, deep in the Rising Star Cave—aside from this mass of hominins there are almost no other vertebrate remains. What was going on here? The brain of *H. naledi* was relatively small,[186] but despite this could they have transported remains to form a deliberate accumulation? If so, what were they thinking? Did they glimpse worlds beyond this one?

Be that as it may, as with any story concerning the search for "missing links" the evolutionary counterpoint to mosaicism is convergence.[187] In tandem these two aspects show that despite the early hominins displaying a variety of evolutionary trajectories there remain common themes. Key in this regard is the capacity to engage in precise manipulation, and as significantly to become fully bipedal. Concerning the former, the employment of opposable digits has evolved multiple times,[188] but one of the closest approximations to the precision grip of the primates is found in the phyllomedusoid frogs.[189] Indeed, and as already noted, the harbinger of a thumb in the sarcopterygian fish is a pointer as to how evolution was going to get a grip on things. Bipedality has also evolved multiple times; although the fully fledged human stance is distinctive, again its evolutionary roots extend more deeply.[190] Variations in the foot structure of early hominins also suggest that bipedal walking came in a variety of flavors.[191] Undoubtedly, the overall theme is a striding out into new worlds, but this does not rule out returns to the trees.[192]

During at least the early history of *Homo* there were various forays into different adaptive zones, each with its missing links.[193] The genealogy that culminated in *H. sapiens* would have been no different, and nobody doubts the story of evolutionary continuity was shadowed by the insensible emergence of increasingly complex cultures.[194] Or do they? Woven into these narratives is a set of astonishing transformations: a capacity to think analogically, to link cause and effect, and to summon intangible realities, not least music and poetry. Contrary to received wisdom, from an evolutionary perspective this entry into a symbolic universe is deeply problematic (chapter 5).

Given the incompleteness of the evidence, this may seem a bold claim. This is because so much of what we would ideally wish to know is archaeologically invisible. When did you last see a sentence fossilize? That said, at first glance the evidence of our acculturation points to it being as gradualistic as our bodily evolution, and the idea of major step-changes, one or more Paleolithic "revolutions," now seems overblown. Pinning down the first stages of this process is certainly not straightforward, but from about 400 kyr onward there is growing evidence of not only fire[195] but symbolic thinking and a sense of deepening awareness.[196] Aside from new materials (such as obsidian) being transported (or exchanged) over more than a hundred kilometers,[197] and evidence of sea crossings of a comparable distance,[198] the most obvious indicators come from the increasingly widespread use of reddish pigments, broadly the iron oxides classed as ocher.[199]

The significance of ocher, however, is complicated by its multiple uses, potentially ranging from treatment of hides, as an ingredient for adhesives (including the hafting of weapons) (figure 4.14), and as (crucially) body ornamentation[200] and displays.[201] Each and every one of these uses implies a corresponding series of cognitive demands.[202] Other evidence, such as from the Qesem Cave in Israel, has revealed sophisticated hunting and cooking as well as butchering, albeit at a relatively crude level[203] (maybe not yet at the level of "Could you pass the salt, please?").

More tellingly, at least to our eyes, are the objects, such as the putative figurines.[204] In contrast to ocher, the snag here is that such objects are extraordinarily rare. Is this simply a failure of the archaeological record or a sign of repeated reinvention? One is grasping at intangibles, but perhaps in these remote ice age worlds the hominins were

FIGURE 4.14 A hallmark of human ingenuity.

FIGURE 4.15 The evolution of hominins is well documented, but where are we now going?

beginning to encounter worlds that are preexistent and abstract, orthogonal to everyday realities, revealing not only previously hidden orders but as importantly the numinous (figure 4.15). Nor should we imagine that this "unfolding" was a monotonic process, with these hidden "landscapes" majestically unfolding. Rather it would be like any other cognitive exploration, with unexpectedly early "breakthroughs" as well as detours that ultimately turned into cul-de-sacs.[205]

One clue in this direction are some puzzling mass occurrences of fossil hominins. Currently, in the day-to-day world of vertebrate paleontology such finds are usually the result of flash floods and similar catastrophic misfortunes, but not necessarily always. In a cave system in Spain (Sima de los Huesos at the Atapuerca site), a number of skeletons (dated to ca. 400 kyr) were found beneath an otherwise inaccessible shaft that has since been labeled a "sepulchral pit."[206] More intriguing still was the previously mentioned mass accumulation of *H. naledi* in an underground chamber that is remote and hard to access (at least today), occurring in sediments that are unlikely to be the product of torrential flooding.[207] That this occurrence could be a Paleolithic charnel house seems at least a possibility, although as noted earlier what weight of "metaphysical assumptions"[208] it can bear is less obvious.

Collectively these lines of evidence point toward hominins at first tentatively dipping their metaphorical toes into a metaphysical ocean, almost infinitely deep and one full of interlayered symbolic meanings. These both deeply resonate as to who we are and invite us into conversations with the unseen. But at this stage, where the links between ourselves and animals have become rapidly more and more tenuous before irrevocably snapping, can the notion of missing links still apply? Today we are unique, but is there a parallel story to be told of other hominins, not least the Neanderthals (figure 4.16)? Their cognitive competence is not in doubt,[209] but did they have the élan, the chutzpah of *H. sapiens*? If not, might that lack of a vital spark in turn explain their demise? The evidence is equivocal,[210] and of course one should be wary of a one-size-fits-all outlook. After all, the Neanderthals occupied various habitats—both open and woody, both near-tundra and clement—stretching across an immense area and subject to the vicissitudes of epic climate change.

Having said that, for the most part the evidence points to a near cognitive equivalence. Consider, for example, the walls built of stalactites located deep within a cave in southwest France, constructed in the form of circles (figure 4.17).[211] Such edifices would be impossible to erect without firelight, but intriguingly with respect to actual evidence for

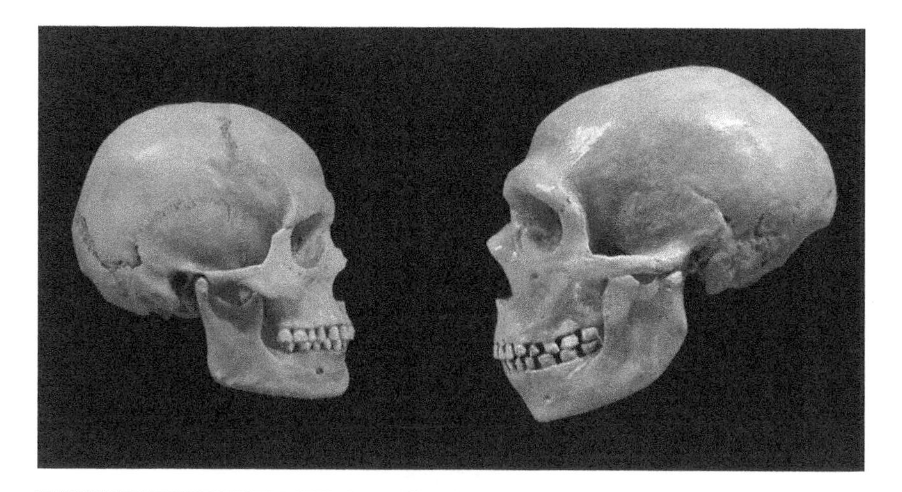

FIGURE 4.16 Similar but intriguingly different: humans (left) and Neanderthals (right) have much in common (and they also interbred), but maybe they looked at the world in subtly different ways? Credit: By hairymuseummatt (original photo), Dr. Mike Baxter (derivative work), https://www.flickr.com/photos/hmnh/3033749380/ (original photo), CC BY-SA 2.0, https://commons.wikimedia.org/w/index.php?curid=21289793.

fires most were placed on the circles themselves. Let us not invoke a Neanderthal Brunhilde; perhaps their construction was no more than a prosaic exercise. Nevertheless, these particular structures, dated at about 175 kyr, are significantly older than much of the other evidence that points to a high degree of cognitive capability among the Neanderthals.

Of all the cognitive benchmarks, the evidence for intentional burial has been particularly fiercely debated (figure 4.18). "Grave shortcomings" was the view of Robert Gargett,[212] yet not only is the evidence strongly indicative but in the famous case of La Ferrassie burials at least five of the skeletons show an east-west orientation.[213] Not necessarily unconnected is the evidence of cannibalism; although some see darker motives, in at least some cases the evidence points not to eating one's grandfather but to part of a ritual.[214] On a lighter note, a musical instrument in the form of a bone flute has been identified,[215] but more secure is the evidence for engravings[216] and perhaps cave art.[217] Some even see

FIGURE 4.17 Deep in the French cavern of Bruniquel are located these astonishing circles constructed by Neanderthals; certainly not accidental, their function(s) remain mysterious. Credit: Luc-Henri Fage/SSAC - Own work, CC BY-SA 4.0, https://commons.wikimedia.org/w/index.php?curid=54235636.

a Neanderthal "face," and maybe the object in question—a piece of flint into which a length of bone is wedged firmly—qualifies?[218] But where the evidence for Neanderthal symbolism has literally taken off is with luckless eagles and avian raptors (as well as corvids and other birds)[219] being stripped of their talons and prominent flight feathers.[220] Irrespective of the cognitive "depths" of this and other types of Neanderthal symbolism, it seems inconceivable that they were mute.[221]

Is, however, the evidence for cognitive equivalence in the Neanderthals quite as secure as it first appears? One bone of contention has been a protracted squabble over whether the cultural innovations shown by the Neanderthals, notably in the Châtelperronian culture, are genuinely independent of *H. sapiens* or the result of either direct contact or diffusion of knowledge.[222] The arguments go back and forth, although overall the case for independent innovation seems the stron-

FIGURE 4.18 Whether or not the Neanderthals actually buried their dead has been contentious, but examples such as these remains from the Shanidar Cave are strongly supportive of such a notion. Much of this picture (scale is 3 cm) shows the crushed skull of an adult, with jaws to the right and the eye sockets above. Further excavation found the upper body of the individual underneath the skull. Image courtesy of Professor Graeme Barker (Cambridge).

ger. But is this enough? One hint that Neanderthals had some sort of cognitive shortcomings concerns their clothing. Here there are two lines of evidence, and both are ingenious approaches. One looks at the not very attractive history of lice—more specifically, the molecular methods that have allowed us to estimate at what time in the past the type of lice that infest clothing diverged from those living in the hair. This gives us a clue as to when *H. sapiens* first started thinking about coats; the estimated divergence could coincide with the most recent ice age (ca. 80 kyr).[223]

The other approach is to consider the standards of Paleolithic couture. Especially when it comes to living in near-tundra conditions, some sort of covering is surely an essential, and there is indeed evidence. The microwear on specialized stone tools (known as lissoirs) is consistent with Neanderthals processing hides to render them supple and waterproof.[224] An added bonus to help protect from hypothermia, especially if one is concealed and waiting for hours for a potential dinner to turn up, is a cozy fur lining. An excellent source of fur comes from such animals as dogs and weasels. Intriguingly, bones of the same animals are common in many *H. sapiens* sites, but far fewer are found in locations where Neanderthals lived, although in other respects the osteological tally (e.g., deer and others) is much the same.[225]

Away from the cooking fires, therefore, Neanderthals may have spent a lot more time shivering. Even so, by itself evidence such as an absence of attractive fur trim on their parkas can hardly be taken as a knock-down argument for cognitive inferiority. How then might we assess the relative minds of ourselves and Neanderthals? Broadly, there are three approaches. The first is that differences exist but our cognitive states are effectively equivalent. The second is that Neanderthals could have had a genuine cognitive deficit,[226] so we should be cautious in automatically assuming equivalences in symbolic capacities. For example, the much-vaunted use of raptor feathers as objects signaling perhaps social status or shamanism might be an overinterpretation of simply an idiosyncratic behavior akin to the magpie pouncing on some glittering object.[227] More generally, even if Neanderthals carried out rituals, perhaps they were more individualistic than collective.[228]

In defining what makes humans cognitively unique (chapter 5), episodic memory and so-called mental time travel may be important markers. If, however, these capacities were only partially developed in the Neanderthals then this might have stymied a fully human cognition.[229] From this perspective, one might go on to suggest that a cognitive deficiency might enforce cultures based on emulation rather than imitation (whereby the logical steps are comprehended), a language but one lacking in recursion, at best rudimentary music, and all in all a stifling of innovation.[230] By no means is everyone persuaded[231] that Neanderthals dwelled in a sort of mental dead-end, and other evidence points to the Neanderthals having language much the same as ours,[232] including types of consonants consistent with conversations.[233] Might one or two of our words even have Neanderthal roots? Either way, aside from what many regard as convincing evidence for symbolic activity, inferences made about their social structure, cooperation, and possible hierarchy[234] are less consistent with the Neanderthals living in a cognitive twilight.

The third possibility is potentially the most interesting: Neanderthals and *H. sapiens* occupied distinct cognitive zones that were not necessarily equivalent.[235] Whether or not the respective organization of the brains was significantly different as a result of different cerebral trajectories is disputed,[236] although the emphasis on visual cortex at the possible expense of other cognitive centers is controversial.[237] The other evidence might cut either way—literally. A recurrent characteristic of some Neanderthal teeth are striations on the incisors and canines. These are consistent with the mouth being used as a "third hand," serving to help to grip animal hide or food. Humans too routinely use their mouth to hold things, but not evidently to the extent of the Neanderthals, where the striations reflect extensive manipulations.[238] The directions of these cut-marks can also be used to infer the preponderance of right-handedness[239] as well as differences in habitats that call on dealing with either hides or chunks of meat (or other foodstuffs).[240] Using the mouth as a third hand might in turn have had an impact on processes of "visuospatial integration" and correspondingly had potential consequences for cognitive architecture. Conceivably this could ultimately determine access to symbolic worlds.[241] From our perspective it has been tempting to view Neanderthals as clumsy cousins, but an alternative tack is to suggest that in reality this behavior demanded refined motor skills that in turn nourished an alternative cognitive world where processes of abstraction were more "virtual" and where external symbols took second place.[242]

Such a view strongly resonates with the overall theme of the myth of missing links. In other words, there is not a monotonic succession of transitional forms but rather an evolutionary "bush" where limbs, wings, and advanced cognition (think respectively of sarcopterygian tetrapods, theropod dinosaurs, and *Homo*) emerge in a variety of configurations and are effectively written into the equations[243] to achieve very much the same set of outcomes. Thus, proto-limbs among sarcopterygians and wings in theropod dinosaurs show variations, but the end points are to walk on land and to fly in the air. The same may apply to the cognitive worlds of *H. sapiens* (and perhaps the Neanderthals), discovering orthogonal abstract realities. From our currently limited perspective such realities may look distinct and the interconnections seem largely intuitive. Individually they appear to represent separate doors that invite us to an infinitely large mental "hyperspace" (figure 4.15). These abstract realities are entirely new worlds, mysteriously closed to animals or at very best the subject of the vaguest intuitions.

As we will see animal vocalizations are not language, their numerosity is not arithmetic, and their music should not be confused with our music. Early in the Paleolithic period, access was most likely highly sporadic, perhaps limited to only a few gifted individuals. Nor should we imagine that there is any limit to this exploration. Correspondingly there may be cultures that are more attuned to this adventure and others who turn away. Either way, on this planet we are the first fruits. In principle the same should apply across the Milky Way and all points beyond. And such may supply an unexpected explanation for the Fermi paradox: there are "extraterrestrials" but not in any way remotely like the ones we would expect.

5

The Myth of Animal Minds

I look into the animal's eyes, and it looks back. We parted company in the trackless forests of preglacial Africa. Millions of years of evolution now separate us, yet our minds appear to meet. Are not the differences paper thin? Was not Charles Darwin a scaled-up ruminating chimpanzee? And what of Rousseau's Noble Savage? Yes, he based his romantic ideal on an inhabitant of the jungle, a pacific fruit-eater with undemanding sexual habits, but as Robert Wokler[1] has explained, too seldom is it appreciated that the role model was not some Arcadian native but the orangutan. Differences exist, but surely they are of degree, not kind? No animal, for example, has language, even if cynics say that the reason they remain mute is lest they be put to work. At any rate, is lack of language such an impediment? Here is Bailey the Yorkshire Terrier, and he is a smart little chap—at a single command from his female owner ("Bailey, fetch Rhino!"; or whatever the toy is called) he rushes off, then triumphantly returns with the correct toy in his chops. And not only that: if we substitute a male voice or a different dialect, Bailey is unfazed. Surely Bailey understands what I mean? As we will learn below, that is unfortunately not the case.

To suggest that our cognitive world is not directly derived from the apes is a tautological no-brainer.[2] If evolution links us so securely to the primates, how can that apply to skeletons and DNA but not the mind? Convergent evolution also shows how remarkably similar cognitive solutions bubble up repeatedly, not only in the crows and dolphins but at least as far afield as the octopus. These observations and a slew of experiments—both among the laboratory cages and deep in the forest—interrogating crows, chimpanzees, dogs, and baboons (and let us not forget the slime molds[3] and mushrooms[4]) have reaffirmed that behind the muzzles and beaks (along with those amoeboid plasmodia and mycelia) reside the nascent stirrings of the human mind. Humans are very smart, but as Darwin insisted their intelligence is one of difference not of kind—or at least this is the mantra repeated endlessly.

AN UNBRIDGEABLE GULF?

On this question of intelligence, however, some would beg to differ. Fancy entitling your paper "Darwin's Mistake," then adding as a subtitle: "Explaining the Discontinuity between Human and Nonhuman Minds." Such an exposition would be career suicide for a young investigator, but in the skilled hands of Derek Penn and colleagues it provided the clarion call.[5] Given our patent evolutionary continuity with other animals, why then are we so utterly different? The responses to the essay by Penn and colleagues were more often tetchy than positively critical, and since then not much has changed. But the evidence is overwhelming: we have permanently abandoned the Darwinian cognitive zone.

Jerome Bruner reminds us that in any taxonomic sense we are no longer a species.[6] Or to move from his particular emphasis on culture, if we are self-designatedly *Homo sapiens*, so too we are many other "species": *academicus, bulla, clausus, credente, deus, discens, docens, duplex, faber, fabricans, fabulator, fabus, ferus, grammaticus, heuristicus, ignorans, imitans, juridicus, luminens, loquens, mimeticus, mortuus, mutans, mysticus, negotiator, noeticus, oneginensis, passiens, prometheus, religiosus, ritualis, sacer, scientificus, sovieticus, symbolicus,* and *ventosus,* not to forget *viator.* But above all, we are *Homo narrans*—only we tell stories.[7] Whatever else this pseudo Linnean taxonomic list hints at, it is how difficult we find it to step out of our skins: humans are exceedingly odd. The remark by Theodosius Dobzhansky, "All species are unique, but the human is uniquest,"[8] finds many echoes. Thus, Thomas Suddendorf pronounced: "We are peculiar indeed."[9]

And it is not only a question of how we stepped into worlds of self-knowledge. The number of human capacities may seem to be relatively small, but they are infinite in their ramifications. If the chimpanzees survive, in a million years' time we can be confident that they will still be termite dipping,[10] equally unaware they have had a history and still with not the smallest inclination to instruct the next generation[11] about of who they are, where they came from, why they matter, or what the future holds. The questions of why humans are so utterly different, when it happened, and under what impetus are as widely debated as they remain obscure. In more candid moments there is agreement that the gap between ourselves and other animals is "mysterious." The actions and capacities that in principle should link us to other animals, be it by common descent or by analogy through evolutionary convergence—and thereby expose our cognitive foundations—turn out on closer inspection to be curiously uninformative.

Just as puzzling, the capacities that make us human are intimately intertwined and very deeply rooted (figure 5.1). It is almost as if we were parachuted onto the planet.[12] We were not, of course, but it is equally difficult to show that one capacity is the direct product of another. Tellingly, it is possible to arrange the list of cognitive acquisitions in almost any historical order you like and end up with an equally compelling series of interdependencies. To complicate matters, the traces of this historical construction (and perhaps in cases such as language there never was a trace) are almost entirely erased. One can find no center in this web of connections; everything seems to be interlinked, without any obvious first cause. Repeatedly we see also that the various capacities are already latent in very young infants.[13]

This holistic framework of cognitive capacities is also remarkably robust to insult. A stroke might impair one or another cognitive

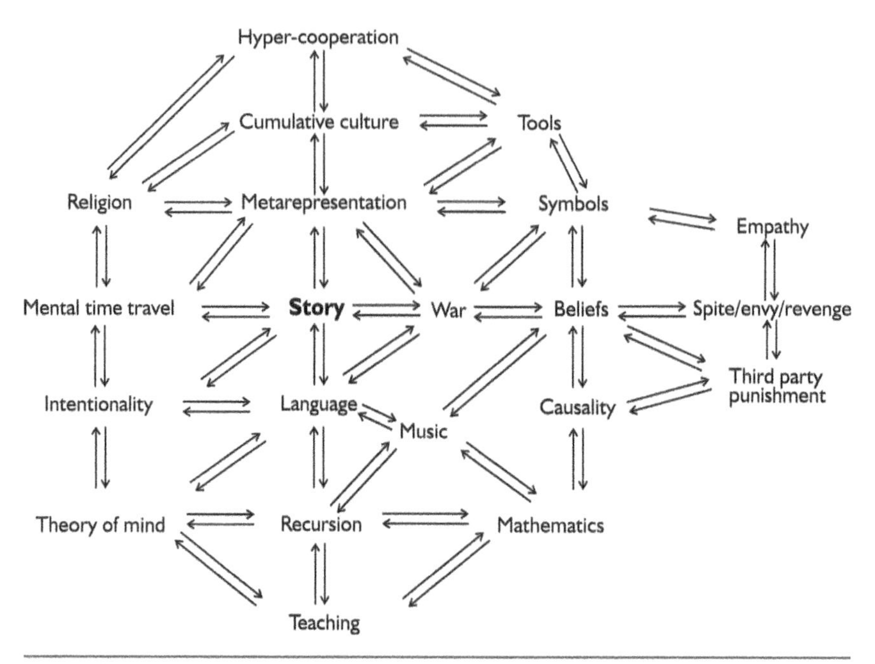

FIGURE 5.1 The cognitive architecture of humans, depicting the range of capacities that define human uniqueness and their interrelationships. Everything depends on everything else, but in cases of damage (such as may follow a stroke), the system remains remarkably robust. Central to this architecture is the capacity to become mythopoeic and tell stories that speak to invisible realities.

function, for example, resulting in a deeply debilitating aphasia (for grammar, or even words). Testing such aphasics is challenging, yet even though the patient can no longer express a key aspect of human cognition—say, the ability to judge what other people's motivations and mental states might be (that is, the theory of mind)—the capacity itself remains intact.[14] Language itself is immensely powerful because it literally articulates ideas, but they and the associated gallimaufry of cognitive capacities that enshrine causal reasoning and theory of mind lie far deeper and stay operative in the face of serious injuries.

And that is why we should be suspicious of explanations that give primacy to one factor (such as tools, language, bipedality, dexterity, or group size) or that appeal to a genetic mechanism that in one way or

FIGURE 5.2 So human aren't they? But as this chapter shows, the gulf between ourselves and chimpanzees (let alone any other animal) is vast, possibly unbridgeable.

another will provide that long-sought-after launch pad.[15] Nor should we regard chimpanzees as watered down humans who just need a bit of beefing up. When Daniel Povinelli and Jennifer Vonk ask if chimpanzees are "suspiciously human"[16] (figure 5.2), their question is more than rhetorical. Equally trying to persuade ourselves that animal A has this nascent cognitive capacity or animal B another one is to miss the point. If beast A or B had language or any other part of our cognitive battery (figure 5.1), it would have *all* of them, either directly expressed or with the obvious potential for expression. And yet the gap remains: we are the "human enigma."[17]

PLAY IT AGAIN, AND AGAIN, AND . . .

Here is another thing that should give us pause for thought. Because animals do not have language, in nearly all cases training is essential. The resultant differences between ourselves and even our nearest kin—such as those that pertain to memorizing sequential information—

are startling.[18] Less often remarked on is that such instructions are not only exhaustive but exhausting. Ironically this tells us much more about ourselves than our animal counterparts: back in the jungle they show not the least inclination to engage in such frivolities. In one case, involving baboons, it took more than 35,000 trials before a satisfactory success rate was achieved (and even then it was still less than 100 percent, and that was only for one task[19]). In a related series of trials,[20] twenty-nine baboons reported for duty, but only six of them got the hang of things, and nearly all were juveniles and only one a female.[21]

One must acknowledge also that from the animals' point of view some setups are decidedly artificial, but in any event one's admiration for the experimentalist can only increase given that repeatedly apparently watertight demonstrations of one or other cognitive Everest turned out to have a fatal flaw. Equally confusing, even slight alterations to a particular setup can radically change the success rate. To make things worse training routines typically take months, with only passing mention of the animal duffers who have been discreetly retired, along with those stalwarts who nevertheless ultimately lose motivation.[22] Indeed, not only are the animal duffers quietly pensioned off, but as often as not there are only a handful of stars.[23] Even Roger (or whatever the animal is called) may shine in one experiment but flounder in the next.[24] Marooned in captivity, these subjects may be rotated from one set of tests to another, year in, year out. What a life.

It is difficult to avoid two conclusions. First, among the welter of experiments it seems uncertain whether there will ever be a definitive, indisputable demonstration of humanlike cognition in any animal. We may see hints and tantalizing similarities, along with weasel words like "proto," "minimal," and "basic," but unequivocal evidence? Second, regardless of what position we hold, we must acknowledge the ideological baggage that comes with any research program. To many researchers, is it not self-evident that humans are the "third chimpanzee"? Is it not blindingly obvious that the basis of mind is material, and accordingly mental abstractions can only be our inventions? In reality, each and every stance comes with a series of default assumptions that usually are tactfully concealed. But they are not always hidden: look out for those chirpy, self-congratulatory pieces that are strident in their assertions and oddly free of self-doubt. Also watch out for similarities being elided into continuities, or a given capacity being taken as a "nascent" expression of humanness that needs no more than a nudge to find full expression.

Overhanging all these endeavors is the ever-present danger of anthropomorphism. If it is not bad enough seeing ourselves as the pinnacle of evolutionary achievement, in addition we risk the grave sin of anthropocentrism. Avoiding that blunder we must then be vigilant not to stumble into an unwitting chimpocentrism[25] as we confuse what characterizes modern-day chimpanzees with our common ancestor lurking in the late Miocene rainforests. Trotting down these winding paths, eventually we may lose all sense of bearing as we become lost in the dense thickets of anthropofabulation.[26]

Nor do the problems end there. For example, we have the story of Clever Hans, the horse who appeared to be an equine mathematician (as well as engaging in other party tricks) by indicating the correct answer by tapping his hoof. The truth was more prosaic; Clever Hans's undoubted skills arose from his being superbly attuned to the almost unnoticeable cues of his owner.[27] So in experiments in animal cognition every effort is made to avoid unintentional cues. Less often remarked on is that animals raised in human company are conspicuously smarter than their native counterparts. That is hardly surprising if one thinks about it, but even as they become more skilled they show no sign of beginning to grasp intentionalities.[28] Nor do they become more human in any other way, let alone more cultural.[29] If they succeed in tasks that their wild compatriots in the jungle show not the least inclination to pursue, what exactly are we learning? Chimpanzee and bonobo "geniuses" like Ai, Kanzi, and Sarah may indeed come up with the goods, but this only comes after years of human company and usually endless training. Yes, the animals have learned something, but more importantly have we?

A CROW CALLED ARCHIMEDES

Perhaps this point of view is much too pessimistic. Dogs are not necessarily animal Einsteins, even though, as Louise Barrett remarked, "I regularly speak to my dog as though she has a degree from Harvard."[30] Surely, however, we cannot dismiss those brainy crows replaying Aesop's ancient fable of the thirsty bird by dropping stones into a half-filled pitcher to raise the water level sufficiently to obtain a drink.[31] Now we fast forward to the laboratory, where instead of a pitcher the crows are provided with a water-filled tube with a tasty morsel out of reach. Dropping in stone after stone, the crow achieves its treat. Surely the bird is putting together two and two, linking cause and effect? And is there not an added bonus that in the wild seldom, if ever, do the birds encounter conveniently located urns half-filled with water with tasty tidbits floating just out of reach? Well, as it happens they are partly primed: many birds habitually carry around pebbles, and conveniently they seem to prefer the larger ones.

With the Aesop's fable experiments, the permutations are many and ingenious. How do the crows behave if the tube is filled with sawdust or if the "stones" are hollow and happen to float? These and other modifications, such as substituting a narrower tube or devising a U-tube configuration,[32] aim to tease out whether what the crows are engaged in is underpinned by a causal understanding. If so, are they not on the very threshold of Archimedean physics? Here, and across the realms of animal cognition, the fundamental question is whether a species can anticipate an action, see in its mind's eye what ought to happen, and set in motion the necessary chain of events. In the case of the Archimedean crows, experimental methodologies and subsequent statistical analyses of the various trials have presented various challenges, but it is clear that the birds have hit the cognitive buffers.[33] Yes, they learn that by dropping stones the water level rises, but they arrive at this result by a trial-and-error basis. Attentive to feedback, they are not thinking things out from first principles. It would, for example, be logical to select the larger stones to accelerate displacement, but they fail to do this. That the causal connections elude them becomes even more obvious if a counterintuitive arrangement is devised that teases out whether the birds grasp the function of the stone and what it actually does, set against their simply being driven by the expectation of the reward. Again there are no prizes for the outcome. If we change the context somewhat, we can now ask the crow to choose stones on the reasonable assumption that the heavier the stone the greater the likelihood of success—we know this, but alas the crow does not.[34] Fundamentally the problem is not simply that success is only achieved by trial-and-error learning but that the crows treat each task as unique and so are incapable of transferring the information that would allow a general solution.[35]

Hope springs eternal: perhaps the deficit is in the experiment rather than the mind? So let us devise a rather different sort of test.[36] Here, we allow the crow to pay close attention to a particular sequence of events that, once successfully achieved, releases a morsel to a nearby crow. In the wild, this sort of observation goes on all the time. After all, if the young did not pay attention they would be in a sorry state. The next step in the experiment, however, is to inquire about whether the observer can then redeploy this knowledge in a new context. Two-year-old infants achieve this easily; by correcting for false moves, they show that they clearly understand the chain of causation. Not so with the crow. Even when the crow observes the same set of actions one hundred times, still it fails in this task of extrapolation. Success is possible, but only if each step is duly rewarded once it has been successfully achieved. The birds are not dim, and they observe what is going on, but they do not have insight or imagination. Crucially, alternative possibilities or counterfactual worlds are literally unimaginable to them.

Other cognitive hurdles point in the same direction. A favorite test is the so-called string-pulling experiment.[37] Here the bird (or other animal) is invited to recover the reward placed in an otherwise inconvenient location, often at the end of a dangling string. Not all species—or individuals—get the hang of the process, but some show an impressive dexterity that is taken to reflect a cognitive prowess. But time to change the setup. Let us impose, for example, a visual barrier[38] or alternatively place the string on a flat board and cunningly disguise the fact that it is now broken. Many other variants can be employed, but the conclusions are much the same. The crow fails to grasp the concept of connectivity; again the notion of cause and effect eludes them.[39] Having no insight they cannot construct a mental scenario.

Again and again, what at first sight looks uncannily like human behavior is actually drawing on simpler mental mechanisms.[40] What applies to crows finds equivalence across a diversity of animals. Repeatedly we infer that their understanding of causalities is generally straightforward so long as it is observed—that is, it is perceptual. Transferring this knowledge to new contexts, however, is a very different proposition. It is far from clear that animals have the least clue as to why things work in the way they do.[41] Free of insight, paradoxically blind to unobservable causal mechanisms, and with the door to analogical reasoning firmly closed, no animal will be joining the metaphorical dots.[42]

THEIR GLASS CEILING

So many and various are the experiments in animal cognition, so ingenious are the attempts to make them watertight, and so diverse are the range of organisms tested, that to argue on the basis of the handful of previously described experiments that humans have access to worlds that are shut to even our nearest relatives or cognitive avatars such as the crows would seem to be frankly incredible. Or is it? Darwin established beyond doubt the evolutionary continuum that links us to all life, both proximate and remote. One escape clause is to argue that the cognitive capacities of animals serve the purposes that match their lives.[43] Each is distinctive,[44] each is different, and that is that. Wings do not make a feathered ape any more than flippers make an aquatic ape. Our bodies are not their bodies; diving a kilometer deep in a submarine canyon is not the same as needing to remember the thousands of locations in a forest where food has been cached for future use.

So we are "just another species"? Emphatically not! Particular species do some things very well indeed, but despite massive encouragement other tasks are either immensely difficult or as likely impossible for them. They are context specific; in contrast, we are context general. The ceiling through which animals can never see—and, correspondingly, the one that to us is transparent—is not a figment of our imagination. We see the world differently. As David Premack[45] has pointed out, a chimpanzee may use a branch but does not see it as part of the adjacent tree, any more than a crow carrying a piece of wire sees the same in the enclosing wire fence. The real challenge is not to explain why we are unique but why humans are alone. Our task would be much assisted if we could seek the advice of an equivalent extraterrestrial (if there are any; see chapter 6). Just because the gulf remains and the supposed bridges are illusory, this does not rule out humans on occasion having recourse to lower level cognitive processing. Nor is it to deny that many animals display considerable behavioral sophistication, albeit usually in very specific contexts.

To give just one example: is not the hoarding of stones and then occasionally hurling them against particular trees a sign of emerging symbolism in chimpanzees?[46] As ever, the challenge is in recognizing genuine intentionality, as compared with an offshoot of a mundane behavior. Thus, chucking rocks at trees may be just part of a typical male display rather than the first step on the royal road to painting and sculpture. The challenges of cognitive interpretation are evident because so many of these actions are easily and readily put into human contexts. For example, it is tempting to test our primate cousins with tokens to be used in exchange for tasty morsels. From our perspective, this, of course, carries all sorts of implications for added value, postponement of enjoyment, and naked greed.[47] And for the primates? As is true so often in the arena of animal cognition, the results are "suggestive" but hardly persuasive. To go on and then plead that the chaps in the cage are suffering a "cognitive overload" simply begs the question.

This is not to deny that animals like chimpanzees can show a considerable degree of cognitive sophistication. To see this, however, as a segue of human capacities looks like special pleading;[48] in reality their worlds are extremely circumscribed. The chimpanzee sees and knows its compatriots also see, but their respective mental states remain closed books. By contrast, our abilities to read other minds can seem almost telepathic.[49] It is not an invariable skill, of course—and thereby lie rich seams of farce and comedy—but more darkly, it permits almost unfathomable descents into a psychological abyss.[50] So too animals may point, but they fail to comprehend it forms a gesture. With us, a single body movement can in different contexts convey a multitude of alternative meanings. Referring to the iconic naked figures on the plaque (figure 5.3) attached to the Pioneer 10 spacecraft, Jonathan Marks amusingly reminds us that an alien might read the raised hand of the man as "No entry to our nudist colony,"[51] while here on Earth the figure's direct stare would go down very badly with an alpha male baboon.

Emphatically this is not to deny that chimpanzees have intelligence and some sort of imagination, but the limits are very severe.[52] Accordingly, any attempt to extrapolate a particular piece of knowledge to an analogous set of circumstances would be doomed to failure. Claims for such capacities in animals have indeed been made, but it is just as likely that the supposed "reasoning" actually depends on perception of present/absent and same/different rather than a genuine *appreciation* of deep-seated analogies. To think prospectively, in other words, to be aware there is not just a causal explanation but one that is "hidden," which may itself be tangible ("Got it! New batteries!") or

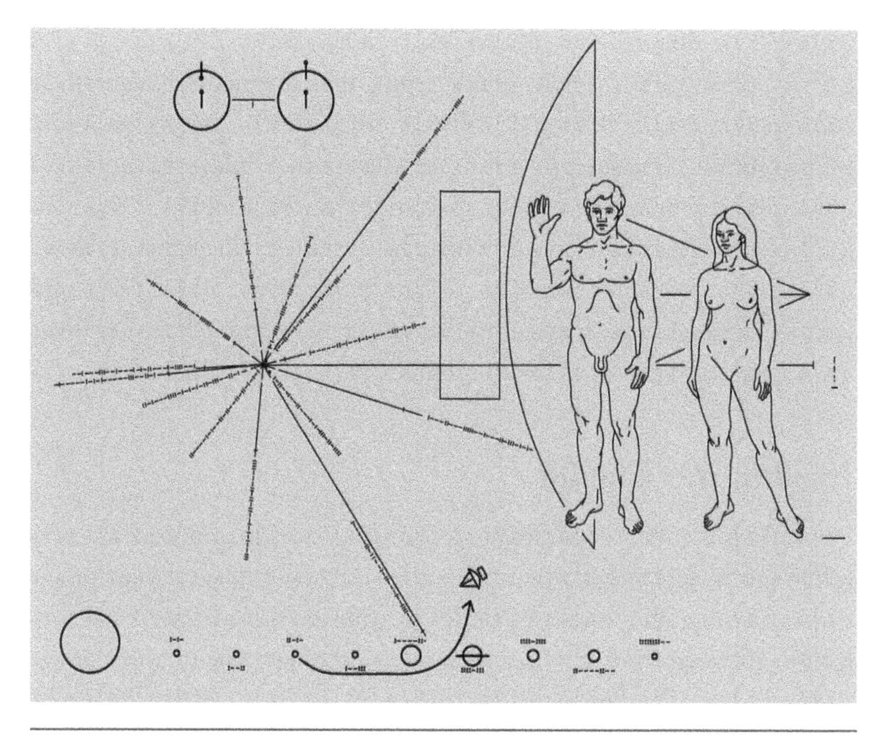

FIGURE 5.3 Now leaving our solar system, this plaque on the Pioneer 10 spacecraft optimistically may be our first greeting to an extraterrestrial, but what conclusions might they draw? Credit: Vectors by Oona Räisänen (Mysid); designed by Carl Sagan & Frank Drake; artwork by Linda Salzman Sagan—vectorized in CorelDRAW from NASA image [1], https://commons.wikimedia.org/w/index.php?curid=1433765.

abstract ("Silly me, try the square root!"), is quite beyond the chimpanzees' grasp.

For us, seeing is more than observing because it is a mental act and potentially is always creative. Moreover, it is our ability "to see the other side of the hill" that allows us to assess and juggle the abstractions that underpin our entire existence, using a range of cognitive capacities that find few counterparts among animals.[53] Other than metaphorically, and to paraphrase Derek Penn and Daniel Povinelli,[54] we can speak but know that we cannot smell an adjective, have eyes but will never see a thought, have fingers but never one that grasps an emotion. This should not be a cause for premature congratulation. Edwin Abbott's *Flatland*,[55] with its two-dimensional inhabitants, is a less whimsical story than it first appears; aside from its gentle irony, it is a somewhat unsettling commentary on the potential reality of other dimensions. What might we be blind to? Suppose, for example, Hermann Minkowski's world of four dimensions (where neither space nor time is absolute) is the reality?[56] Here, rather than time flowing past us with its uncatchable "now," it is we who encounter events as we "travel" along our respective world lines. Minkowski's geometric insights might open doors that to animals are not so much closed as for all intents and purposes nonexistent.

THE ONE-WAY MIRROR

What then is it that separates our cognitive world from that of every other species? The usual mantra is that the self-evident evolutionary continuity demands a correspondingly unbroken connectivity in cognition. However rudimentary in expression or specific in application, the harbingers of advanced cognition (not least tool use, vocalizations, planning, apparent prosociality, and even culture) prefigure the human condition. Darwin was in no more doubt about our cognitive antecedents than he was that human predominance was a direct result of our intellectual faculties. Two cardinal problems, however, are usually overlooked. Given these adaptive advantages, then why has this capacity only evolved once, even though advanced intelligence is rampantly convergent? The other argument for cognitive continuity is the claim that we all inhabit the same world. We do not—or rather, we and all life occupy one world, but we alone dwell in others. Can this correct? After all, are there not cognitive strands that bridge the gap between beast and man?

Do not chimpanzees cooperate (figure 5.2)? For example, they patrol their boundaries, warn the unaware of clear and present danger (snakes[57]), and will recruit the most effective collaborator[58] (but we will return with other cautionary tales). And to give the most famous of such examples, is not the formation of male coalitions a convincing sign of cognitive competence? To be sure, among primates generally there is little evidence for such a link,[59] but the case of hunting by chimpanzees appears to be an open-and-shut case of humanlike intelligence. Indeed, given our evolutionary proximity, how could this activity (or, as noted later, the use of stone tools to crack open nuts) be anything but a pointer to how we became fully human? In the case of hunting, what is not in dispute is that in their natural habitat chimpanzees routinely coordinate their activities, nor does anyone question that the sharing of meat may have more to do with coalition building (in the Gombe chimpanzees) than a nutritional top-up (as in fact seems to be the case in the Tai populations).[60] But the fundamental question remains: are the chimpanzees genuinely cooperating, communicating their intentions ("No, no! Fred! Over there! Pass me the gun . . ."), and putting their heads together to agree on the best strategy? Alternatively, is it simply opportunistic—merely a mob of adventurers, for whom the greater the numbers, the higher the chance of success, falling into line behind whoever happens to take the initiative? All the evidence suggests the latter.[61] There is no plan, no concept of a plan, and no execution of a plan. The contrast with humans could not be starker. With us, cooperation, reciprocity, and shared intentions—not only employing language but also the apparently mundane process of pointing[62]—allow common goals, with an unexpected consequence of allowing us to entirely rewrite the laws of biology.[63]

Nor is this all in the head. Our uniqueness is also embodied, not least in our movements (think of dance;[64] not "dancing" bears, who do no such thing,[65] let alone the "waggle dance" of bees[66]) and technologies that find no counterparts in animals. Our roots are Darwinian, but ironically (and only humans are capable of irony) we find ourselves alone. So at odds is this view with the received wisdom that it is essential to stress that animals are not automatons. They have minds but not our sort of mind; or to be more precise, our minds entirely encompass and transcend theirs. They are capable of abstractions, but only if it is based on what they perceive.[67] So their world is free of concepts, beliefs, and propositions.[68] A mouse knows full well there are cats (or owls), but it does not have a clue that there is the category "cats" (or Persians or Manxes, let alone lions). Without concepts there can be no extended thinking, let alone recognition of analogies.

To be sure, the claims for animals possessing some sort of capacity for analogical representation exist, but typically these abilities only manifest themselves after intensive training.[69] Animals know and understand, but neither know that they know nor that others understand. That is, they cannot "mind read": be aware that the other is an independent agent with a mind of its own.[70] So too we should be skeptical of so-called mirror self-recognition,[71] not least when even cognitive stars such as the kea fail to make the grade.[72] Suspicion deepens when we find that the success rate among enculturated animals is greater than among their "ignorant" counterparts in the wild, but even here the failure rate is high.[73] If animals had notions of self-concept or were capable of mind reading, they would no longer be beasts. They would be us.

Animals may appear beneath the proscenium, but because they lack any sense of agency never can they be actors, never learn to take a bow. Tom Suddendorf and Michael Corballis[74] used the theater as an apt analogy for the scope of human cognition, be it the imaginative range of the stage and its changing sets (incidentally a word with more than fifty meanings), the narrative, the rehearsals, the director's tantrums, and of course the audience. By contrast, animals reside (or from our perspective are trapped) in the functional domain. They can use cues to take an appropriate action, but because the causal significance is invisible there can never be a thing we call "explanation." Remove the cues and the chimpanzee is lost. Abstractions exist, but these are irretrievably linked to what is perceptual. To extend these capacities to a level of representations, to employ concepts and then redeploy them in the unlimited ways, what we choose to do is literally inconceivable to animals.

Chimpanzees will communicate,[75] but when it comes to working together they fall silent.[76] Fred the chimpanzee can certainly cooperate in all sorts of other ways but only with his chums—if the other is unfamiliar, then its welfare is of no concern.[77] Correspondingly, the female chimpanzee will be attentive to her offspring but never grasp that she is a mother. Equally, a chimpanzee may learn how a chunk of rock is a more effective nut-cracking tool than a piece of wood (and the heavier the stone, the better[78]), but never will they conceive of putting this notion into the context of force. An animal may observe an apple fall from a tree, but never will it infer that a fruit has yielded to the invisible force of gravity. Animals see but do not reason,[79] and most likely the distinction between accidental and intentional eludes them. No animal could ever become a scientist. It is not only that they have very limited access to rationality, but they lack the imagination that is essential for any scientist. Only then is there the counterintuitive realization that so-called common sense may be nothing of the kind. Would that some social scientists realized this.

Apes might engage in deception (although much of the evidence is anecdotal[80]) and be adept at employing a false negative. But a false positive is almost entirely beyond any animal's grasp. Their lives are rule bound, but they themselves could never articulate the rules. Thus, much is made of false belief, the notion that not only is there a reality but one that we can represent and as crucially misinterpret. Such is patently present in young children.[81] In contrast, attempts to demonstrate false belief in animals have been an endless, and for the most part unconvincing, saga.[82] It is not that animals are ignorant of their neighbors, but as Kristin Andrews notes, "Apes track false beliefs but might not understand them,"[83] a thesis echoed earlier by Juliane Kaminski and colleagues, who noted, "Chimpanzees know what others know but not what they believe."[84] So a chimpanzee will realize that its neighbor has a different perspective but is quite incapable of stepping into his companion's shoes and so adopting this perspective.[85] So not surprisingly they can observe and evidently understand a human action but are incapable of any sort of imitation.[86]

In a nutshell, and the metaphorical formulations of "stepping into shoes" or "a nutshell" underlines this gulf in cognitive apprehensions, we alone are capable of higher-order representations. Crucial to this enterprise is first the emergence of "analogical abstraction in three-month" infants,[87] followed in due course by the construction of intricate analogies. Here we see a virtuoso performance of melding, blending, fusing, and undoing apparently disparate categories, with the most gifted seeing not just the analogies that all of us employ but going deeper still, to find analogies between analogies. As Douglas Hofstadter and Emmanuel Sander remarked, analogies are "the fuel and fire of thinking."[88] Such thoughts can be completely independent of perceptual realities, so they are strictly unobservable and to any animal literally remain inconceivable.[89] So too they can be framed as counterfactuals, allowing our imaginations to reconfigure worlds. By the end of their first year, infants are already grasping the possibility of counterfactual alternatives.[90] Nor is it any accident that the pretend play of children finds no counterpart in animals, other than (and most suspiciously) in the "linguistic" enculturated apes.[91] And this is one reason to think that although animals could hardly function without memories or manage without a capacity of foresight, it is much less likely they know that they too are actors in the drama, capable of an alternative course of actions and devising scenarios—exercising free will, if you like. Such depends on the human cognitive scaffolding.

Why our cognitive scaffolding is equipped with ladders, if not elevators, that animals lack is much less clear. One clue comes from the poet Robert Browning, who in his characteristically staccato line proclaims, "What's time? Leave Now for dogs and apes! Man has Forever!"[92] That animals have memories is beyond dispute,[93] but what is it that they actually remember?[94] The usual mantra is phrased as: "What? Where? When?" and so far as the great apes are concerned the evidence of such episodic memory is lacking.[95] Much attention has also been given to the highly specific capacity of some birds to cache their food, perhaps in hundreds of separate locations. Clearly a sort of memory is at a premium, and to recover the goodies they must know the precise locations and draw on what is known as semantic memory. In that sense they anticipate, but do they remember the individual episodes? Do they recall being there themselves? Do they have any concept of time?[96] "Gosh, it seems like yesterday . . ." (and so falls within the category of episodic memory and thus appears to be somewhat different from its semantic equivalent).

Nor is it clear that by drawing on the processes of associative learning[97] animals would ever be in a position to envisage other outcomes,[98] let alone construct a history of events.[99] For that matter, does a squirrel concealing hazelnuts have the least notion of winter?[100] Could it summon up the concept of "General Winter" or sigh nostalgically while recalling the words "deep and crisp and even"?[101] Only then would they draw on past experience, project to an undefined future, think back to the past, and if needed generate an alternative future scenario.[102] Again and again, what looks like future-thinking falls at the hurdle of metarepresentation, that strange ability not only to anticipate a future event but understand it as that future event.[103]

So perhaps also our memories only exist because we tell stories.[104] If animals could do that, they would tell us about it.[105] In the trade, this ability is known as mental time travel or more generally a metarepresentation.[106] Animals might have foresight,[107] but only if they possessed intentions, concepts, and counterfactuals could they then enter this new world. Seemingly they cannot,[108] and for us it would be equally impossible unless we knew that we were persons, owning our thoughts and engaged in endless self-reflection. Only by stepping outside ourselves can we intuit our own mortality, but at the same time realize that both the dead and the unborn have their parts to play.

THE KINDNESS OF ANIMALS?

Just how fundamental is the divide becomes equally apparent when we misattribute mental states to animals, reading their often noisy reactions into moral equivalents, especially altruism.[109] If, for example, animals could be demonstrated to show empathy, realizing the predicament of a trapped companion, then surely this would help to narrow the gap? That rats (and ants[110]) will engage in very similar rescue operations is not in doubt, but it is far less clear that at least the rodents understand the emotions of the victim, reading the mind of their struggling companions.[111] There are good Darwinian reasons for rescue being adaptive, but the intentionality that has the well-being of the victim at heart is a very different matter. This is very much in doubt.[112]

In humans, the stranger in distress is to be helped, but animal empathy remains a minefield. Appeasement and reconciliation fall far short because fully fledged empathy must involve not only stepping into the other's shoes (thus begging the question of a theory of mind), but a third party providing metaphorical handkerchiefs. Much is suggestive, but hard evidence is in much shorter supply.[113] The chimpanzees ostensibly provide the gold standard for genuine consoling, but oddly, in striking contrast to their captive counterparts, in the wilds of Uganda not only is reconciliation much less common but consolation is effectively unknown.[114] And the jungle has other dangers, but the responses are hardly consistent. So in the Tai chimpanzees the "first aid" offered is of an exemplary standard, but far to the east in Gombe the wounds provoke repulsion. Why the difference? Most likely the explanation is Darwinian; in Tai the leopards represent a constant threat but much less so in Gombe.[115]

It is far from clear that our nearest cousins are anywhere near a moral dimension. What about fairness? Can we discern any rudiments or even glimmerings of an aversion to inequity? Chimpanzees can certainly help, but then at the same time are they seeing the other chimpanzee's point of view? Such propensities have been more widely identified, but the evidence is hardly overwhelming.[116] One such trial invited New Caledonian crows to collaborate in order to release a reward.[117] In itself that presented no problem, but the acid test was whether one of them would alter its behavior depending on whether the partner was in attendance. That did not happen because the vital connection between assistance and knowing what the other requires cannot be made. The domains remain separate, whereas in humans social awareness and technical skills meld effortlessly.

Other tests look to the chimpanzees and other apes. Much is made of food sharing, but the overall picture is hardly encouraging. In the wild, mothers never spontaneously offer food to their infants,[118] but in captivity they only dole out the unpalatable food.[119] In marked contrast to children,[120] in chimpanzees spontaneity and active involvement are in very short supply. The overwhelming sense is one of almost complete indifference.[121] Even when collaborating, this has nothing to do with sharing but only the aim of obtaining a bigger share of the goodies.[122] Unsurprisingly then, in cases where two chimpanzees can share food, whether or not one was the actual collaborator makes no difference to the outcome.[123] Chimpanzees can certainly be alert to the actions of others, but they remain "heartless." They are fundamentally competitive, lack common goals, and remain unaware of what (let alone who) a partner really is or how to see another's perspective.[124] To be charitable (which to animals is a closed book), one might label chimpanzees sturdy individualists. In reality they are more like sociopaths,[125] whose only echo among humans would be those feral children such as the boy of Aveyron.[126]

One might think that bonobos, often regarded as more tolerant (a trait that in any event is probably exaggerated[127]), might come off better in a concern for strangers. But that is not the case:[128] perversely (from our perspective), in third-party interactions they prefer a neighbor who throws a wrench in the works over one who might help.[129] From our perspective these are alien worlds, free of a meaningful context, where intentions can never be shared, and negotiation is a contradiction in terms. When it comes to kindness, their cognitive avatars the ravens fare no better.[130] Similar disappointments will await the orphaned chimpanzee; the maternal siblings might look after the little waif, but without social bonds any sort of adoption is very rare, will take months, and even then is pretty half-hearted.[131] And what about when the infant chimpanzee is dead? The interactions between the corpse and its mother can be pretty gruesome,[132] but do the "mourners" have any proper concept of death, or are we wearing exceptionally thick anthropomorphic spectacles?[133]

From our perspective, theirs is a bleak world, so in one sense we can be grateful animals are innocent when it comes to crime and punishment. Yes, given their generally self-regarding nature it is unsurprising that a chimpanzee can retaliate against a thief, but despite this there is no evidence for spite.[134] Nor can they conceive of a third-party punishment.[135] In this moral vacuum, the concept of ownership draws yet another blank.[136] In contrast to our worlds, colored as they are by guilt, envy, regret (gentle or otherwise), injustice, and indignation, the animal mind will never show a flicker of recognition.[137] And why should we ever expect these animals to be so capable? Margaret Visser was surely correct when she wrote, "Justice, [is] a wholly supernatural concept that is not to be found in nature."[138]

Even in a more immediate context it is only because of cultural norms, often symbolic and where conformity is not repressive but embedded in a complex social matrix,[139] that the ironies of Jane Austen will delight us and the savage ruminations of C. S. Lewis's Screwtape appall us. Unable to step out of their minds, animals will never step into a moral universe. Fairness is far beyond their ken,[140] and they will never learn to take turns.[141]

GIVE US THE TOOLS

How, and possibly why, we find ourselves in such strange circumstances will be addressed in due time. The immediate task is to build on the already mentioned examples of animal cognition and tease out why what at first sight appears to be the same activities, albeit simpler, are actually oceans apart. Although it remains convenient to talk about categories—say, "toolmaking," "mathematics," or "language"—the frontiers between them are fluid and mutable (figure 5.1). If there is an overarching theme, it is that we are irreversibly and utterly enculturated. There is, therefore, no obvious boundary between ourselves and the surrounding world. The importance of this becomes more obvious if we consider toolmaking. Tool use by animals is very widespread (including, somewhat implausibly, snails[142]) but highly sporadic,[143] and it is far from clear that braininess and toolmaking have much of a connection.[144] Even among equally cognitively competent species most of them never employ tools.[145] And among the former, no animal uses one tool to make another. Admittedly, in the laboratory those cognitive aristocrats, the New Caledonian crows, can use one tool to access the one they actually need,[146] but given there is a "'captivity bias' in animal tool use,"[147] once again we see how enculturated animals enjoy a premium that finds no equivalent in the wild.

Yet when it comes to toolmaking their achievements are certainly impressive. Is it, however, all that it is cracked up to be? Take the New Caledonian crow Betty, who is alas no longer available for interviews: in an Oxford laboratory her star turn was to take a piece of straight wire and bend it into a hook so that it was in a configuration suitable to retrieve a recessed bucket of food.[148] Nobody was claiming that Betty was thinking through a strategy, but it still seemed to be a spectacular demonstration of avian innovation. But repeatedly these apparently heroic examples of animal cognition crumble in the face of what Sara Shettleworth has called "killjoy explanations."[149] It transpires that Betty's feat was less remarkable because in the wild tool bending by New Caledonian crows is part of their natural repertoire.[150]

In the young crows, the toolmaking is not only honed by lengthy intervals of observation and is a painfully slow learning process,[151] but its foundation is at least in part innate.[152] Nor is that the end of the story, because the parents are highly tolerant, and the young enjoy an extended period of feeding.[153] Nobody questions what in the adult is an impressive capacity to manipulate the tools, but it is no coincidence that, unlike other crows whose beaks recurve, those of New Caledonia have a remarkably straight and stout bill.[154] Combined with pronounced

FIGURE 5.4 Dipping for termites, the chimpanzee seems to be on the path to technology; the reality is very different.

binocular vision,[155] this ensures the tool is held securely and manipulated precisely. Moreover, in general comparisons between these birds and chimpanzees when it comes to comparing tool use, the latter emerge as the all-round better performers.[156]

Does this not give us confidence that by watching an ape crack open a nut using a hammer and anvil stone (and even using supporting wedges[157]) or use sticks to dip for nutritious termites or ants (figure 5.4) we are revisiting the first stages of human technology?[158] Such a view is reinforced by tool selection, such as selecting a certain material for the hammer,[159] or altering the length of a stick according to the aggressiveness of the insects,[160] or even converting the termite probe into a sort of brush.[161] For us, tool use is so customary that we take for granted that assessment of both form and function (even when they are far from obvious) are essential prerequisites. Naturally chimpanzee tool kits are much less sophisticated, but the default position is that they too must draw on the same mental parameters for effective use. Other evidence appears to point in the same direction. Chimpanzees in the open forests of Senegal spear bush babies (albeit with a lamentable success rate),[162] while deeper in the jungle of the Congo another group uses one tool to open up the termite nest and a more slender stick to dip for the insects.[163] These and other regional variants, including chimpanzees that dig up plant roots[164] or employ "pestle-pounding,"[165] represent local traditions—in other words, cultures. But as Thibaud Gruber and coworkers[166] point out, the chimpanzees themselves appear not to know that they have cultures.

As with all other aspects of animal cognition, the skills apes have are strangely one-dimensional, strongly stereotyped, devoid of flair and panache, and tellingly are not in any way cumulative.[167] For example, female chimpanzees ultimately leave their home troop and join nearby bands, bringing with them skills such as nut cracking. Potentially they could transmit new methods, but this never happens.[168] For one reason or another, social conformity[169] is adopted (or as likely imposed) on the immigrants. To be sure, the chimpanzees may be able to extend their techniques, such as when the investigators provide honey but of varying accessibility,[170] but there is little to suggest that a local "tradition" involves experimentation, let alone signs of evolving. By and large chimpanzees are intensely conservative. By implication so too are their cultures (and in the case of nut cracking, some groups are patently less efficient[171]), and their persistence is more likely to be due to variations in the environment than any internal social dynamic.

Related lines of evidence are equally telling. Let us take a group of chimpanzees already familiar with various sorts of tools, but today they will be equipped with a sort of rake, ideal for sweeping up food.[172] There is only one snag: the rakes come in two varieties. Both are identical in appearance, but only one type actually functions (the other has a hidden self-collapsing mechanism). Unsurprisingly, the animals quickly learn which is which. When, however, they are given the opportunity to test the rakes, they are just as likely to abandon the usable rake and cling to the dud version. At this juncture it is time for a breather. About three years later, and reassured that chimpanzees have indeed excellent memories, the trials restart, and once again with prior training. Now they are provided with further assistance: color-coded cues to distinguish the functional rakes from their collapsing counterparts. The chimpanzees quickly get the hang of things, and the next step is to remove the cues and see what happens. Sad to say, we are back to square one. They can see the tool, and they know how to use it, but they are incapable of interrogating its function. This is not the case with fully enculturated chimpanzees, but they are beneficiaries of a rich and stimulating environment with ever-present human encouragement. All of this is par for the course, and tellingly their semi-enculturated counterparts fail to grasp the causal connections.[173]

GRASPING THE INFINITE

Our tools are completely different from those of chimpanzees or any other animal. Not only are the former capable of endless redeployment, but these tools become direct extensions of ourselves. We live with them in a cognitive symbiosis, and happily give tools nicknames. We make the object, but in turn the object makes us:[174] the body and its motor actions, the brain and its cognitive flexibility, and the artifact are all intertwined and engaged in a constant conversation that molds who we are and who we think we are (figure 5.5). Ironically our tools are physical, but they would be impossible to make without abstractions[175]—and as products of our intelligence, they return the compliment.[176]

This intentionality seems to go back to the very dawn of our toolmaking.[177] Such objects are necessarily functional (I write these words with my fountain pen), but so too a tool can be a joke, a social statement, or just beautiful. All the world is now seen through the prism of what is made. Lambros Malafouris[178] gives as examples the stone knap-

FIGURE 5.5 The potter makes the pot, but both the process and end product are richly cognitive.

per and the potter, although just as applicable would be a painter, a novelist, or, come to think of it, a scientist. In any of the creative acts the processes are fluid and transformative; the outcome is often unexpected, and in hindsight also remains provisional.

To emphasize only the geniuses and the most talented among us is to miss the point. Being creative is central to being human, and it effectively draws on a "collective brain."[179] Even among children there is an ungovernable desire to share goals and intentions, to take turns, to not only imitate but imitate to imitate, to draw attention to things, and to enter a world of shared assumptions and a willingness to find social norms—in short, to form a psychological collective and to step into the cultural ocean and swim with increasing confidence.

Identification of animal cultures has become a commonplace, but some are deeply skeptical that there is a true correspondence to their human counterparts.[180] When hunter-gatherers are asked if their cultures are equivalent to the behaviors of the animals, whose lives they know inside out, their response is one of indignation.[181] Not only that, but in complete contrast, human cultures are cumulative and endlessly inventive. Without our unique capacity for mental time travel, such cultures would be stillborn.[182] Equally, cumulative cultures could never emerge without an openness to ritual,[183] a capacity to be altruistic, and crucially embedded in a moral framework where reward and blame are not fictions.[184] However propitious the circumstances and whatever encouragement is offered, the animal can never see such matters as a joint endeavor.[185] And if they cannot see into each other's minds, should we be surprised? They can only emulate—that is, they appreciate the end product rather than the necessary steps (along with the pitfalls of not infrequent blind alleys) of how to get there.[186]

By contrast, we are exceptionally skilled at imitation,[187] which is both cognitive and social.[188] This not only allows us to follow another's actions but to understand what the demonstrator intends. Tellingly, the "facial correlates of determination"[189] that we use find no counterpart in a chimpanzee. This is especially important when cause and effect are not immediately apparent—when they are cognitively "opaque," so to speak.[190] Without these gifts, cumulative culture would for us be an impossibility. Not only does imitation help to facilitate successive improvements but, along with hypercooperation and cultural transmission, it ensures a self-ratcheting culture.[191]

BACK TO SCHOOL

Intimately linked to imitation is another capacity, which is entirely familiar but actually exceedingly odd indeed: teaching.[192] Here too we see a profound shift in mental worlds. As ever, the apparent similarities with animals are misleading. Take the example of meerkats. At first glance they offer an object lesson in animal pedagogy, albeit this and other examples nearly always involve food rather than some abstract thirst for knowledge. In any event, scorpions are an important part of the meerkat diet, but they come with self-evident risks. So there is a progression, from the young being presented first with dead scorpions. As the pups mature, the prey are now alive but are conveniently defanged until they are competent to tackle the real McCoy.[193] The stages of instruction, however, are governed on the types of vocal call the pup makes. Fool the "teacher" with the call of a more experienced pup, and in principle the juvenile will face an extremely painful breakfast. Meerkats are not alone. Cats also can demonstrate hunting techniques to their kittens, but never in a way that is remedial. In other words the kittens learn but they are not taught

We recognize the animal duffers, but animals themselves have no knowledge of dunces. And how could they? They are incapable of mind reading, so they will never grasp whether the "pupil" is ignorant or simply misinformed. The young will certainly observe the actions of the adult, and they may even be handed a tool, but never will they receive instruction.[194] As a consequence, their road to competence is usually painstakingly slow. Even in the laboratory, after repeated demonstrations of a particular technique, the chimpanzee usually fails to attend.[195] Ironically, the nearest approach to teaching among animals is probably the ants with their so-called tandem running,[196] but it seems vanishingly unlikely that the leading ant has any insight into the mental state of her pupil. In humans, it is the exact reverse.[197]

So familiar is this capacity that too often we fail to appreciate just how odd human pedagogy actually is. Encouragement, dialogue, and collaboration[198] are its hallmarks, and they are only possible if minds can read each other's intentions.[199] And this mind reading is much more closely aligned to print reading than might first appear. Teaching and text are now inseparable[200] and thereby there might also be a deeper link to language and iconography.[201] In any event, the consequences are self-evident, but in one sense teaching is so deeply enculturated that it is the species rather than the individual that is the genius. Infants have an ungovernable and spontaneous desire to instruct and inform, to meld observation and judgment with correction, and to applaud an aesthetic solution. It is no accident that successful teachers are often charismatic. Their words nevertheless fall on deaf ears unless we are able to glean their intentions and, just as importantly, communicate back at a shared and deep level of understanding.

Teaching is also one more reflection for our capacity for hyperco-operation, and here too this is vital to our success. Coaching and one-to-one teaching may be popular (and cripplingly expensive), but our cognitive universe is polysemous, linked by a vast reservoir of cultural information and learning. Big brains are necessary, but they are very far from sufficient.[202] Genetics plays little if any part in this process (after all, we only started reading a few thousand years ago), so cultural evolution is both the product and manufacturer: "Grist and mills," as Cecilia Heyes[203] declares. Tellingly, in a subsequent work querying "natural pedagogy," she goes on to point out how "children are taught to be teachable."[204]

SPEAKING UP

None of these developments would be possible without language, which is often taken to be a hallmark of human uniqueness. Maybe it is, but is there not a continuum between animal vocalizations and speech? Do not all sounds pour forth from the same type of larynx, syrinx, or similar sound box? "Speech-ready" was the term Tecumseh Fitch and colleagues used.[205] Did not Cardinal de Polignac say of the ape, "Speak and I will baptize you"?[206] And even if animals do not speak, is not their grasp of words significant?

At the beginning of this chapter, the terrier Bailey made his debut. Before we whistle him back into the arena, let us look at some other canine word sleuths. Meet the border collies Chaser and Rico, who also have built up impressive vocabularies, only needing to hear the new word a single time to associate it with a particular toy (figure 5.6).[207] To explain this skill, known as "fast mapping," the general assumption has been that by connecting word to object these dogs are effectively in the realm of names. Nobody is claiming that this is language, but

FIGURE 5.6 Dogs love their toys, and at first sight the name will elicit an immediate response; dig a little deeper, however, and it is obvious the dog has no concept of names.

hovering in the background are a host of assumptions revolving around comprehension, referents, concepts, identity, and so forth. In many ways the abilities of Bailey match those of Chaser and Rico, and in its own way they arguably are the more impressive given that the collies are sheepdogs, so for them commands go paw-in-paw. The key test, however, is not only recognition of existing toys, but how the dog deals with a new name. Typically in such experiments there are a series of stages; for example, first is making sure that, having been introduced to two new toys, Bailey can find one of them in a collection of old favorites. Then we have to ensure that the dog does not get confused when she has to find her new toy in a pile of not only familiar toys but also of entirely new ones. All is going swimmingly, but suddenly the applause dies down. The final trial should be a cinch—after all, Bailey has clearly shown she knows how to find the requested toy. Now, going back to the two new toys, all Bailey has to do is hear the name "Triceratops" (or whatever) then promptly trot off and choose the correct toy.

Consternation! Bailey is hopeless; evidently she has no idea that one of the toys actually has a name.[208]

In her own way Bailey is far from stupid; she and the other dogs are all ears when we speak to them or use a variety of other cues.[209] But Bailey and her friends (and this applies to any other animal) will never understand what we desire, what we are actually thinking, what motivations drive us, or what alternatives we are weighing.[210] In fact, in many ways when it comes to assessing animal cognition, dogs are the worst possible example. Whether you regard them as symbiotic partners that played a key role in our own domestication (or less charitably, as social parasites), dogs are superbly attuned to our gestures, including pointing (where chimpanzees struggle[211]). Dogs are, with the possible exception of spaniels, eminently trainable. But bring in an actor to fake a heart attack or pretend to be pinned down by a massive weight, then a dog is at sea.[212] They can be trained to watch out for people who are clutching their chests or being buried by debris—or who need a message delivered in the trenches—and then for us they become more animals but our companions.

Well, if the dogs are a bit of disappointment, surely our near-relatives will furnish those linguistic missing links? After all, at least some animals would seem to have a protolanguage, thus providing the long-sought evidence for a continuum. With their specific alarm calls for different predators, the vervet monkeys (figure 5.7) have taken center stage. But the argument that such sounds convey specific meanings and so count as words does not stand up to examination. However adept they are at social relationships, they seem scarcely self-aware, and the minds of their companions are shuttered.[213] Not only that but in these and other animals the sounds are almost never context specific[214] but are employed, for example, as a warning against an external threat (say, a leopard) and during aggressive encounters within the group. This is not to trivialize animal communications; what they do, they do it very well, but it is hardwired, and the sole aim is to convey information. So it is hardly surprising that in, for example,

FIGURE 5.7 Many animals vocalize, but vervet monkeys epitomize the use of what sound like protowords, until, that is, you listen more carefully.

Campbell's monkeys the vocal repertoire of the females[215] is not only much richer than that of the males, but the latter almost never join in these "conversations."[216] A cynic might raise an eyebrow, but in any event what these monkeys cannot do is to convey mental states.[217] "You look like a million bucks" is music to our ears but just noise to an animal.

If the much cited examples of primate vocalizations transpire to be quite different from any sort of protolanguage (if, of course, it was ever spoken) then perhaps we should turn for inspiration to birdsong (figure 5.8). At first glance this would seem to lead to much more fertile comparisons.[218] Not only are there apparent similarities to language, but given that these vocalizations are patently convergent, we might avoid the sort of circularities that arise when we try to persuade ourselves that the hoots and cries of monkeys are in any way instructive. It is true that in many birds the songs must be learned. Tellingly, at an

FIGURE 5.8 Birdsong is not only strikingly melodious but appears to have striking parallels to human language, but only at first sight.

early stage there is a fluidity of expression that strongly recalls the babbling stage of infants.[219] To parrot a phrase does a disservice to the skilled mimicry of many birds, and some birds have repertoires that involve duets and even antiphonal four-part songs.[220] Add to this the popular suggestion that human language first emerged as singing, and it might appear that the gap between birdsong and language is encouragingly narrow. Birdsong can be highly ordered, with sets of notes forming a syllable and these in turn arranged in a motif. Such might be construed as a syntax should they be rearranged in a particular order to thereby convey a meaning. Closer inspection, however, indicates that even hints at a rudimentary language are entirely misplaced.

Such is evident from human languages. These show an immense diversity, but plant any infant in a different setting and shortly afterward he will be chattering away in Mandarin or she will acquire a thick Glaswegian accent peppered with words such as *glaikit* (= foolish) or *stravaiging* (= wandering). Birdsong is fundamentally different, entailing as it does a mixture of "come hither" (to females) and "bugger off" (to unwanted rivals). Birdsong is learned, but there is a spectrum between the species that imitate their "tutor" (who emphatically is not a teacher) and those who are receptive to many sounds and so join the ranks of accomplished mimics. Some can even dispense with instruction and still emerge as accomplished songsters. As with humans, some birds do appear to have a critical interval, what is known as a sensitive period, when the song must be acquired; but in contrast, other species can learn throughout their life. Humans can as well, but we must learn our first tongue at an early stage of infancy or otherwise remain seriously impaired. Why there is such a diversity of birdsongs and modes of acquisition, ranging from a handful of songs in the great tit to a thousand plus in the brown thrasher, is also a somewhat open question.[221] Any solution, however, is going to look at not only an adaptive explanation but one that is ultimately sexual.

Once again, the paradox is that despite having so much in common, the gulf between even birdsong and language remains.[222] To be sure, one thing in common is how the production of precisely modulated sounds involves neural and motor feedback between hearing and the processes of articulation. These sounds can be quite complex and although usually stereotyped can show a degree of flexible utterance, not to mention "dialects." The crucial question, however, is does birdsong ever involve semantics, with the implication that the bird can juggle concepts, find a meaning, or convey a concept—in other words, engage in a rational discourse? It seems not.

The sounds birds make can involve a sort of abstraction. For example, familiar sounds can be redeployed into new combinations. So they are not for the cognitively faint-hearted. Yet, on the other hand, there is no obvious correlation between "song learning and cognitive ability."[223] Ironically, only we can term these animal noises as "language" and identify a "grammar," but their sounds are neither our sort of language (with its recursive structure) nor do they ever employ a context-free grammar.[224] Notably, when redeploying sounds these sorts of generalizations cannot be transformed into a genuine syntax.

So much is evident from the inability of birds to distinguish the same song but now with unfamiliar sounds incorporated.[225] If they had this competence, they could extract those abstract rules that even newborns show.[226] Even if birds had what might pass as a "syntax," crucially it is phonological rather than conceptual.[227] Language is much more than stringing words together, and the rules of engagement are not the same.[228] Accordingly the hierarchy of animal communication has almost no depth. These noises are uniformly imperative. In complete contrast, human speech is declarative,[229] laden with information and at heart a cooperative and social venture.[230]

A further indication that birdsong and language cannot possibly be equivalent is that in the former different songs do not refer to different things.[231] Birdsong is never redeployed to other types of communication, let alone other cognitive domains. As with everything else in animal cognition, it is monotonic whereas human cognitive capacities are intricately interlaced.[232] Language may be based on relatively simple rules, but its scope is effectively unlimited because of its combinatorial complexity. Not only are the very rare cases of combinatorial communication among animals (and, let us not forget, bacteria[233]) a very pale echo of human language, but in the latter case our sounds only make sense because they are ostensive and embedded in a theory of mind.[234]

Also key to language is recursion, a capacity that emerges at a very early age.[235] Recursion represents our ability to embed phrase within phrase and still retain the overall meaning, in principle indefinitely. Consider David Premack's example: "Women think that men think that they think that men think that women's orgasm is different."[236] Surely the answer is obvious? Claims, for example, that starlings can be trained to embed recursive "phrases" in their songs have failed to stand up to scrutiny.[237] In the original research study that aimed to demonstrate starling recursion, the investigators had to run a vast number of trials (up to 50,000), and even then two of the eleven starlings never got the hang of it.[238] This alone should raise a cautionary eyebrow (a metaphorical gesture that only we would understand). More importantly it deflects us from the realization that if it takes months of sustained effort with these birds (in a sort of cognitive archaeology), what exactly are we trying to establish?

In language, as with all the ramifications of human cognition, we see interdependences and parallels. Correspondingly, metarepresentations, false beliefs, or our episodic memories are necessarily recursive, again raising the suspicion that language is inextricably united to an entire cognitive architecture (figure 5.1).[239] Only in this new world can we tell fictional/factional/counterfactual stories about almost anything, traveling as far back in time as we wish and constructing innumerable future alternatives. Language, mental time travel, and all else that makes us human are again interdependent.[240]

Such is equally evident from the various attempts to teach chimpanzees to communicate by using either signing or tokens. Predictably the enterprise is exceptionally laborious. And in the final analysis it is not clear it was worth the effort. The discontinuities with language remain as sharp as ever.[241] These apes show no sign of employing a rule-based system that might echo a grammar, and no sign of rearranging the "words" to impart new, let alone amusing, meanings.[242] It may be a novel way for the chimpanzee to communicate, but it echoes the point that birdsong or any other type of vocalization (along with all the other channels for communication) is profoundly uninteresting, except of course to the animal itself. Here it matters vitally: territory demarcation, mate attraction, and the other exigencies of living in a Darwinian universe circumscribe their narrow existences. From our perspective it is a deeply monotonous world, but only through our eyes and ears. In any other circumstance we would be in exactly the same boat—and, in passing, this might be one explanation for the apparent absence of extraterrestrials (chapter 6).

Not, of course, that many of our intentions cannot be conveyed nonverbally. Yet language, often taken as the defining aspect of being human, is intermeshed with all that makes our cognitive world unique. It is integral to our being; as is evident from very young children, long before full articulation is possible the fundamentals of intentionality are being mapped out.[243] So too language becomes embedded in a world packed with metaphorical meanings. It is, therefore, much more than just words, articulating as it does symbolic and abstract formulations.

The gulf with animals remains unbridgeable.[244] Implicit to human language but absent in any animal is a conveyed world of an expected context and potentially implications that speaks to the mind of the hearer.[245] Not only is human language remarkably flexible, but it is also subtle (think of jokes or sarcasm, not to mention the pitfalls of ambiguity); by contrast, animal communication is more similar to Morse code. Animals can certainly follow our commands, but language itself is an entirely closed book. Wittgenstein's lion may roar, but otherwise the cat is eternally mute. *Per impossibile* if they were to understand language animals would regard it as akin to telepathy.

And if our language should fail, would we just regress to making animal sounds? Far from it. As Paul Oppenheimer noted, it is a much more sinister journey, a dismal path to gibberish, incoherence, and a surrender to entirely malevolent forces.[246] Just as language is a deeply cognitive process, so it also has inescapable metaphysical consequences. In orthodox circles much has been made of Broca's areas being central to language, but cases of drastic brain surgery and hydrocephaly have hinted at a much more puzzling story.[247] And there are other mysteries, such as Marc Liblin, an inhabitant of eastern France.[248] In ways that make no ordinary sense, as a boy he acquired an unknown language, and only a chance encounter in the town of Rennes resulted in its identification as an archaic version of the tongue spoken in an exceedingly remote Pacific island of Rapa Iti.

In the absence of anything like a credible protolanguage, how language emerged is almost entirely conjectural. Much has been made of how many of the building blocks necessary for language, including memory and vowel-like articulations, lurk in the primate brain.[249] One can, however, as easily turn this on its head: given this, then why do they not have even the remotest rudiments of language? Not surprisingly, many posit a gradual sequence of events, ones in which, for example, gestures or pantomime could have provided the initial impetus.[250] Others see a much more rapid transition, potentially achieved by the interlocking of hitherto disparate functions. From an evolutionary perspective this is equally reasonable, but I suspect either approach misses a deeper point. Rather than treating language as the *sine qua non* of human uniqueness—be it as some sort of gold standard or the fuse that detonated an ongoing series of cultural explosions—we must turn the story inside out. Just as with the apparent absence of extraterrestrials (chapter 6), we need to ask ourselves what sort of universe we are actually living in, one that rather oddly we can understand.

From this perspective, oxymoronically language is the articulation of meaning. Rather than being an ever more complicated series of vocalizations that were recruited to deal with larger tribes, or to meet the needs of teaching or instruction, or whatever hypothesis you might prefer (and it is telling that each and every one is equally compelling and equally untestable), language is not only integral to our understanding the world around us, both visible and invisible, but also, as the Word, it is the primal expression of existence. It can certainly be likened to a mirror, but with the constant proviso that we can see through it darkly. Its infinite depths are revealed more by hints, as poetry insists, than by direct demonstration, as science allows, but only then in an extremely circumscribed manner.

THE SQUARE ROOT OF MINUS ONE

The invitation is open to us, but mysteriously remains closed to animals. What is true of language applies equally to mathematics. The parallels are exact, but again this cognitive capacity is holistic rather than step-like. The capacity for numerical abstraction is evident in newborns.[251] Correspondingly, just as grammatical aphasics can still display

FIGURE 5.9 Numerosity, the capacity to assess the relative number of objects, in this case foxhounds, is very widespread among animals, but to imagine numerosity as a precursor for even simple arithmetic is misleading.

causal reasoning, so too similarly disadvantaged patients remain competent in mathematics.[252] Maybe mathematics and language share some deeper basis, but it is just as likely that although the latter routinely articulates the former they are autonomous. What about animals? Certainly many species are adept at approximation of relative numbers in the process known as numerosity (figure 5.9). This capacity has deep evolutionary roots[253] and is supposed to be the immediate precursor to mathematics.[254] But the apparent proximity hides an unbridgeable gulf. Numerosity is part and parcel of the Darwinian world, and the numerosity of the guppy fish compares very favorably with that of an average undergraduate.[255] But in complete contrast no animal can add or subtract, let alone understand a square root. Mathematics is different. Not only, as Eugene Wigner insisted, is it unreasonably effective,[256] but it delineates realities orthogonal to everyday experience.

To explain why numerosity cannot be conflated with any sort of mathematics,[257] it is necessary to outline the basis for numerosity. In essence, it can be divided into numerical distance and numerical magnitude, either of which can confer considerable adaptive advantages. For a gazelle, being able to distinguish eleven leopards from three is considerably more important than the distinction between ten and eleven leopards (or for a fox, foxhounds; figure 5.9). Correspondingly, with numerical magnitude a social fish that habitually lives in shoals, such as the guppy, will find it more useful to discriminate eight from nine compared with thirty-five from forty. There is a subtext here inasmuch as discrimination of very small numbers, one to about four, is known as subitization, and it seems to involve a neural process distinct from the operation of numerosity.[258]

With respect to numerosity, what matters is that the capacity to distinguish numerical magnitudes and distances follows a famous psychophysical principle known as the Weber-Fechner law.[259] Crucially, this revolves around perception, and it might be more easily explained by considering the parallel example of weights. Place a feather in one hand and a small rock in the other. Obviously the difference is self-evident, but now you ask how much you need to add to (or for that matter subtract from) either object for you to perceive a difference. Because both of your hands are full, your companion obliges you; it transpires that she has to add disproportionately more weight to the rocks than the feather. In other words, the perception is not a linear one but logarithmic. This is also the case with numerosity, and thereby it too must be a sensory process.

Yet this is patently not the case with even the mental process of addition, let alone conjuring up something like a complex number (figure 5.10).[260] Such processes are not just cognitive but entail abstractions that may be intangible yet are no less real than that feather. Yet just as animal vocalizations are elided into language, so the default

FIGURE 5.10 Mathematics: a challenge for many humans but forever a closed book to any animal.

Darwinian assumption is that mathematics is no more than an extension of numerosity—more sophisticated and certainly deeper, but not fundamentally different. To explain this continuity there is an appeal to so-called supramodality, which is a process that transcends specific sensory inputs.[261] This then is deemed to involve abstractions. How any such abstraction emerges is no more obvious than the analogous processes that generate language. In directly comparable ways, each comparison conflates two entirely different operations. Vocalizations are assumed to flow into language, and numerosity is seen to meld into mathematics.[262] It all seems so reasonable, but in each case this same caesura emerges. On the one side is a world of perception, context, and the demands of day-to-day living. On the other is a very different world that is equally real and capable of infinite expression, rich in abstractions, and mysteriously visible only to us.

As with other areas of animal cognition, there is a natural temptation to conflate capacities, to bestow on at least our nearest cognitive cousins the bronze medal, the accolade of "well done, missed by a whisker, almost there." That animals possess the capacity to judge approximate numbers has been amply documented, nor is it disputed that in principle this might confer concrete adaptive advantages. Yet the minute-by-minute life of any animal is ever demanding, with the situation rapidly and often unpredictably shifting. No doubt estimating relative numbers has a role, but to imagine the beast peering out of the jungle as a nascent mathematician is wide of the mark. Animals need to quantify, at least sometimes (but how often outside the laboratory is a moot point), but never do they calculate.

When we consider the wider field, this disjunction between numerosity and mathematics becomes more obvious. Outside a cultural milieu, mathematics is unimaginable,[263] depending as it does on symbolic representations.[264] Echoing Eugene Wigner, Richard Hamming remarked, "I have tried, with little success, to get some of my friends to understand my amazement that the abstraction of integers for counting is both possible and useful."[265] This sense of the extraordinary is all the greater when we realize that this abstraction may have then led to writing.[266] Yes, a handful of species can arrive at a simulacrum of calculation, but do not expect any of them to call for paper and pencil. And such "mathematics" is only possible with intensive training, telling us almost as much about human perseveration as it does about an animal's slender grasp of ordinality (that is, the number four lies between three and five), let alone cardinality (that is, I can have three dogs, three oranges, or three imaginary beings).

The star-turn in this department was the chimpanzee Ai, who after several years' training got to number nine; if one was being charitable, Ai displayed a sort of cardinality[267] but never in the spontaneous fashion that children show.[268] In humans such capacities are effectively spontaneous. This is not to say the playing field is quite level; unsurprisingly, different cultures, say Chinese and European, show respective strengths and weaknesses in how they undertake the equivalent mathematical procedures.[269] Equally important is that many isolated tribes can only count a few numbers, and they get along perfectly well using only approximations.[270] But any human, anywhere, can be taught what a complex number is, how to solve a quadratic equation, or how to simply engage in long division.

In trying to assess exactly how animals engage with their numerosities leads to further pitfalls. First, and to return to a worn theme, the degree of training is typically exhaustive, routinely involving thousands of trial runs and even years of unstinting work, before the response is deemed competent.[271] Even then failure is rarely reported, and a success rate well below the 100 percent mark is deemed to be acceptable. More problematic is that, because the capacity for numerosity has a sensory basis (as is evident from the logarithmic scaling of the Weber-Fechner law), how can we be sure that rather than engaging in the approximation of numbers the brain is actually assimilating the sensory cues but assessing them according to their relative magnitude and then integrating the result?[272] This may itself involve a sophisticated process, but it falls far short of any sort of abstraction.

Given that the only conduit into the animal mind are sense data (and thereby hangs a long philosophical debate), disentangling whether numerosity is anything more than a process of sensory integration is something of a minefield. Too often the experimental approach is burdened with prior assumptions, and the results are often conflicting. The animal automatically sees and hears (or employs some other sensory modality), and somehow one has to devise an experiment that takes into account the potentially confounding role of the various sensory cues, which might vary, for example, in terms of their relative size or clarity.

A strong hint that numerosity has little, if anything, to do with processes of mental abstraction has arisen from studies that showed that when the sensory cues are altered there is a corresponding shift in numerical performance.[273] And this tallies with everything else we know about animal cognition. As with toolmaking or vocalizations, animals deliver what their species requires but no more. All are cognitive dead ends, free of absolute values and devoid of abstractions, without meaning yet sufficient unto the day. When they change, they track how the environment has changed. Some can lead to remarkably sophisticated behaviors—think of the "cunning" of the salticid spiders—but the rules they employ are simple.[274] What is irrelevant is ruthlessly filtered out. Their brain ensures survival, not deep insights into the way of the world. It is difficult to say which is the more peculiar: that no other animal has bothered to step outside its natural frame of reference, or that only we have.

THE MUSIC OF THE SPHERES

Well, we cannot expect this view to win warm applause in North Oxford or other refuges of lost causes, such as the university faculty club. For what it is worth, my own epiphany in terms of animal cognition can be traced, I think, to a report by Patricia Gray and her colleagues.[275] They review the many interesting convergences between the music of animals, be it generated by cetaceans or songbirds, and by evolutionary inclusion humans as well. They argue that music is not just an ordered series of sounds but sounds that we can listen to. The poet John Keats's nightingale pulls at our very heartstrings ("Was it a vision, or a waking dream?"[276]), while others find the oceanic cadences of whale song strangely moving. In emphasizing how the structures of music—be they harmonies, melodies, inversions, or riff sessions—link beasts and humans, Gray and her colleagues appear to provide us with compelling evidence for a cognitive continuity. They then, however, take not so much a step as an imaginative leap. Conceding that these musical similarities are hardly surprising, given both the methods of sound production and the neural interconnections (such as between hearing, modulation, and motor control), they went on to ask if there might not be a much deeper reason for such similarities. Suppose, they suggested, that each species, in its own way, is beginning to access the universal music?

In his *Life of Pythagoras,* the Neoplatonist philosopher Porphyry related how this extraordinary individual could hear the harmony of the Universe, a theme also taken up by Plato (in the *Republic* and *Cratylus*) and still later by Philo (as in *On Dreams*). Even today it can be heard, at least according to the entomologist Evelyn Cheesman. A remarkable and undersung individual, during her time on the Vanuatu island of Erromanga (and then later on New Guinea) Cheesman heard the Music of the Spheres. To the locals, the fact that the stars were singing was a commonplace; she also insisted that these sounds were compellingly real, even though they were without any structure one might associate with music. As Cheesman remarked, "We can only speak of it as celestial music beyond our language or song."[277]

To the scientific mind, or any other card-carrying materialist, claims of celestial music would be taken as compelling evidence for an unhinged mind. That, however, would be difficult to reconcile with the steely individualism of Cheesman. The music she heard came from the empyrean realms; but at a more mundane level, music, language, and mathematics are all human expressions, but they collectively open portals to transcendental realities. And just as language is sundered from vocalizations and mathematics divorced from numerosity, so there is an equal caesura between the "music" of animals and even the simplest of human tunes.[278]

Given the earlier case studies of language and mathematics, this should hardly be a cause for surprise. Consider the evidence. Among humans music is a universal (figure 5.11), and a world without music would be almost as desolate as one free of language. Indeed, it is far from obvious that one could exist without the other. Parallels exist, not least in its apparently unlimited combinations. In marked contrast to language, however ironic in tone, music has pervasive and unresolved

FIGURE 5.11 In contrast to any animal music, that made by humans is deeply enculturated.

ambiguities. Yes, that tortured genius Ivor Gurney wrote of how among the trees one could see "the beech with its smooth A major trunk, its laughing E major foliage."[279] So too evocative pieces such as Honegger's *Pacific 231* might summon up images of these magnificent steam locomotives, and more immediately Vaughan Williams' *The Lark Ascending* immediately transports me to the remoteness of a Dorset chalk down on a cool spring day, sitting on an ancient barrow with the infinity of England stretching before me. At a more abstract level, Wagner's leitmotifs delineate the entire Ring cycle, and when decoded they immeasurably enhance what by any standard is a towering philosophical achievement of how we now stare into the existential abyss but remain human.

Being unaware of Valkyries, Tarnhelms, or Notungs does little to erode the transcendental power of this astonishing music. And the animal counterparts? Not only Vaughan Williams but composers as diverse as Vivaldi and Messiaen might find inspiration from goldfinches and blackbirds, and far from being mimicry it is deeply creative. But the traffic is entirely one way. Animal "music" is always context specific, and like their other vocalizations revolves around imperatives and not declaratives; as already noted with birdsong, as often as not the demands are those of territorial demarcation or invitations to sexual possibilities. Even with the maddeningly elusive functions of humpback whale song there is no indication that the sounds carry specific meanings,[280] any more than does birdsong. By contrast, our music is entirely open-ended and deeply enculturated (figure 5.11), weaving together our shared identity and often central to ritual. In just the same way as our tools and everything else that makes us human.

In its own way music exemplifies the paradoxes of undoubted evolutionary continuity versus the paradox of cognitive caesuras. Even more strongly than with language, an air of puzzlement descends on exactly where the differences lie. As with so many facets of animal cognition, the foundational necessities are apparently present but never find the appropriate combinatorial resolution. Tecumseh Fitch, for example, identifies four key components in human music: singing, dancing, drumming, and social synchronization.[281] In each case we can find counterparts among the animals, and so too music that employs, for example, the regularities known as perfect consonances or temperate scale.[282] For the most part, however, the comparisons turn out to be imprecise, and the occurrences are curiously sporadic. Chimpanzees, for example, can drum, and one such, Barney, gave an impressively rhythmical display.[283] But he "performed" only once; it was never to be repeated by him nor any of the other members of his group. Crucial in our music is the process of entrainment, an exact synchronization to an external rhythm or beat. From the simplest tapping of feet to the almost lunatic River Dances, entrainment in humans is deeply embedded. But turning to animal examples, as Margaret Wilson and Peter Cook noted, "Fireflies can't help it, pet birds try, and sea lions have to be bribed," whereas "humans want to."[284]

And thereby, once again, lies the difference. Our music may be meaningless, but it is deeply intentional and full of meaning. A vital clue might come from studies of rhesus monkeys and the perhaps incongruous introduction of a sort of metronome.[285] The distinction one seeks is whether in their response to the regular tick-tocks (faster and slower) these monkeys are being merely reactive, compared with predictive. The latter case obviously applies to us and presupposes a mental model of time. One can almost guess the result. For the rhesus, even to get there was a huge challenge, and the training took a year of sustained effort. Even then, although glimmers of a capacity to anticipate the action of the metronome were emerging, it still appeared that a truly predictive skill would be forever unattainable for them. Similar work with chimpanzees has been hardly more encouraging.[286]

In captivity other animals may do much better, with Snowball the cockatoo as the star performer,[287] but there is no reason to think that the musical beat is seen as an intentional representation.[288] As with mathematics and language, human music transpires to be equally unique. Ignore the repeated litany of those who insist on cognitive continuities, that somehow we are "setting the bar too high" and should dumb down our expectations of animal musicality. We share common roots, of course, but mysteriously this capacity is not only transformed but entirely transcends its apparent equivalents. Thus, to regard music as an incidental to our existence, a mere bagatelle (so to speak) or most lamentably "auditory cheesecake," is to miss the point spectacularly.[289] Just as the supposed links of mathematics to numerosity (and language to vocalizations), it is only natural to ask what might have triggered the generation of music in our minds, what adaptive advantages it might have conferred. As likely as not, one will end up with yet another just-so story. Sexual selection is hardly likely, but maybe (as some would argue) the murmurs between mother and infant were the crucial starting point? If that were so, then it would be reassuring to find crooning counterparts among the apes or even the songbirds, quietening their restive charges who fear the cradle may drop.

CRYSTALLIZING THOUGHT

The natural response in all these cases of apparent cognitive disjunction is to reaffirm the principle of evolutionary continuity, even if that means a frantic papering over the cracks. Alternatively, if it is grudgingly conceded that the differences are sufficiently important to demand an explanation, then the investigator will pinpoint a factor—say, bipedality, cooperative breeding or helpless infants, increasing population size, a "simple" genetic change, climate change, even a change in diet—that separated us from every other species as we were propelled into an ascent that may have left us dizzy but with a unique vantage point. Simply to argue that being bigger brained makes us smarter misses the point.[290] In contrast with the animals, our intelligence is of a completely different order.

Our task is not made any easier by the various adaptive explanations that are routinely wheeled out to explain what ignited the human fuse. They typically suffer from at least three conceptual problems. First, they cannot all be correct,[291] and some areas of received wisdom may be much more shaky than popularly imagined.[292] Second, let us say it was trigger X. Why then did no other species also rise to this challenge and self-ratchet the process with continuous feedback? Third, and most problematic, you say it is the cause, but I suggest it is the result. Did a bigger brain drive our trajectory or vice versa? Given its plasticity, the latter is just as likely. Our brains adapt to our culture. And then there are the problems of evolutionary continuity and turning to our closest relations. We and apes are proximal in many ways, but as often as not if one seeks convincing similarities they must be imported from much more remote groups. For enthusiasts of convergence this is all grist to the mill, but it becomes decidedly less convincing when one has to cherry-pick one example from an orangutan, another from a crow, a third from a dolphin, and the last from an ant.

I suggest we are approaching the question in an entirely back-to-front manner. It involves much more than a shift in perspective. This is not to dispute that cognitive capacities evolved to make sense of the world. Neither should we be surprised that such general capabilities extend far deeper into the tree of life than those handful of species that happen to possess a nervous system. Indeed, in one way or another all life—bacteria,[293] ciliates,[294] plants[295]—are cognitive. Yet even among the invertebrates what is often identified as cognitive sophistication may conceal "an inconvenient truth"[296] that the supposed similarities with vertebrates are much less precise than imagined. On the other hand, what humans have managed (or achieved, or more likely been granted) is radically different and truly astonishing. As exemplified by our tools and artifacts, the material world has become part of us—not just as an extended phenotype but as part of our cognitive architecture. Concurrently we now encounter realities, if you like Platonic forms, that are intangible to any other species, and ones that paradoxically predate any life, here or anywhere else. They are the eternal verities, if you will. Such is a familiar trope among mathematicians; although the expert sees great beauty in the equations, this is hardly a commonplace experience. More accessible are music, poetry, painting, sculpture, and anything else that in one way or another distills reality and in doing so invites transcendence. Each seems to be a different avenue to the same objective, where meaning becomes valid and immutable.

From this perspective the archaeological record is both helpful and tantalizing, if not sometimes downright frustrating. For example, are the earliest known stone tools of hominins a direct extension of chimpanzee lithics or qualitatively different?[297] Nobody doubts that our first lithics were crude in comparison to succeeding cultures, but there still seems to step change between what the chimpanzees can achieve and what our immediate ancestors began to grasp, both mentally and physically.[298] Chimpanzees, for example, might produce sharp-edged flakes as they bash stones together, but there is no evidence that these are anything more than accidental by-products.[299] And was it at this point,[300] or with Acheulean hand axes,[301] or even later that such tools became coextensive with our minds, if you like reflecting a crystallization of thoughts? In one sense, it is a "chicken and egg" question as motor skills and cognitive chutzpah fed on each other.[302] Either way, and as noted earlier, these tools can only be crafted by a species adept at solving problems, engaged in analogical reasoning, and open to instruction, a situation utterly alien to any animal.[303] Equivalent arguments apply to the use of fires[304] (including cooking) and employment of ocher, body ornamentation and clothing, weaponry, inhumation, and art.

Rather than seeking a mysterious factor X, a grafting on of some strange capacity, or trying to maintain the fiction that whatever separates us from animals is only a matter of degree, we need to take a metaphysical perspective. Entirely alien to the materialist, this viewpoint insists that humans have entered new realms. These are ones of ideas and imagination[305] where the unobservable is real, a world of concepts and generalizations working in easy harmony.[306] We are the way the universe becomes self-aware, but crucially and more akin to a voyage of discovery to immanent realities, ones that presuppose the world to be truly meaningful and open to interrogation at many levels. Needless to say the process can only be interpreted through our bodies, our senses, and our cultural milieu. The last in particular will strongly determine what we are capable of "seeing." The Pirahã people of Brazil, for example, are extremely unusual in that they have a language without recursion.[307] The consequences appear momentous because theirs is a cognitive world that is almost amnesiac, lacking a history or folklore and at best an indeterminate future. But rather than arguing their reality is incommensurable with those of others, it would be more fruitful to say that all are perspectives, all true, all slanted, but some at least more open to infinite investigation. So we open one door, and then having entered a still richer cognitive landscape find in turn it beckons us to yet more portals.

An alternative construction, which I prefer, is as follows. It is not so much that as *Homo narrans* we began to tell each other stories

FIGURE 5.12 Books are closed to all animals, but for us they are the doorways to the imaginative spirit of *Homo narrans*, the writer and the written.

(figure 5.12), but to our very considerable surprise we found ourselves in a story that was not of our making but seemed to expect us to be part of the narrative. As discussed elsewhere, we have become both the writer and the written. So the musicologist Jean Molino wrote of his ideas, "Have I told fictions here? If so, I would at least have been faithful to the founding act of hominization: the telling of stories."[308] Crucially, however, these worlds of meanings were not invented or imposed but discovered, and apply not just for our fellow species or even all life but for the entire universe. As Owen Barfield intuited with respect to language,[309] we became mythopoeic creatures and have never looked back. Language "evolves" (as, in Barfield's view, does consciousness) and, to simplify greatly, words once spanned metaphysical and material meanings and were anchored in a mythos. Now they have become "splintered," leading to precision of expression but at the disastrous expense of meaning. In this context, I am struck by what are said to be the last words of the Levantine historian and intellectual enthusiast John Freely: "I am going to find out where the words come from."[310]

This view, of course, is orthogonal to mainstream Darwinian thinking. But unless one is a subscriber to the bizarre (and self-contradictory) view that the only prism through which the world can be seen is a Darwinian squint, we are inviting evolutionary theory to make contact with metaphysical ideas that enjoy their own validity. Rather than thinking of Mind being trapped in our skulls, we intuit how Mind can be accessed and interpreted by our brains rather than being a derivative of neural processing. Of course, as a sort of "filter" (a woefully inadequate metaphor[311]), the brain struggles to register the ineffable, the indescribability of existence, the constant effort to see beyond the mundane and to balance precision with generalities and accept absolutes even when everything is provisional. But this is what makes us, and paradoxically only us, human. We are the oddest thing that ever existed.

6

The Myth of Extraterrestrials

The expedition from Threga IX could not have arrived on a hotter day. As the shore of the lagoon shimmered, the scientific teams were swiftly dispatched, fanning across the Carboniferous world to study goniatites and lycopods, synapsids and seed-ferns. They had strict rules, mind you: only the absolute minimum to be collected, and absolutely nothing, and I mean *nothing*, to be left behind. Well, rule books were made for bureaucrats, safely writing memos many light years away. The captain, a grizzled space-dog of the old and fast-vanishing breed, had other ideas.

The captain's aquarium had a small breeding colony of Grendels, vaguely fishlike creatures, with the oddest tail you have ever seen and more to the point ferocious appetites. Their amusement value had long since passed, however. So as the rest of the crew snoozed, he strolled down unobserved to the water's edge to release the critters. Off they swam! The Grendels duly created temporary mayhem, but in the Darwinian theater they only played a walk-on part and in due course entered the fossil record. More than 300 million years later, I christened them *Typhloesus*,[1] and on account of their bizarre appearance I jokingly referred to them as my "alien goldfish."[2] For the time being these strange fossils remain in phylogenetic limbo, but not for a moment do I imagine them to be genuine aliens.

But is the idea of some such visiting card so ridiculous? Odds might militate against it, but the next time you visit a dinosaur trackway site why not keep your eyes open? There will be the usual footprints of tridactyl hooligans and much deeper imprints made by dimmer quadrupedal leviathans, but just suppose there were much smaller footprints: size nine and (just to drive the message home) the prod-marks made by a tripod supporting the video equipment. Nothing like keeping an open mind!

Would a scrutiny of the fossil record for signs of an alien visitation be as complete and utter a waste of time as scouring the sky with radio telescopes that are ever hopeful of receiving those strangely elusive messages? Perhaps instead we should be looking for artificial capsules, suitably shielded and containing a staggering 10^{19} bits of information for every gram of payload,[3] or indeed some other sort of artifact[4] dispatched by benevolent aliens. Surely we have every reason to be optimistic?[5] Like those legions of animal behaviorists who are convinced that only a hair's breadth separates our mental states from those of animals (chapter 5), so enthusiasts for extraterrestrials have an almost messianic conviction as they intone, "They Will Be Found."

SO WHERE ARE THEY?

They ought to be easy to spot, but as Enrico Fermi (figure 6.1) famously asked during that lunch in Los Alamos, "Where are they?"[6] There has been no shortage of proposed solutions to solve this quandary. In the second edition of Stephen Webb's informative overview, he lists an impressive seventy-five—twenty-five more than when the book first appeared in 2002.[7] In the end, only a handful seem to offer anything remotely like a plausible suggestion, and in the meantime what David Brin aptly called "The Great Silence"[8] casts an almost theological uncertainty over our deliberations (figure 6.2). Discounting the famous "Wow" signal,[9] it is oddly quiet out there—no signals, no visitations,[10]

FIGURE 6.1 The brilliant Italian physicist Enrico Fermi (1901–1954) provocatively asked: Where are they?

no equivalent to the *Mary Celeste* spinning slowly through the Kuiper Belt (or perhaps even closer as one of James Benford's "Lurkers"[11]), no buried monoliths on the moon.[12] Fermi's question is coming home to roost with a vengeance, and something does not add up, big time.[13] As Dr. Johnson stated, "All argument is against it, but all belief is for it." He, of course, was talking about ghosts, but to invert his remarks the

FIGURE 6.2 Received wisdom is that there must be extraterrestrials, and probably in abundance. But Fermi's paradox naggingly suggests we are alone. Credit: Sheila Terry/Science Source.

widespread intuition[14] is that despite there not being a shred of evidence there *must* be extraterrestrials.

The vital statistics appear to be on the side of the believers: in the Milky Way alone the number of potentially habitable worlds (not to mention companion moons) may be a tiny fraction of the total inventory, but the numbers are still stupendously big.[15] One estimate[16] argues that about half of potentially habitable planets house primitive life, and that the number of advanced civilizations exceeds two thousand, with our nearest neighbors perhaps approximately thirty light years away. And that is just in the Milky Way. In the so-called Local Cluster, we have two immediate neighbors of comparable size, and within a hundred million light years are thousands more galaxies—and that is still in our backyard. Not only that, but our Milky Way and Andromeda are familiar spiral galaxies, but colossal elliptical galaxies (such as Maffei 1, some nine million light years away) may be far more propitious for planets and may potentially contain orders of magnitudes more homes.[17]

Who knows? But the total number of terrestrial planets[18] in the visible universe might be on the order of 10^{20} (with the actual number being potentially infinite). Remember, however, that as we are the only yardstick; there is no reliable metric for comparison. Even if there are gadzillions of planets, perhaps we would need to throw the dice almost as many times for a fluke-like emergence of life. And if that event was not stupendously against the odds, maybe nearly all planets turn out to be permanent disaster zones. In either case, this seems to be unlikely, but with all matters pertaining to the Fermi paradox nobody knows.[19]

It is the most peculiar subject because the absence of evidence may mean everything or absolutely nothing. Some hypotheses may appear more plausible and others less so, but it would be very unwise to rank them. For the foreseeable future (i.e., until tomorrow), even the most outlandish idea might turn out to be the correct solution. Just as importantly our expectations are hamstrung by our own technological outlook. Getting men to the moon was just the first step to Andromeda, wasn't it? One small step for humanity—only a few more jumps to the stars and all points beyond. And if the necessary technology is almost within our grasp, why not envisage an alien traveling in the opposite direction?

Things move on, and if our near-future appears to be one of a world dominated by self-replicating robots, whose cognitive prowess will be outstripping the smartest human, then won't it be their interstellar equivalents that should be forging their way toward us[20] from that home of the Grendels, on dear, old Thega IX? In this context it is difficult to think out of the box, and no doubt in a hundred years hence the focus will be reset to provide a new set of explanations. These will seem to be just as plausible—and most likely just as wrong.

At any rate, let us not get ahead of ourselves. The Milky Way is a big enough space, and even in our Local Cluster there will be hundreds of billions of planets, some surely home to our counterparts. So is it best to sit tight and wait? By all means we can do that, but there are a few straws in the stellar wind that suggest, sad to report, we are the only people looking at the skies.[21] Skeptics will point out that our chances of success using existing technologies are still very low, analogous perhaps to looking for a single straw-colored, inch-long needle in a million large haystacks[22] (or something like that). But even if the haystack really does contain a needle, I still find the current perspective sobering. Thus far, scanning the Milky Way and also directed searches such as those aimed at nearby stars and neighborhood galaxies (which in their own way are easier targets)[23] have collectively drawn a complete blank.

Well, let us say for the sake of argument that civilizations are indeed extremely rare and that on average a galaxy only houses one at a time. At the moment, that is us. Given, however, the immense plentitude of galaxies, if only a tiny handful serve as nurseries for advanced intelligences then sooner rather than later they should move from appropriating the energy of their parent sun (as the so-called Dyson sphere[24]), and knuckle down to harvesting that of the entire galaxy. In the trade, these are known as Kardashev Type III civilizations, and the starlight captured must be still ultimately released as waste heat. In other words, they would release waste in the form of infrared radiation, which astronomers are skilled at detecting. So the search is on for a galactic equivalent to a Dyson sphere (figure 6.3). In one such study,[25] some 30,000 candidate galaxies (culled from an initial target of approximately 100,000) were surveyed, between them housing perhaps 10^8 planets capable of evolving intelligence (working on the assumption—again more or less arbitrary—that approximately one planet in every hundred thousand has such a capacity[26]). Sure enough, this search revealed that some of these galaxies are suspiciously infrared rich (and correspondingly underluminous). Could at least one of these be the long-sought-after Kardashev Type IIIs? Almost certainly not.[27] The subsequent winnowing process undertaken by the astronomers suggests that none of these galaxies is truly anomalous.

Our prospects are not looking too good, are they? Yet the chances get even worse. Planets did not form as soon as stars crystallized out of the sequel to the Big Bang, although they date back to perhaps eleven billion years.[28] More importantly, Earth-like planets require much of the periodic table to be built, function, and perhaps nurture biospheres.

FIGURE 6.3 Alien megastructures such as a Dyson sphere would soak up stellar energy and re-radiate it in the infrared, thereby potentially providing an astronomical target in the search for extraterrestrial life.

These "metals" (the term astronomers employ, rather quaintly, for all the elements heavier than hydrogen and helium) are the product of supernovas and other extremely energetic processes. So several cycles of star formation and destruction are required, but metal-rich stars have already formed less than a billion years after the Big Bang.[29] More importantly still, it transpires that our solar system is actually a great deal younger than many of its equivalents. Thus, in the journey to intelligence it would seem that the latter would have enjoyed a considerable head start, on the order of at least a couple of billion years.[30] This in turn might place us at something of a disadvantage, or maybe not. At the moment we simply do not know.

Already it is obvious that solar systems come in all shapes and sizes, and by no means all are likely to be propitious locations.[31] The influential rare earth hypothesis[32] posits that all sorts of peculiar and special circumstances need to come together not just to make a habitable planet—home to pond scum—but one to serve as an incubator for intelligence and produce people like Enrico Fermi. In a sense, our neighbors

FIGURE 6.4 An early image from the Mars Pathfinder site with the so-called Twin Peaks less than a mile away: a bleak and probably lifeless scene. Credit: NASA/JPL. Last updated: September 2, 2020. https://www.nasa.gov/image-feature/marss-twin-peaks.

Venus and Mars (figure 6.4) show what can go wrong: in their early days they (and even the moon[33]) may have housed primitive biospheres, but today both are almost certainly uninhabitable.[34] And possibly those trillions of exoplanets far beyond our solar system followed the same fate as Mars and Venus: all have been derailed or at best shunted into dead ends[35] where only the most hardy of life-forms maintain a desperate metaphorical toehold.

Maybe only Earth developed the desiderata necessary for not only a healthy biosphere but ultimately the emergence of intelligence. High on this wish list might be things such as plate tectonics[36] and in due course an oxygenated atmosphere.[37] Alternatively, perhaps life readily evolves but almost invariably fails at the eukaryotic hurdle or never manages to make a nervous system (in themselves unlikely suppositions).[38] In one way or another, is it only the Earth that manages to reach the finishing line? A variant on this idea revolves around long-term planetary maintenance in a sort of Gaia-like homeostasis.[39] Here, life is assumed to pop up all over the place, but it almost always fails to evolve fast enough so as to gain control of the geochemical feedbacks that ensure long-term habitability. This is another neat idea, and one example of a selection of constriction points or "bottlenecks" that are wheeled out to explain the sheer improbability of our being here.

But in this particular context what does "fast enough" mean? The last I read about life, it likes nothing better than an adaptive challenge—one of the games it plays best. Convergence comes a close second and has the added benefit of ensuring that if one set of contenders falls out of the race, something else similar has a very good chance of reaching the finishing tape.

SENSIBLE, UNLIKELY, MAD

So despite all these escape clauses, and irrespective of the fact that we seem to be relative newcomers, once interstellar travel is the norm then a galactic diaspora should not take more than about 100 Ma[40] (and neither does intergalactic colonization seem to be beyond the realms of possibility[41]). The paradox remains that we should not be here to talk about the absence of extraterrestrials. Our ancestors, unassuming little beasts that looked something like the Cambrian fish *Metaspriggina*,[42] would have ended up in a neat line on a piece of toast, accompanied by a crackingly good Chablis (or whatever high-class intoxicant our carbon-based visitor[43] would prefer). Indeed, as has been pointed out, were we to find even a trace of extraterrestrial life—a Martian microbe will do—we should be very worried indeed. The problem is not the risk of imminent invasion by squads of laser-cannoning goons but that extraterrestrials present a future existential challenge. As Nick Bostrom points out from that perspective, "It would be good news if we find Mars to be sterile. Dead rocks and lifeless sands would lift my spirit" (figure 6.4).[44]

One crumb of comfort might be that we and those elusive aliens could all be running the same race—and, as it happens, facing the same risks. In other words, until about now each and every biosphere has been repeatedly knocked into the long grass by galactic-wide catastrophes, notably those resulting from immensely energetic gamma-ray bursts (GRB).[45] In principle these and similar high-energy events might be very bad news for a terrestrial biosphere.[46] Thus, a GRB or some such similar "phase transition"[47] might explain why only around now people are popping up all across the galaxy and saying, "That's odd. Where is everybody?" It is a neat solution, but from our perspective it is far from clear that any such galaxy-wide event is a likely threat.[48]

So notwithstanding Webb's excellent *tour d'horizon* and his seventy-five alternatives, when it comes to the Fermi paradox these can be reduced to three explanations: "Sensible, unlikely, mad."[49] There are no prizes for guessing the correct answer, but fairness demands we address briefly the first two. *Sensible* is home to all those *bien-pensants,* their reassuring murmur arising from the university faculty clubs, North Oxford, and other zones of intellectual rectitude. The softest version of this argument is that we will probably never make contact, but no matter: scattered across the galaxy is a necklace of civilizations, all probably short-lived, and none surmounting the challenges of interstellar travel.[50] There's the end to it. Variants of this idea exist, of course, such as the "zoo hypothesis."[51] Alternatively, perhaps the extraterrestrials are very shy or at least go to extraordinary lengths to conceal themselves.[52] Or maybe, knowing the very long-term outlook is not looking too rosy for the time being, they have decided to hunker down in a state of suspended animation (the aestivation hypothesis[53]) until the thermodynamic situation is looking decidedly more favorable, as in the ever-cooling universe over time the currency of energy dramatically increases.

The second explanation, *unlikely,* is that to everybody's surprise (except me and a few other curmudgeons), we are alone. That is a bit of a disappointment because there I was, all ready to tell you how convergent evolution (chapter 2) gave a *vade mecum* to alien life.[54] "Alone" is, however, relative. What we seek, wisely or otherwise, is somebody to talk to—maybe to discuss the merits of gin and tonic, or perhaps get a few hints as to the solving of the Riemann hypothesis or the Birch and Swinnerton-Dyer conjecture. But alternatively what if the universe is indeed awash with life, familiar and unfamiliar, but only we humans have moved beyond the animal sphere? "Out there" there is the usual cacophony, but we can leave them to howl and chatter for eternity because never will they tell stories, discover true myths, encounter analogical reasoning, and in their spare time get around to building rocket ships, self-replicating probes, or whatever else carries the message to all those other worlds.

So is it only us? Now we are dangerously close to my final explanation for Fermi's paradox: *mad.*

Not that I am first to write about the ridiculous: we should remember Niels Bohr's dictum that the hypothesis is certainly fascinating but it is also crazy. And, as Bohr then pointed out, only one problem remains—it is not crazy enough. First in line in terms of outlandish explanations is that there certainly is a universe but it happens to be virtual (figure 6.5). One proponent of this intriguing, if chilling, proposal is Stephen Baxter. Drawing on the so-called Bekenstein bound, his calculations suppose that to make a believable virtual universe the total

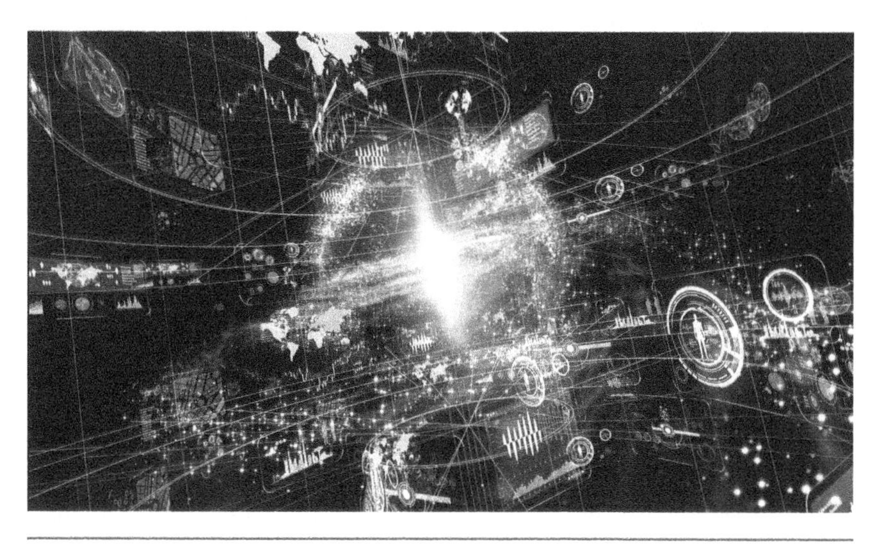

FIGURE 6.5 The worst of solutions to the Fermi paradox: we live in a virtual universe.

computational power available would not permit anything much larger than about a hundred light years across.[55] So when the Neanderthals glanced at the moon or Galileo swung his telescope to its counterparts orbiting Jupiter, all these bodies did not actually exist but were part of a virtual reality. Cosmic wallpaper, if you like. So long as we remain earthbound, all remained well. However, as our probes extend outward—first to the moon, and now in the guise of Voyager 1 and 2 to the very edges of the solar system—so necessarily the planets and whatnot must shift from being virtual to becoming "real."

Such things cannot go on forever, and sooner or later the game will be up. In Baxter's superb novella "Touching Centauri,"[56] the plot revolves around the dispatch of a laser signal that will be bounced off a distant object a few light years from Earth. The time of return was known to a very high degree of precision, but at the appointed moment there was no reflection, no "echo." What could be "seen" those few light years away was an illusion, part of the virtual reality of the skies that the Bekenstein bound forbids to ever become real. Once the bluff of our "masters" had been called, events unfolded in an eschatology that is both terrifying and oddly hopeful. Not surprisingly, Baxter's ideas have attracted attention,[57] and in principle they are testable.[58] In the end Baxter's ideas probably cannot work, but as I will discuss they may be not so far from the correct explanation. Moreover, one sideline that might be consistent with living in a virtual reality is that the "construction" is unlikely to be perfect so there will be odd "glitches"—what from our perspective we might call paranormal occurrences.

Living in a virtual universe is by no means the only mad idea. How about living in black holes,[59] and thus safely out of sight from ordinary folk such as you and me? In the spirit of explaining Fermi's paradox, we should not rule out what seems intuitively implausible. A related theme has been developed by John Smart with his transcension hypothesis.[60] Here, the idea is that life-forms not so very much more advanced than ourselves slip into other dimensions, flickering from our mundane realities into alternative universes. This transfer process might only take a matter of decades. Such beings are de facto invisible to us—until we take the same route and presumably join them. It is perhaps a more credible idea than Baxter's virtual universe, and again has the virtue of not pussyfooting around when it comes to radical explanations for the Fermi paradox. Again I suspect the transcension hypothesis is on the right track, but paradoxically it is too remote, too extraterrestrial. But either way let us keep an open mind.

A UNIVERSE BUILT ON IMAGINATION

So what is the real solution? In a sense it might be a variant on the suggestions of Stephen Baxter, John Smart, and others.[61] It is the territory of zen-like paradoxes, where matters slip between the fingers of comprehension, where things are both opposites and identical, where documentation is both elusive and also crystal clear. Any approach will seem tangential, always struggling to articulate what sounds, at least to orthodox ears, absurdly arcane if not impossible. Or paradoxically it is so obvious as scarcely to need articulation. In essence, we need to decide what sort of universe we are living in and as importantly exactly who we are. In an earlier chapter, I partially unpacked this idea when I suggested that despite our evolutionary continuity with other animals we (and maybe also the Neanderthals) have now begun to enter completely new worlds. And these regions have many features that will not please the materialists. I refer, of course, to the various paranormal phenomena.

Scientifically, the paranormal is almost completely out of bounds, and for good reason. First, these phenomena are an anathema to any materialist agenda, in which the most charitable view would point to the limitless capacity for human delusion. They also stand against the scientific method, in that replication of such events is pretty well out of the question, although persistent sites of haunting do exist.[62] So too the types of events, happenings, occurrences, and situations are dizzyingly different, so again I hear materialists complain that this is exactly why they are so improbable. That many of the descriptions of these strange events are relayed by people who in all other respects appear to be unexceptional is a striking feature, but again the critic would reply that in legal cases witnesses are often highly unreliable.

Nevertheless, the literature is vast. We can take our pick, but one aspect that perhaps receives less attention are time shifts. What, for example, are we to make of a Dr. Clay, who in 1927 found his drive across Cranborne Chase in England to be accompanied by an Iron Age horseman?[63] That is almost normal compared with other, genuinely eccentric, occurrences.[64] It is easy to appreciate why Charles Fort referred to all these examples as "damned."[65] This is not, of course, in terms of religious salvation (although some instances are genuinely terrifying), but in reference to such events being orthogonal to any normal set of expectations. Such occurrences have little obvious typology,[66] and one might wonder what on earth the paranormal has to do with the Fermi paradox.

What matters is that these events, however sporadic and unexpected they may be (and we are willing to grant that they are not *all* a result of human credulity), serve to indicate that the universe is very much more than a brutal materialist fact. Central to any progress, we must risk entering what we might call Prospero's Maze and asking: Who really are we? What are the wellsprings of our imagination? What is the nature (so to speak) of consciousness? Perhaps that last question enfolds everything else. We would do well to follow Jeff Kripal's dictum (and before him many others): "Consciousness is the fundamental ground of all that we know, or ever will know."[67] To conflate this "ground" with any sort of materiality is a massive misapprehension. Consciousness has no other yardstick, has no commonality with anything else.

It is hardly surprising that many find this view vertiginous and feel that somehow it must be coaxed back into its epiphenomenal coffin. Such a stance is self-defeating, and the collateral damage immense. If consciousness (or mind), rather than quarks, dark matter, or any other physical agency, is the "substrate" for our existence then things begin to take on a very different perspective. What, for example, are the real roots of our imagination? Elsewhere I reprised the strange stories of Ramanujan's uncanny access to mathematical worlds,[68] and such stories of people drawing on deep wellsprings of imagination and invisible helpers could be multiplied many times.[69]

Another aspect of this access is the mysterious process of authorship, or in the view of J. R. R. Tolkien, a process of subcreation, our ability to develop a narrative that somehow comes true. What, for example, are we to make of an account provided by the British writer Theodore Powys's adopted daughter (figure 6.6) concerning her father's 1927 novel *Mr. Weston's Good Wine?* This somewhat neglected masterpiece

FIGURE 6.6 Theodora Gay Scutt, the adopted daughter of Theodore and Violet Powys, here posing with Lady Ottoline Morrell. In the daughter's memoirs she relates a very curious story concerning her father. Credit: Philip Edward Morrell, August 1936, Violet Rosalie Powys (née Dodds); Theodora Gay Scutt (née Powys); Theodore Francis Powys; Lady Ottoline Morrell NPG Ax143988. Used by permission National Portrait Library, London.

is about a wine merchant (Mr. Weston) and his assistant Michael, who travel in an old Ford through the lanes of Dorset. In reality, Michael is the archangel—so you may guess the true identity of Mr. Weston. In his book Powys skillfully unpacks the eucharistic themes woven into rural life of death, redemption, and damnation. It is a cracking good read, and a convincing one—but is it pure fiction? Apparently not: about ten years after *Mr. Weston's Good Wine* was published, Powys went out for his customary walk. It was a snowy day and as he was walking along a remote lane, Powys slipped and broke his foot, immobilizing himself. Fortunately an ancient Ford motorcar pulled up, driven by a wine merchant whose assistant was away on nearby business. The merchant's destination was the Fox Inn in Folly (a tiny hamlet about a mile or so from Powys's home in Mappowder), where in prewar rural Dorset not surprisingly very little wine was drunk. He lifted Powys, a heavy man, into his car and conveyed him home. Powys's later inquiries into the identity of the wine merchant drew a blank. Powys's "invention" somehow had come to life—imagination had become reality.[70]

Nor is this the only such example. Many writers are familiar with the creation of a character who then takes charge of the novel. Even for a minor scribbler like myself, I have found it difficult in my previous efforts to avoid the conclusion that in some odd way the books write themselves. In this context, Kripal rightly makes much of the science fiction writer Philip K. Dick.[71] This deeply enigmatic figure insisted, in what was surely much more than a metaphor, that we are not the artist but the drawing, not the writer but the story itself, where seemingly distinct outer and inner worlds become each other.[72] My knowledge of Dick's work is relatively slender although his *The Man in the High Castle* (1962) is widely agreed to be among the most remarkable. Here and elsewhere in Dick's work I have repeatedly encountered a disconcerting "reality," a sense of alternative worlds that really exist. In *The Man in the High Castle*, a final twist is how the principal characters realize that it is they who are in the "wrong" world, compared with ours where the Allies did win the Second World War.

A string of thinkers, from Blaise Pascal to G. K. Chesterton, have insisted that imagination is more than fantasy, that it can be the portal to new worlds. They and many others have articulated, or more often struggled, to explain how it is that their "invented" worlds are very far from fictions, that the impingement of their worlds on this world is far from delusional. J. R. R. Tolkien is an interesting case in point, not least because on the surface he was conservative in outlook and led a life that was unimaginable without a deep Catholic faith. That is only part of his story, however: like all the others in the Inklings literary group, including Charles Williams, Tolkien's wellspring of imagination ran very deep, and he was heterodox about what constituted reality. As related by Verlyn Flieger,[73] Tolkien had an overwhelming impression (albeit usually kept carefully under wraps and often accompanied by careful dissimulation) that one really could step into other worlds or perhaps tunnel one's way there.

And what may all this have to do with the Fermi paradox? There are two approaches. From one perspective, the answer is already widely intuited but flies in the face of received wisdom. That is, so far as the "visible" universe is concerned, we are completely alone. For those of a materialist disposition, there are abundant "natural" explanations. "Rare Earth" or this or that particular evolutionary "bottleneck" will serve perfectly well. The optimists will continue their searches, and the lack of evidence will act as a further spur to endless, and endlessly futile, investigations. Good luck to them.

And the alternative? From this perspective the question of the Fermi paradox becomes effectively irrelevant. In an odd way, the universe at large really does not matter very much now. It brought us to where we are, but now we are moving to a different sort of business. Repeatedly there has been a central intuition that it is not consciousness per se that is slippery and elusive but rather the material world, which is made of shadows. It is these encounters with, for want of a better term, orthogonal realities that show that in contrast to those evanescent extraterrestrials, we are very far from being alone. From our current perspective, these intersections are almost strange and (if only in hindsight) also unnerving. We seem to be both the actors and the audience, powerful and helpless, guides and tourists, articulate and mute, facing a door wide open or scrabbling at one firmly locked, sensible and deeply chaotic, but all utterly believable. Tolkien and his co-Inklings invited us into a series of wondrous worlds. So too did Shakespeare: in his *The Tempest*, when Prospero's daughter proclaimed the brave new world she was exactly right, and her father should have never broken his staff.

CODA

Darwin was right, of course. Evolution happens. Less gratifyingly, in other respects he was gloriously wrong—or to be more charitable, utterly out of his depth. First, the former: as has been repeatedly observed, he was neither the first to intuit evolution nor alone in assembling data in support of the theory. But where he was unique was not only his capacity to marshal the information, but his ability to draw together webs of enquiry that aimed to entrap the intellect. He had another advantage as well: the time was propitious. Yet even if his body had been washed ashore on some remote Patagonian cove after the loss with all hands of the HMS *Beagle*, or if Wallace had succumbed to some loathsome tropical disorder, several hands would have penned very much the same *Origin* sooner rather than later. Crucially, Darwin's masterpiece and subsequent works (not least *The Descent of Man*) insisted on both evolutionary continuity and the apparent nascent stirrings of human mentality in animal minds, noting "the difference in mind between man and the higher animals, great as it is, certainly is one of degree and not of kind."[1]

Such remains the received wisdom to the present day, but in correspondence Darwin was distinctly more cagey. Think of his letter to the philosopher and political economist William Graham.[2] Having dismissed the suggestion that natural laws imply purpose, a few lines later Darwin stated his inner conviction "that the Universe is not the result of chance. But then with me the horrid doubt always arises whether the convictions of man's mind, which has been developed from the mind of lower animals, are of any value or at all trustworthy. Would any one trust in the convictions of a monkey's mind, if there are any convictions in such a mind?" We need not be overly anachronistic, but in this letter we might anticipate Einstein's nervousness of a dice-throwing god, as well as Dennett's solipsistic concept of a Darwinian "universal acid."

In any event, Darwin's nervousness is all the more striking given his panopticonical vision of the world. Given his immense sweep of evolutionary insights, it is curious how his thinking quickly runs into the sands when it encounters the metaphysics of mind. If monkeys give him shudders even in the more domestic circumstances with a dog snoozing at his feet, the gulf appears. In his letter to the American botanist Asa Gray,[3] he first appealed to designed laws without a designer, but noted how out of his intellectual depth he felt; he then commented, "A dog might as well speculate on the mind of Newton." Well, as Bernard of Chartres reminded us,[4] we all stand on the shoulders of giants. This is a powerful conceit, but we can switch metaphors and think not so much of open vistas, but rather of a figure who beckons us into a deeper order. Welcome to Prospero's Maze. Despite the many twists and turns, we can rely on a strong thread, at least to begin with.

First, the facts of evolution are not in dispute. Yet not all paths are possible, and underlying constraints predispose the navigation to a limited number of outcomes, not least intelligence. Above the high hedges of the Maze swings our Milky Way, and we are encouraged to think that "up there" much the same evolutionary story is not only unfolding but innumerably often. Our eyes then return to the ground, and there in front of us stands the enigmatic figure of Mortimer, who from now on will be our guide. Glancing back at the sky, he murmurs that Enrico Fermi was correct. His paradox holds that against all logic and common sense we are indeed alone. But in what sense?

Now we are *Homo narrans*, finding ourselves embedded in a story with which we are not just observers but vital participants. We are both the writers and the written, but in an endless loop that descends into even deeper realities that for the most part we scarcely intuit.[5] To the materialist, these must all be absurdities, shadow games we play before sliding into the utter darkness of irrecoverable extinction. But the evidence suggests otherwise. Although firmly in the materialist camp, J. B. S. Haldane is famous for his saying, "Now, my suspicion is that the universe is not only queerer than we suppose, but queerer than we *can* suppose."[6] We would be wise to add as our coda "and so are we."

ACKNOWLEDGMENTS

My principal debt of thanks must be to the hundreds of scientists and writers whose work has formed the foundation of this book. Nevertheless, to have their years of endeavors summarized in a few bald sentences (or less) hardly counts as recognition, not least if I may come to conclusions markedly at odds with the cherished world pictures of my very diverse sources. More immediately, I thank my colleagues, notably Jen Hoyal Cuthill, Nick Butterfield, Emily Mitchell, Tim Jenkins, Arik Kerschenbaum, Jeff Kripal, and Rob Asher, for conversations and often encouragement. Both Tim and Rob, along with Dan Field, kindly reviewed some of the chapters. As ever, libraries across Cambridge were extraordinarily helpful, and I also particularly acknowledge the support from the Department of Earth Sciences as well as my College, St John's. Vivien Brown coped heroically with the primary inky scribbles and the endless redrafts, and to her I offer my best thanks. The same is extended to my agent, Barbara Levy, as well as Susan Arellano in Templeton Press and Helen Wheeler of Westchester Publishing Services. Finally, my wife, Zoë, provided continual encouragement of the creative atmosphere, not least as corks were pulled and stiff G&Ts came to hand.

NOTES

CHAPTER 1

1. The quotation (p. 135) is from Simon Conway Morris, "Life: The Final Frontier for Complexity?" in *Complexity and the Arrow of Time,* ed. Charles H. Lineweaver, Paul C. W. Davis, and Michael Ruse (Cambridge: Cambridge University Press, 2013), 135–61.
2. As remarked by William Bains and Dirk Schulze-Makuch, "The Cosmic Zoo: The (Near) Inevitability of the Evolution of Complex, Macroscopic Life," *Life* 6, no. 3 (2016), 25, https://doi.org/10.3390/life6030025.
3. Exemplified by the highly influential book by John Maynard Smith and Eörs Szathmary, *The Major Transitions in Evolution* (New York: W. H. Freeman, 1995).
4. For a refreshingly balanced overview of eukaryogenesis and in many ways its unremarkable origins (notwithstanding profound consequences), see Austin Booth and W. Ford Doolittle, "Eukaryogenesis, How Special Really?" *Proceedings of the National Academy of Sciences of the United States of America* 112, no. 33 (2015): 10278–85; also see the discussion of their article by Nick Lane and William F. Martin, "Eukaryotes Really Are Special, and Mitochondria Are Why," *Proceedings of the National Academy of Sciences of the United States of America* 112, no. 35 (2015), E4823, https://doi.org/10.1073/pnas.1509237112, and the authors' reply: "Reply to Lane and Martin: Being and Becoming Eukaryotes," *Proceedings of the National Academy of Sciences of the United States of America* 112, no. 35 (2015), E4824, https://doi.org/10.1073/pnas.1513285112.

5. Marilee A. Ramesh, Shehre-Banoo Malik, and John M. Logsdon Jr., "A Phylogenomic Inventory of Meiotic Genes: Evidence for Sex in *Giardia* and an Early Eukaryotic Origin of Meiosis," *Current Biology* 15, no. 2 (2005): 185–91; see also Paulo G. Hofstatter, Giulia M. Ribeiro, Alfredo L. Porfírio-Sousa, and Daniel J. G. Lahr, "The Sexual Ancestor of All Eukaryotes: A Defense of the 'Meiosis Toolkit': A Rigorous Survey Supports the Obligate Link between Meiosis Machinery and Sexual Recombination," *BioEssays* 42, no. 9 (2020), e2000037, https://doi.org/10.1002/bies.202000037.
6. Eugene V. Koonin, "Intron-Dominated Genomes of Early Ancestors of Eukaryotes," *Journal of Heredity* 100, no. 5 (2009): 618–23; see also Miklos Csuros, Igor B. Rogozin, and Eugene V. Koonin, "A Detailed History of Intron-Rich Eukaryotic Ancestors Inferred from a Global Survey of 100 Complete Genomes," *PLoS Computational Biology* 7, no. 9 (2011), e1002150, https://doi.org/10.1371/journal.pcbi.1002150.
7. Romain Derelle, Philippe Lopez, Hervé Le Guyader, and Michaël Manuel, "Homeodomain Proteins Belong to the Ancestral Molecular Toolkit of Eukaryotes," *Evolution and Development* 9, no. 3 (2007): 212–19.
8. Such as the so-called mini-chromosome maintenance proteins; see Yu Liu, Emma T. Steenkamp, Henner Brinkmann, Lise Forget, Hervé Philippe, and B. Franz Lang, "Phylogenomic Analyses Predict Sister-group Relationship of Nucleariids and Fungi and Paraphyly of Zygomycetes with Significant Support," *BMC Evolutionary Biology* 9 (2009), e60, https://doi.org/10.1186/1471-2148-9-272. For when RNA takes an interest in gene expression (so-called RNA interference), see, for example, Eugene V. Koonin, "Origin of Eukaryotes from within Archaea, Archaeal Eukaryome and Bursts of Gene Gain: Eukaryogenesis Just Made Easier?" *Philosophical Transactions of the Royal Society of London, Series B: Biological Science* 370, no. 1678 (2015), 20140333, https://doi.org/10.1098/rstb.2014.0333.
9. See, for example, Lael D. Barlow, Eva Nývltová, Maria Aguilar, Jan Tachezy, and Joel B. Dacks, "A Sophisticated, Differentiated Golgi in the Ancestors of Eukaryotes," *BMC Biology* 16, no. 1 (2018), 27, https://doi.org/10.1186/s12915-018-0492-9; also the correction in *BMC Biology* 16, no. 1 (2018), 35, https://doi.org/10.1186/s12915-018-0510-y.

10. See, for example, Bill Wickstead, Keith Gull, and Thomas A. Richards, "Patterns of Kinesin Evolution Reveal a Complex Ancestral Eukaryote with a Multifunctional Cytoskeleton," *BMC Evolutionary Biology* 10 (2010), 110, https://doi.org/10.1186/1471-2148-10-110.
11. Joel B. Dacks, Pak P. Poon, and Mark C. Field, "Phylogeny of Endocytic Components Yields Insight into the Process of Nonendosymbiotic Organelle Evolution," *Proceedings of the National Academy of Sciences of the United States of America* 105, no. 2 (2008): 588–93; Joel B. Dacks, Andrew A. Peden, and Mark C. Field, "Evolution of Specificity in the Eukaryotic Endomembrane System," *International Journal of Biochemistry and Cell Biology* 41, no. 2 (2009): 330–40. Also with specific respect to SNARE proteins, see Tobias H. Kloepper, C. Nickias Kienle, and Dirk Fasshauer, "An Elaborate Classification of SNARE Proteins Sheds Light on the Conservation of the Eukaryotic Endomembrane System," *Molecular Biology of the Cell* 18, no. 9 (2007): 3463–71.
12. Xavier Grau-Bové, Arnau Sebé-Pedrós, and Iñaki Ruiz-Trillo, "The Eukaryotic Ancestor Had a Complex Ubiquitin Signaling System of Archaeal Origin," *Molecular Biology and Evolution* 32, no. 3 (2015): 726–39.
13. Alex de Mendoza, Arnau Sebé-Pedrós, and Iñaki Ruiz-Trillo, "The Evolution of the GPCR Signaling System in Eukaryotes: Modularity, Conservation, and the Transition to Metazoan Multicellularity," *Genome Biology and Evolution* 6, no. 3 (2014): 606–19.
14. For an overview, see V. Lila Koumandou, Bill Wickstead, Michael L. Ginger, Mark van der Giezen, Joel B. Dacks, and Mark C. Field, "Molecular Paleontology and Complexity in the Last Eukaryotic Common Ancestor," *Critical Reviews in Biochemistry and Molecular Biology* 48, no. 4 (2013): 373–96.
15. See, for example, Sven B. Gould, Sriram G. Garg, and William F. Martin, "Bacterial Vesicle Secretion and the Evolutionary Origin of the Eukaryotic Endomembrane System," *Trends in Microbiology* 24, no. 7 (2016): 525–34.
16. See, for example, articles by Anja Spang et al., "Complex Archaea That Bridge the Gap between Prokaryotes and Eukaryotes," *Nature* 521, no. 7551 (2015): 173–79; and Katarzyna Zaremba-Niedzwiedzka et al.,

"Asgard Archaea Illuminate The Origin of Eukaryotic Cellular Complexity," *Nature* 541, no. 7637 (2017): 353–58; with commentaries by James O. McInerney and Mary J. O'Connell, "Microbiology: Mind the Gaps in Cellular Evolution," *Nature* 541, no. 7637 (2017): 297–99; Caner Akil and Robert C. Robinson, "Genomes of Asgard Archaea Encode Profilins That Regulate Actin," *Nature* 562, no. 7727 (2018): 439–43; and Laura Eme and Thijs J. G. Ettema, "The Eukaryotic Ancestor Shapes Up," *Nature* 562, no. 7727 (2018): 352–53. One should also draw attention to an odd microorganism (*Parakaryon*) from deep-sea hydrothermal vents off Japan that seems to be intermediate between prokaryotes and eukaryotes; see Masashi Yamaguchi et al. "Prokaryote or Eukaryote? A Unique Microorganism from the Deep Sea," *Journal of Electron Microscopy (Tokyo)* 61, no. 6 (2012): 423–31. Could it be a convergent proto-eukaryote? A hint of this is given by Masashi Yamaguchi and Cedric O'Driscoll Worman, "Deep-sea Microorganisms and the Origin of the Eukaryotic Cell," *Japanese Journal of Protozoology* 47, no. 1–2 (2014): 29–48.

17. See, for example, Bill Wickstead and Keith Gull, "The Evolution of the Cytoskeleton," *Journal of Cell Biology* 194, no. 4 (2001): 513–25.
18. Quotation is from George Gaylord Simpson, *The Meaning of Evolution: A Study of the History of Life and Its Significance for Man*, rev. ed. (New Haven, CT: Yale University Press, 1967), 17.
19. Simpson, *Meaning of Evolution*, 286.
20. Although discussing the genome of the amoeboid *Naegleria*, the article by Lillian K. Fritz-Laylin et al., "The Genome of *Naegleria gruberi* Illuminates Early Eukaryotic Versatility," *Cell* 140, no. 5 (2010): 631–42, repeatedly underscores these points.
21. See, for example, the article by Maureen A. O'Malley, Jeremy G. Wideman, and Iñaki Ruiz-Trillo, "Losing Complexity: The Role of Simplification in Macroevolution," *Trends in Ecology and Evolution* 31, no. 8 (2016): 608–21; see also the case of gene loss reviewed by Ricard Albalat and Cristian Cañestro, "Evolution by Gene Loss," *Nature Reviews Genetics* 17, no. 7 (2016): 379–91.
22. See the overview by Yuri I. Wolf and Eugene V. Koonin, "Genome Reduction as the Dominant Mode of Evolution," *BioEssays* 35, no. 9 (2013): 829–37.

23. See, for example, the case among cell lineages documented by Ricardo B. R. Azevedo et al., "The Simplicity of Metazoan Cell Lineages," *Nature* 433, no. 7022 (2005): 152–56.
24. Eugene V. Koonin, "The Biological Big Bang Model for the Major Transitions in Evolution," *Biology Direct* 2 (2007), 21, https://doi.org/10.1186/1745-6150-2-21.
25. See the abstract of Nicolas Glansdorff, Ying Yu, and Bernard Labedan, "The Last Universal Common Ancestor: Emergence, Constitution and Genetic Legacy of an Elusive Forerunner," *Biology Direct* 3 (2008), 29, https://doi.org/10.1186/1745-6150-3-29. The same applies to the associated virome; see Mart Krupovic, Valerian Dolja, and Eugene V. Koonin, "The LUCA and Its Complex Virome," *Nature Reviews Microbiology* 18, no. 11 (2020): 661–70.
26. Florian Busch, Chitra Rajendran, Kristina Heyn, Sandra Schlee, Rainer Merkl, and Reinhard Sterner, "Ancestral Tryptophan Synthase Reveals Functional Sophistication of Primordial Enzyme Complexes," *Cell Chemical Biology* 23, no. 6 (2016): 709–15.
27. See the article on imidazole glycerol phosphate synthase by Bernd Reisinger et al., "Evidence for the Existence of Elaborate Enzyme Complexes in the Paleoarchean Era," *Journal of the American Chemical Society* 136, no. 1 (2014): 122–29.
28. See, for example, the discussion of thiroredoxins in Raul Perez-Jimenez et al., "Single-Molecule Paleoenzymology Probes the Chemistry of Resurrected Enzymes," *Nature Structural and Molecular Biology* 18, no. 5 (2011): 592–96; and Irene Kaganman, "Resurrected Enzymes," *Nature Methods* 8, no. 6 (2011), 452, https://doi.org/10.1038/nmeth0611-452.
29. See, for example, Matija Harcet, Masa Roller, Helena Cetković, Drago Perina, Matthias Wiens, Werner E. G. Müller, and Kristian Vlahovicek, "Demosponge EST Sequencing Reveals a Complex Genetic Toolkit of the Simplest Metazoans," *Molecular Biology and Evolution* 27, no. 12 (2010): 2747–56; Ana Riesgo, Nathan Farrar, Pamela J. Windsor, Gonzalo Giribet, and Sally P. Leys, "The Analysis of Eight Transcriptomes from All Poriferan Classes Reveals Surprising Genetic Complexity in Sponges," *Molecular Biology and Evolution* 31, no. 5 (2014): 1102–20; Ilya Borisenko, Marcin Adamski, Alexander Ereskovsky, and Maja Adamska,

"Surprisingly Rich Repertoire of *Wnt* Genes in the Demosponge *Halisarca dujardini*," *BMC Evolutionary Biology* 16, no. 1 (2016), 123, https://doi.org/10.1186/s12862-016-0700-6; and Federico Gaiti, Katia Jindrich, Selene L. Fernandez-Valverde, Kathrein E. Roper, Bernard M. Degnan, and Miloš Tanurdžić, "Landscape of Histone Modifications in a Sponge Reveals the Origin of Animal *cis*-Regulatory Complexity," *eLife* 6, e22194, https://doi.org/10.7554/elife.22194; with commentary by Veronica Hinman and Gregory Cary, "Multicellularity: The Evolution of Gene Regulation," *eLife* 6 (2017), e27291, https://doi.org/10.7554/eLife.27291.

30. See, for example, Werner E. G. Müller and Isabel M. Müller, "Origin of the Metazoan Immune System: Identification of the Molecules and Their Functions in Sponges," *Integrative and Comparative Biology* 43, no. 2 (2003): 281–92; Glen R. D. Elliot and Sally P. Leys, "Evidence for Glutamate, GABA and NO in Coordinating Behaviour in the Sponge, *Ephydatia muelleri* (Demospongiae, Spongillidae)," *Journal of Experimental Biology* 213, no. 13 (2010): 2310–21; Eve Gazave et al., "NK Homeobox Genes with Choanocyte-Specific Expression in Homoscleromorph Sponges," *Development Genes and Evolution* 218, no. 9 (2008): 479–89; Sally P. Leys, "Elements of a 'Nervous System' in Sponges," *Journal of Experimental Biology* 218, no. 4 (2015): 581–91; Gemma D. Richards, Elena Simionato, Muriel Perron, Maja Adamska, Michel Vervoort, and Bernard M. Degnan, "Sponge Genes Provide New Insight into the Evolutionary Origin of the Neurogenic Circuit," *Current Biology* 18, no. 15 (2008): 1156–61; and Sandie M. Degnan, "The Surprisingly Complex Immune Gene Repertoire of a Simple Sponge, Exemplified by the NLR Genes: A Capacity for Specificity?" *Developmental and Comparative Immunology* 48, no. 2 (2015): 269–74.
31. Nick Lane, *Journal of Theoretical Biology* 434, no. 58–67 (2017): 62; less plausibly he links this to the acquisition of mitochondria.
32. Stephen Jay Gould, *Full House: The Spread of Excellence from Plato to Darwin* (Cambridge, MA: Belknap Press, 1996); see also the critique by Kim Sterelny, "Bacteria at the High Table," *Biology and Philosophy* 14, no. 3 (1999): 459–70.

33. Birgit Luef et al., "Diverse Uncultivated Ultra-small Bacterial Cells in Groundwater," *Nature Communications* 6 (2015), 6372, https://doi.org/10.1038/ncomms7372; Lydia-Ann J. Ghuneim, David L. Jones, Peter N. Golyshin, and Olga V. Golyshina, "Nano-Sized and Filterable Bacteria and Archaea: Biodiversity and Function," *Frontiers in Microbiology* 9 (2018), 1971, https://doi.org/10.3389/fmicb.2018.01971.
34. Andy Knoll and Dick Bambach, "Directionality in the History of Life: Diffusion from the Left Wall or Repeated Scaling of the Right?" Supplement, *Paleobiology* 26, no. 4 (2000): 1–14.
35. Sarah Adamowicz, Andy Purvis, and Matthew A. Wills, "Increasing Morphological Complexity in Multiple Parallel Lineages of the Crustacea," *Proceedings of the National Academy of Sciences of the United States of America* 105, no. 12 (2008): 4786–91.
36. For a helpful overview, see D. W. McShea, "Three Trends in the History of Life: An Evolutionary Syndrome," *Evolutionary Biology* 43, no. 4 (2016): 531–42.
37. Eva Bianconi et al., "An Estimation of the Number of Cells in the Human Body," *Annals of Human Biology* 40, no. 6 (2013): 463–71.
38. Ricardo Cardoso Neves, Kristine J. Kürstein Sørensen, Reinhardt Møbjerg Kristensen, and Andreas Wanninger, "Cycliophoran Dwarf Males Break the Rule: High Complexity with Low Cell Numbers," *Biological Bulletin* 217, no. 1 (2009): 2–5; also Ricardo Cardoso Neves and Heinrich Reichert, "Microanatomy and Development of the Dwarf Male of *Symbion pandora* (Phylum Cycliophora): New Insights from Ultrastructural Investigation Based on Serial Section Electron Microscopy," *PLoS One* 16, no. 4 (2015), e122364, https://doi.org/10.1371/journal.pone.0122364.
39. See, for example, Renate Czaker, "Dicyemid's Dilemma: Structure versus Genes. The Unorthodox Structure of Dicyemid Reproduction," *Cell and Tissue Research* 343, no. 3 (2011): 649–58.
40. See the overview by F. G. Hochberg Jr., "The 'Kidneys' of Cephalopods: A Unique Habitat for Parasites," *Malacologia* 23 (1982): 121–34.
41. Tsai-Ming Lu, Miyuki Kanda, Noriyuki Satoh, and Hidetaka Furuya, "The Phylogenetic Position of Dicyemid Mesozoans Offers Insights into

Spiralian Evolution," *Zoological Letters* 3 (2017), 6, https://doi.org/10.1186/s40851-017-0068-5.

42. See, for example, George S. Slyusarev, Viktor V. Starunov, Anton S. Bondarenko, Natalia A. Zorina, and Natalya I. Bondarenko, "Extreme Genome and Nervous System Streamlining in the Invertebrate Parasite *Intoshia variabili*," *Current Biology* 30, no. 7 (2020): 1292–98.e3; with commentary Gonzalo Giribet, "Genomes: Miniaturization Taken to Extremes," *Current Biology* 30, no. 7 (2020): R314–16.
43. Philipp H. Schiffer, Helen E. Robertson, and Maximillian J. Telford, "Orthonectids Are Highly Degenerate Annelid Worms," *Current Biology* 28, no. 12 (2018): 1970–74.e3.
44. Hidetaka Furuya, Mitsunori Ota, Ritsuko Kimura, and Kazuhiko Tsuneki, "Renal Organs of Cephalopods: A Habitat for Dicyemids and Chromidinids," *Journal of Morphology* 262, no. 2 (2004): 629–43.
45. Jonathan Foox and Mark E. Siddall, "The Road to Cnidaria: History of Phylogeny of the Myxozoa," *Journal of Parasitology* 101, no. 3 (2015): 269–74; also the overview by Stephen D. Atkinson, Jerri L. Bartholomew, and Tamar Lotan, "Myxozoans: Ancient Metazoan Parasites Find a Home in Phylum Cnidaria," *Zoology (Jena)* 129 (2018): 66–68.
46. See the review by E. Sally Chang, Moran Neuhof, Nimrod D. Rubinstein, Arik Diamant, Hervé Philippe, Dorothée Huchon, and Paulyn Cartwright, "Genomic Insights into the Evolutionary Origin of Myxozoa within Cnidaria," *Proceedings of the National Academy of Sciences of the United States of America* 112, no. 48 (2015): 14912–17.
47. See, for example, Pavla Bartošová et al., "*Sphaerospora Sensu Stricto*: Taxonomy, Diversity and Evolution of a Unique Lineage of Myxosporeans (Myxozoa)," *Molecular Phylogenetics and Evolution* 68, no. 1 (2013): 93–105.
48. Jonathan Ben-David, Stephen D. Atkinson, Yulia Pollak, Gilad Yossifon, Uri Shavit, Jerri L. Bartholomew, and Tamar Lotan, "Myxozoan Polar Tubules Display Structural and Functional Variation," *Parasites and Vectors* 9, no. 1 (2016), 549, https://doi.org/10.1186/s13071-016-1819-4.
49. E. A. Adriano and B. Okamura, "Motility, Morphology and Phylogeny of the Plasmodial Worm, *Ceratomyxa vermiformis* n. sp. (Cnidaria: Myxozoa: Myxosporea)," *Parasitology*, 144, no. 2 (2017): 158–68.

50. See, for example, Peter C. Andersen, Brent V. Brodbeck, and Russell F. Mizell III, "Metabolism of Amino Acids, Organic Acids and Sugars Extracted from the Xylem Fluid of Four Host Plants by Adult *Homalodisca coagulata*," *Entomologia Experimentalis et Applicata* 50, no. 2 (1989): 149–59.
51. See, for example, M. Malone, R. Watson, and J. Pritchard, "The Spittlebug *Philaenus spumarius* Feeds from Mature Xylem at the Full Hydraulic Tension of the Transpiration Stream," *New Phytologist* 143, no. 2 (1999): 261–71.
52. See, for example, Laura J. Crews, Margaret E. McCully, Martin J. Canny, Cheng X. Huang, and Lewis E. C. Ling, "Xylem Feeding by Spittlebug Nymphs: Some Observations by Optical and Cryo-scanning Electron Microscopy," *American Journal of Botany* 85, no. 4 (1998): 449–60; E. A. Backus, K. B. Andrews, H. J. Shugart, C. L. Greve, J. M. Labavitch, and H. Alhaddad, "Salivary Enzymes Are Injected into Xylem by the Glassy-winged Sharpshooter, a Vector of *Xylella fastidiosa*," *Journal of Insect Physiology* 58, no. 7 (2012): 949–59; and Wonjung Kim, "Mechanics of Xylem Sap Drinking," *Biomedical Engineering Letters* 3 (2013): 144–48.
53. See, for example, Peter C. Andersen, Brent V. Brodbeck, and Russell F. Mizell III, "Assimilation Efficiency of Free and Protein Amino Acids by *Homalodisca vitripennis* (Hemiptera: Cicadellidae: Cicadellinae) Feeding on *Citrus sinensis* and *Vitis vinifera*," *Florida Entomologist* 92, no. 1 (2009): 116–22.
54. See, for example, Peter C. Andersen, Brent V. Brodbeck, and Russell F. Mizell III, "Metabolism of Amino Acids, Organic Acids and Sugars Extracted from the Xylem of Four Host Plants by Adult *Homalodisca coagulata*," *Entomologia Experimentalis et Applicata* 50 (1989): 149–59.
55. See, for example, Richard A. Redak, Alexander H. Purcell, João R. S. Lopes, Matthew J. Blua, Russell F. Mizell III, and Peter C. Andersen, "The Biology of Xylem Fluid–Feeding Insect Vectors of *Xylella fastidiosa* and Their Relation to Disease Epidemiology," *Annual Review of Entomology* 49 (2004): 243–70.
56. See, for example, Daniela M. Takiya, Phat L. Tran, Christopher H. Dietrich, and Nancy A. Moran, "Co-cladogenesis Spanning Three Phyla:

Leafhoppers (Insecta: Hemiptera: Cicadellidae) and Their Dual Bacterial Symbionts," *Molecular Ecology* 15, no. 13 (2006): 4175–91.

57. See, for example, Shuji Shigenobu and Alex C. C. Wilson, "Genomic Revelations of a Mutualism: The Pea Aphid and Its Obligate Bacterial Symbiont," *Cellular and Molecular Life Sciences* 68, no. 8 (2011): 1297–309.
58. John P. McCutcheon, Bradon R. McDonald, and Nancy A. Moran, "Convergent Evolution of Metabolic Roles in Bacterial Co-symbionts of Insects," *Proceedings of the National Academy of Sciences of the United States of America* 106, no. 36 (2009): 15394–99; John P. McCutcheon and Nancy A. Moran, "Functional Convergence in Reduced Genomes of Bacterial Symbionts Spanning 200 My of Evolution," *Genome Biology and Evolution* 2 (2010): 708–18.
59. Part of the title of James T. van Leuven, Russell C. Meister, Chris Simon, and John P. McCutcheon, "Sympatric Speciation in a Bacterial Endosymbiont Results in Two Genomes with the Functionality of One," *Cell* 158, no. 6 (2014): 1270–80.
60. Ed Yong, "Snug as a Bug in a Bug in a Bug," *National Geographic,* June 20, 2013, https://www.nationalgeographic.com/science/phenomena/2013/06/20/snug-as-a-bug-in-a-bug-in-a-bug; reporting on Filip Husnik et al., "Horizontal Gene Transfer from Diverse Bacteria to an Insect Genome Enables a Tripartite Nested Mealybug Symbiosis," *Cell* 153, no. 7 (2013): 1567–78.
61. See, for example, Gitta Szabó, Frederik Schulz, Elena R. Toenshoff, Jean-Marie Volland, Omri M. Finkel, Shimshon Belkin, and Matthias Horn, "Convergent Patterns in the Evolution of Mealybug Symbioses Involving Different Intrabacterial Symbionts," *ISME Journal* 11, no. 3 (2017): 715–26.
62. This refers to a phenacoccinid mealybug; see Rosario Gil, Carlos Vargas-Chavez, Sergio López-Madrigal, Diego Santos-García, Amparo Latorre, and Andrés Moya, "*Tremblaya phenacola* PPER: An Evolutionary Beta-Gammaproteobacterium Collage," *ISME Journal* 12, no. 1 (2018): 124–35.
63. Along with input from the host; see, for example, Qiong Rao et al., "Genome Reduction and Potential Metabolic Complementation of the Dual Endosymbionts in the Whitefly *Bemisia tabaci,*" *BMC Genomics*

16, no. 1 (2015), 226, https://doi.org/10.1186/s12864-015-1379-6; also John P. McCutcheon and Nancy A. Moran, "Parallel Genomic Evolution and Metabolic Interdependence in an Ancient Symbiosis," *Proceedings of the National Academy of Sciences of the United States of America* 104, no. 49 (2007): 19392–97.

64. See, for example, John P. McCutcheon and Nancy A. Moran, "Extreme Genome Reduction in Symbiotic Bacteria," *Nature Reviews Microbiology* 10, no. 1 (2012): 13–26.
65. See, for example, Sergio López-Madrigal, Amparo Latorre, Manuel Porcar, Andrés Moya, and Rosario Gil, "Mealybugs Nested Endosymbiosis: Going into the 'Matryoshka' System in *Planococcus citri* in Depth," *BMC Microbiology* 13 (2013), 74, https://doi.org/10.1186/1471-2180-13-74, although these authors also emphasize that in some ways the intricate symbiosis between the bacteria is more akin to "a new composite organism rather than a bacterial consortium" (p. 9); also see Sergio López-Madrigal, Séverine Balmand, Amparo Latorre, Abdelaziz Heddi, Andrés Moya, and Rosario Gil, "How Does *Tremblaya princeps* Get Essential Proteins from Its Nested Partner *Moranella endobia* in the Mealybug *Planoccocus citri?*" *PLoS One* 8, no. 10 (2013), e77307, https://doi.org/10.1371/journal.pone.0077307; and Gordon M. Bennett and Nancy A. Moran, "Small, Smaller, Smallest: The Origins and Evolution of Ancient Dual Symbioses in a Phloem-Feeding Insect," *Genome Biology and Evolution* 5, no. 9 (2013): 1675–88.
66. See, for example, Nancy A. Moran, "Accelerated Evolution and Muller's Rachet in Endosymbiotic Bacteria," *Proceedings of the National Academy of Sciences of the United States of America* 93, no. 7 (1996): 2873–78.
67. See, for example, Ryuichi Koga and Nancy A. Moran, "Swapping Symbionts in Spittlebugs: Evolutionary Replacement of a Reduced Genome Symbiont," *ISME Journal* 8, no. 6 (2014): 1237–46; and Filip Husnik and John P. McCutcheon, "Repeated Replacement of an Intrabacterial Symbiont in the Tripartite Nested Mealybug Symbiosis," *Proceedings of the National Academy of Sciences of the United States of America* 113, no. 37 (2016): E5416–24.
68. Meng Mao, Xiushuai Yang, and Gordon M. Bennett, "Evolution of Host Support for Two Ancient Bacterial Symbionts with Differentially

Degraded Genomes in a Leafhopper Host," *Proceedings of the National Academy of Sciences of the United States of America* 115, no. 50 (2018): E11691–700.

69. See, for example, Husnik et al., "Horizontal Gene Transfer," 1567–78.
70. Shuji Shigenobu and David L. Stern, "Aphids Evolved Novel Secreted Proteins for Symbiosis with Bacterial Endosymbiont," *Proceedings of the Royal Society of London, Series B: Biological Sciences* 280, no. 1750 (2013): 20121952, https://doi.org/10.1098/rspb.2012.1952.
71. For *Prochloroccus,* see F. Partensky, W. R. Hess, and D. Vaulot, "*Prochlorococcus,* a Marine Photosynthetic Prokaryote of Global Significance," *Microbiology and Molecular Biology Reviews* 63, no. 1 (1999): 106–27; and for *Pelagibacter,* see Stephen J. Giovannoni et al., "Genome Streamlining in a Cosmopolitan Oceanic Bacterium," *Science* 309, no. 5738 (2005): 1242–45.
72. Pedro Flombaum et al., "Present and Future Global Distributions of the Marine Cyanobacteria *Prochlorococcus* and *Synechococcus,*" *Proceedings of the National Academy of Sciences of the United States of America* 110, no. 24 (2013): 9824–29.
73. J. Jeffrey Morris, Richard E. Lenski, and Erik R. Zinser, "The Black Queen Hypothesis: Evolution of Dependencies through Adaptive Gene Loss," *mBio* 3, no. 2 (2012), e00036-12, https://doi.org/10.1128/mbio.00036-12.
74. See, for example, Samay Pande and Christian Kost, "Bacterial Unculturability and the Formation of Intercellular Metabolic Networks," *Trends in Microbiology* 25, no. 5 (2017): 349–61.
75. See, for example, C. Kevin Boyce, Maciej A. Zwieniecki, George D. Cody, Chris Jacobsen, Sue Wirick, Andrew H. Knoll, and N. Michele Holbrook, "Evolution of Xylem Lignification and Hydrogel Transport Regulation," *Proceedings of the National Academy of Sciences of the United States of America* 101, no. 50 (2004): 17555–58.
76. Maciej A. Zwieniecki, Howard A. Stone, Andrea Leigh, C. Kevin Boyce, and N. Michele Holbrook, "Hydraulic Design of Pine Needles: One-Dimensional Optimization for Single-Vein Leaves," *Plant, Cell and Environment* 29, no. 5 (2006): 803–9.
77. Jean-Christophe Domec, Barbara Lachenbruch, Frederick C. Meinzer, David R. Woodruff, Jeffrey M. Warren, and Katherine A. McCulloh,

"Maximum Height in a Conifer Associated with Conflicting Requirements for Xylem Design," *Proceedings of the National Academy of Sciences of the United States of America* 105, no. 33 (2008): 12069–74.

78. Among extant floras, the maximum in both conifers and eucalypts is approximately one hundred meters, and among the latter this gigantism has evolved at least seven times; see D. Y. P. Tng, G. J. Williamson, G. J. Jordan, and D. M. J. S. Bowman, "Giant Eucalypts—Globally Unique Fire-Adapted Rain-Forest Trees?" *New Phytologist* 196, no. 4 (2012): 1001–14.
79. Howard C. Berg, "Marvels of Bacterial Behavior," *Proceedings of the American Philosophical Society* 150 (2006): 428–42; also Ferris Jabr, "How to Swim in Molasses," *Scientific American* 309, no. 2 (2013): 90–95.
80. E. M. Purcell's sparky article "Life at Low Reynolds Number," *American Journal of Physics* 45, no. 3 (1977): 3–11.
81. Stephen Puleo, *Dark Tide: The Great Boston Molasses Flood of 1919* (Boston: Beacon Press, 2003).
82. Charles Darwin, *On the Origin of Species by Means of Natural Selection, or the Preservation of Favoured Races in the Struggle for Life* (London: John Murray, 1859).
83. M. R. Drost, M. Muller, and J. W. M. Osse, "A Quantitative Hydrodynamical Model of Suction Feeding in Larval Fishes: The Role of Frictional Forces," *Proceedings of the Royal Society of London, Series B: Biological Sciences* 234, no. 1276 (1988): 263–81.
84. Victor China and Roi Holzmann, "Hydrodynamic Starvation in First-Feeding Fishes," *Proceedings of the National Academy of Sciences of the United States of America* 111, no. 22 (2014): 8083–88.
85. See, for example, Christian Jørgensen, Sonya K. Auer, and David N. Reznick, "A Model for Optimal Offspring Size, Including Live-Bearing and Parental Effects," *American Naturalist* 177, no. 5 (2011): E119–35.
86. See, for example, John T. Huber and John W. Beardsley, "A New Genus of Fairyfly, *Kikiki,* from the Hawaiian Island (Hymenoptera: Mymaridae)," *Proceedings of the Hawaiian Entomological Society* 34 (2000): 65–70. For only slightly larger examples among the beetles, see A. A. Polilov, "Anatomy of the Smallest Coleoptera, Featherwing Beetles of the Tribe Nanosellini (Coleoptera, Ptiliidae), and Limits of Insect Miniaturization," *Entomological Review* 88 (2008): 26–33.

87. Henrik K. Farkas, "Über die Eriophyiden (Acarina) Ungarns II. Beschreibung einer neuen Gattung and zwei neuer Arten," *Acta Zoologia Academiae Scientiarum Hungaricae* 7 (1961): 73–76; and S. Mahunka, "Aethiopische Tarsonemiden (Acari: Tarsonemida). II," *Acta Zoologia Academiae Scientiarum Hungaricae* 22 (1976): 69–96.
88. For an overview of arthropod miniaturization, see Alessandro Minelli and Giuseppe Fusco, "No Limits: Breaking Constraints in Insect Miniaturization," *Arthropod Structure and Development* 48 (2019): 4–11.
89. See, for example, John T. Huber and John S. Noyes, "A New Genus and Species of Fairyfly, *Tinkerbella nana* (Hymenoptera, Mymaridae), with Comments on Its Sister Genus *Kikiki,* and Discussion on Small Size Limits in Arthropods," *Journal of Hymenoptera Research* 32 (2013): 17–44.
90. D. Weihs and E. Barta, "Comb Wings for Flapping Flight at Extremely Low Reynolds Numbers," *AIAA Journal* 46, no. 1 (2008): 285–88; G. Davidi and D. Weihs, "Flow around a Comb Wing in Low-Reynolds-Number Flow," *AIAA Journal* 50, no. 1 (2012): 249–52; and Shannon K. Jones, Young J. J. Yun, Tyson L. Hedrick, Boyce E. Griffith, and Laura A. Miller, "Bristles Reduce the Force Required to 'Fling' Wings Apart in the Smallest Insects," *Journal of Experimental Biology* 219, no. 23 (2016): 3759–72.
91. Alexey A. Polilov, Natalia I. Reshetnikova, Pyotr N. Petrov, and Sergey E. Farisenkov, "Wing Morphology in Featherwing Beetles (Coleoptera: Ptiliidae): Features Associated with Miniaturization and Functional Scaling Analysis," *Arthropod Structure and Development* 48 (2019): 56–70.
92. Cathal Cummins, Madeleine Seale, Alice Macente, Daniele Certini, Enrico Mastropaolo, Ignazio Maria Viola, and Naomi Nakayama, "A Separated Vortex Ring Underlies the Flight of the Dandelion," *Nature* 562, no. 7727 (2018): 414–18.
93. Emma van der Woude, Hans M. Smid, Lars Chittka, and Martinus E. Huigens, "Breaking Haller's Rule: Brain-Body Size Isometry in a Minute Parasitic Wasp," *Brain, Behavior and Evolution* 81, no. 2 (2013): 86–92.
94. See also the report on navigational abilities in miniaturized ants by Ravindra Palavalli-Nettimi and Ajay Narendra, "Miniaturisation Decreases Visual Navigational Competence in Ants," *Journal of Experimental Biology* 221, no. 7 (2018), jeb177238, https://doi.org/10.1242/jeb.177238.

95. Anastasia A. Makarova, V. Benno Meyer-Rochow, and Alexey A. Polilov, "Morphology and Scaling of Compound Eyes in the Smallest Beetles (Coleoptera: Ptiliidae)," *Arthropod Structure and Development* 48 (2019): 83–97.
96. Allan W. Snyder, "Acuity of Compound Eyes: Physical Limitations and Design," *Journal of Comparative Physiology* 116 (1977): 161–82.
97. K. Kirschfeld, "The Resolution of Lens and Compound Eyes," in *Neural Principles in Vision*, ed. F. Zettler and R. Weiler (Berlin: Springer, 1976), 354–70.
98. For an excellent (and neglected) overview, see Hansjochem Autrum, "Performance Limits of Sensory Organs," *Interdisciplinary Science Reviews* 13, no. 1 (1988): 27–39.
99. Zuzanna Błaszczak, Moritz Kreysing, and Jochen Guck, "Direct Observation of Light Focusing by Single Photoreceptor Cell Nuclei," *Optics Express* 22, no. 9 (2014): 11043–60.
100. Eric J. Warrant, Almut Kelber, Anna Gislén, Birgit Greiner, Willi Ribi, and William T. Wcislo, "Nocturnal Vision and Landmark Orientation in a Tropical Halictid Bee," *Current Biology* 14, no. 15 (2004): 1309–18; also commentary by Michael F. Land, "Nocturnal Vision: Bees in the Dark," *Current Biology* 14, no. 15 (2004): R615–16.
101. See, for example, Hema Somanathan, Renee M. Borges, Eric J. Warrant, and Almut Kelber, "Visual Ecology of Indian Carpenter Bees. I: Light Intensities and Flight Activity," *Journal of Comparative Physiology, A* 194, no. 1 (2008): 97–107.
102. Hema Somanathan, Renée M. Borges, Eric J. Warrant, and Almut Kelber, "Nocturnal Bees Learn Landmark Colours in Starlight," *Current Biology* 18, no. 21 (2008): R996–97.
103. Almut Kelber, Anna Balkenius, and Eric J. Warrant, "Scotopic Colour Vision in Nocturnal Hawkmoths," *Nature* 419, no. 6910 (2002): 922–25; and Almut Kelber and Lina S. V. Roth, "Nocturnal Colour Vision—Not as Rare as We Might Think," *Journal of Experimental Biology* 209, no. 5 (2006): 781–88.
104. Anna Balkenius and Almut Kelber, "Colour Constancy in Diurnal and Nocturnal Hawkmoths," *Journal of Experimental Biology* 207, no. 19 (2004): 3307–16.

105. Eric Warrant, Magnus Oskarsson, and Henrik Malm, "The Remarkable Visual Abilities of Nocturnal Insects: Neural Principles and Bioinspired Night-Vision Algorithms," *Proceedings of the IEEE* 102, no. 10 (2014): 1411–26.
106. For an incisive overview, see Michael F. Land and Dan-Eric Nilsson, *Animal Eyes* (New York: Oxford University Press, 2002).
107. Hema Somanathan, Almut Kelber, Renée M. Borges, Rita Wallén, and Eric J. Warrant, "Visual Ecology of Indian Carpenter Bees. II: Adaptations of Eyes and Ocelli to Nocturnal and Diurnal Lifestyles," *Journal of Comparative Physiology, Series A: Neuroethology, Sensory, Neural, and Behavioral Physiology* 195, no. 6 (2009): 571–83.
108. See, for example, D. A. Baylor, T. D. Lamb, and K. W. Yau, "Responses of Retinal Rods to Single Photons," *Journal of Physiology* 288 (1979): 613–34.
109. Eric J. Warrant, "Vision in the Dimmest Habitats on Earth," *Journal of Comparative Physiology, Series A: Neuroethology, Sensory, Neural, and Behavioral Physiology* 190, no. 10 (2004): 765–89; David C. O'Carroll and Eric J. Warrant, "Vision in Dim Light: Highlights and Challenges," *Philosophical Transactions of the Royal Society of London, Series B: Biological Sciences,* 372, no. 1717 (2017), 20160062, https://doi.org/10.1098/rstb.2016.0062.
110. Johan Pahlberg and Alapakkam P. Sampath, "Visual Threshold Is Set by Linear and Nonlinear Mechanisms in the Retina That Mitigate Noise: How Neural Circuits in the Retina Improve the Signal-to-Noise Ratio of the Single-Photon Response," *BioEssays* 33, no. 6 (2011): 438–47.
111. See, for example, Anna Honkanen, Jouni Takalo, Kyösti Heimonen, Mikko Vähäsöyrinki, and Matti Weckström, "Cockroach Optomotor Responses below Single Photon Level," *Journal of Experimental Biology* 217, no. 23 (2014): 4262–68.
112. See also Anna Lisa Stöckl, David Charles O'Carroll, and Eric James Warrant, "Neural Summation in the Hawkmoth Visual System Extends the Limits of Vision in Dim Light," *Current Biology* 26, no. 6 (2016): 821–26; and commentary by Petri Ala-Laurila, "Visual Neuroscience: How Do Moths See to Fly at Night?" *Current Biology* 26, no. 6 (2016): R231–33.

113. Petri Ala-Laurila and Fred Rieke, "Coincidence Detection of Single-Photon Responses in the Inner Retina at the Sensitivity Limit of Vision," *Current Biology* 24, no. 24 (2014): 2888–98; and commentary by Paul R. Martin, "Neuroscience: Who Needs a Parasol at Night?" *Current Biology* 24, no. 24 (2014): R1164–66.
114. Jonathan N. Tinsley, Maxim I. Molodtsov, Robert Prevedel, David Wartmann, Jofre Espigulé-Pons, Mattias Lauwers, and Alipasha Vaziri, "Direct Detection of a Single Photon by Humans," *Nature Communications* 7 (2016), 12172, https://doi.org/10.1038/ncomms12172; and commentary by Davide Castelvecchi, "People Can Sense Single Photons," *Nature* (2016), https://doi.org/10.1038/nature.2016.20282.
115. Nicolas Brunner, Cyril Branciard, and Nicolas Gisin, "Possible Entanglement Detection with the Naked Eye," *Physical Review A* 78, no. 5 (2008), 052110, https://doi.org/10.1103/PhysRevA.78.052110.
116. N. David Mermin, "Physics: QBism Puts the Scientist Back into Science," *Nature* 507, no. 7493 (2014): 421–23.
117. J. Ebert and G. Westhoff, "Behavioural Examination of the Infrared Sensitivity of Rattlesnakes (*Crotalus atrox*)," *Journal of Comparative Physiology, Series A: Neuroethology, Sensory, Neural, and Behavioral Physiology* 192, no. 9 (2006): 941–47; they calculate a threshold value as 3.35×10^{-3} mW/cm^2.
118. Helmut Schmitz and Herbert Bousack, "Modelling a Historic Oil-Tank Fire Allows an Estimation of the Sensitivity of the Infrared Receptors in Pyrophilous *Melanophila* Beetles," *PLoS One* 7, no. 5 (2012), e37627, https://doi.org/10.1371/journal.pone.0037627; like extreme vision, they invoke a sort of neural summation.
119. In, for example, sharks: Carl G. Meyer, Kim N. Holland, and Yannis P. Papastamatiou, "Sharks Can Detect Changes in the Geomagnetic Field," *Journal of the Royal Society Interface* 2, no. 2 (2005): 129–30; and in birds: Michael Winklhofer, Evelyn Dylda, Peter Thalau, Wolfgang Wiltschko, and Roswitha Wiltschko, "Avian Magnetic Compass Can Be Tuned to Anomalously Low Magnetic Intensities," *Proceedings of the Royal Society of London, Series B: Biological Sciences* 280, no. 1763 (2013), 20130853, https://doi.org/10.1098/rspb.2013.0853.

120. P. Semm and R. C. Beason, "Responses to Small Magnetic Variations by the Trigeminal System of the Bobolink," *Brain Research Bulletin* 25, no. 5 (1990): 735–40, who report in the ophthalmic nerve of the bobolink (a migratory bird) a response in the ophthalmic nerve of 200 nanoteslas, equivalent to 0.4 percent of the Earth's magnetic field.
121. See, for example, J. D. Pettigrew, P. R. Manger, and S. L. Fine, "The Sensory World of the Platypus," *Philosophical Transactions of the Royal Society of London, Series B: Biological Sciences* 353, no. 1372 (1998): 1199–210.
122. Nicole U. Czech-Damal et al., "Electroreception in the Guiana Dolphin (*Sotalia guianensis*)," *Proceedings of the Royal Society of London, Series B: Biological Sciences* 279, no. 1729 (2012): 663–68.
123. J. D. Pettigrew, "Electroreception in Monotremes," *Journal of Experimental Biology* 202, no. 10 (1999): 1447–54.
124. See, for example, Stephen M. Kajiura and Kim N. Holland, "Electroreception in Juvenile Scalloped Hammerhead and Sandbar Sharks," *Journal of Experimental Biology* 205, no. 23 (2002): 3609–21; and Laura K. Jordan, Stephen M. Kajiura, and Malcolm S. Gordon, "Functional Consequences of Structural Differences in Stingray Sensory Systems. Part II: Electrosensory System," *Journal of Experimental Biology* 212, no. 19 (2009): 3044–50.
125. D. Petracchi and G. Cercignani, "A Comment on the Sensitivity of Fish to Low Electric Fields," *Biophysical Journal* 75, no. 4 (1998): 2117–18; see also James Glanz, "Physicists Advance into Biology," *Science* 272, no. 5262 (1996): 646–48.
126. Tian Yow Tsong, "Exquisite Sensitivity of Electroreceptor in Skates," *Biophysical Journal* 67, no. 4 (1994): 1367–68; Rob C. Peters, Lonneke B. M. Eeuwes, and Franklin Bretschneider, "On the Electrodetection Threshold of Aquatic Vertebrates with Ampullary or Mucous Gland Electroreceptor Organs," *Biological Reviews of the Cambridge Philosophical Society* 82, no. 3 (2007): 361–73. The latter also draw attention to the influence of the geomagnetic field.
127. Nor should it be assumed that all species are equally sensitive. Yellow stingrays, for example, outstrip cownose rays, but the former seek out moving prey whereas the schooling habits of the cownose rays may

lead to interference and require supplementation by vision and smell; see Christine N. Bedore, Lindsay L. Harris, and Stephen M. Kajiura, "Behavioral Responses of Batoid Elasmobranchs to Prey-Simulating Electric Fields Are Correlated to Peripheral Sensory Morphology and Ecology," *Zoology* 117, no. 2 (2014): 95–103. Body shape is also a factor, and the iconic shape of the skate is not only linked to its swimming but to the distribution of the electrosensory array; see, for example, Marcelo Camperi, Timothy C. Tricas, and Brandon R. Brown, "From Morphology to Neural Information: The Electric Sense of the Skate," *PLoS Computational Biology* 3, no. 6 (2007), e113, https://doi.org/10.1371/journal.pcbi.0030113. Yet despite all this, the limits of detection do indeed seem to be fantastically small; see Christine N. Bedore and Stephen M. Kajiura, "Bioelectric Fields of Marine Organisms: Voltage and Frequency Contributions to Detectability by Electroreceptive Predators," *Physiological and Biochemical Zoology* 86, no. 3 (2013): 298–311.

128. Angel Caputi, "How Do Electric Eels Generate a Voltage, and Why Don't They Get Shocked?" *Scientific American* 294, no. 3 (2006): 104.
129. For an overview, see Gary J. Rose, "Insights into Neural Mechanisms and Evolution of Behaviour from Electric Fish," *Nature Reviews Neuroscience* 5, no. 12 (2004): 943–51.
130. Sara K. Tallarovic and Harold H. Zakon, "Electric Organ Discharge Frequency Jamming during Social Interactions in Brown Ghost Knifefish, *Apteronotus leptorhynchus*," *Animal Behaviour* 70, no. 6 (2005): 1355–65.
131. See, for example, Gary Rose and Walter Heiligenberg, "Temporal Hyperacuity in the Electric Sense of Fish," *Nature* 318, no. 6042 (1985): 178–80.
132. Walter Heiligenberg, C. Baker, and J. Matsubara, "The Jamming Avoidance Response in *Eigenmannia* Revisited: The Structure of a Neuronal Democracy," *Journal of Comparative Physiology* 127 (1978): 267–86.
133. See, for example, Masashi Kawasaki, "Sensory Hyperacuity in the Jamming Avoidance Response of Weakly Electric Fish," *Current Opinion in Neurobiology* 7, no. 4 (1997): 473–79.

134. In the region of the brain known as the electrosensory lateral line lobe; see Masashi Kawasaki, "Temporal Hyperacuity in the Gymnotiform Electric Fish, *Eigenmannia*," *American Zoologist* 33 (1993): 86–93.
135. Richard A. Altes, "Ubiquity of Hyperacuity," *Journal of the Acoustical Society of America* 85, no. 2 (1989): 943–52.
136. See, for example, G. Westheimer, "Editorial: Visual Acuity and Hyperacuity," *Investigative Ophthalmology and Visual Science* 14, no. 8 (1975): 570–72.
137. Some of this work is based on cats, but humans show much the same tolerances; see, for example, N. V. Swindale and M. S. Cynader, "Vernier Acuity of Neurones in Cat Visual Cortex," *Nature* 319, no. 6054 (1986): 591–93; and Kathryn M. Murphy and Donald E. Mitchell, "Vernier Acuity of Normal and Visually Deprived Cats," *Vision Research* 31, no. 2 (1991): 253–66.
138. See, for example, Ying Zhang and R. Clay Reid, "Single-Neuron Responses and Neuronal Decisions in a Vernier Task," *Proceedings of the National Academy of Sciences of the United States of America* 102, no. 9 (2005): 3507–12.
139. S. P. McKee and G. Westheimer, "Improvement in Vernier Acuity with Practice," *Perception and Psychophysics* 24, no. 3 (1978): 258–62.
140. C. Bushdid, M. O. Magnasco, L. B. Vosshall, and A. Keller, "Humans Can Discriminate More Than 1 Trillion Olfactory Stimuli," *Science* 343, no. 6177 (2014): 1370–72.
141. Anna Maria Angioy, Alessandro Desogus, Iole Tomassini Barbarossa, Peter Anderson, and Bill S. Hansson, "Extreme Sensitivity in an Olfactory System," *Chemical Senses* 28, no. 4 (2003): 279–84.
142. See the remark (p. 134) in the review by Karl-Ernst Kaissling, "Chemo-electrical Transduction in Insect Olfactory Receptors," *Annual Review of Neuroscience* 9 (1986): 121–45; see also Anna Menini, Christiana Picco, and Stuart Firestein, "Quantal-like Current Fluctuations Induced by Odorants in Olfactory Receptor Cells," *Nature* 373, no. 6513 (1995): 435–37.
143. Trese Leinders-Zufall, Andrew P. Lane, Adam C. Puche, Weidong Ma, Milos V. Novotny, Michael T. Shipley, and Frank Zufall, "Ultrasensitive

Pheromone Detection by Mammalian Vomeronasal Neurons," *Nature* 405, no. 6788 (2000): 792–95, correction: *Nature* 408 (2000): 616.

144. M. Sumper, E. Berg, S. Wenzl, and K. Godl, "How a Sex Pheromone Might Act at a Concentration below 10^{-16} M," *EMBO Journal* 12, no. 3 (1993): 831–36.

145. See, for example, T. Strünker, L. Alvarez, and U. B. Kaupp, "At the Physical Limit—Chemosensation in Sperm," *Current Opinion in Neurobiology* 34 (2015): 110–16, with an emphasis on the extraordinary sensitivity of sea-urchin sperm.

146. See the essay by John P. McGann, "Poor Human Olfaction Is a 19th-Century Myth," *Science* 356, no. 6338 (2017), eaam7263, https://doi.org/10.1126/science.aam7263 (and summary on p. 597).

147. Flavia Mancini, Chiara F. Sambo, Juan D. Ramirez, David L. H. Bennett, Patrick Haggard, and Gian Domenico Iannetti, "A Fovea for Pain at Fingertips," *Current Biology* 23, no. 6 (2013): 496–500.

148. Eva K. Sawyer and Kenneth C. Catania, "Somatosensory Organ Topography across the Star of the Star-Nosed Mole (*Condylura cristata*)," *Journal of Comparative Neurobiology* 524, no. 5 (2016): 917–29.

149. See, for example, Kenneth C. Catania, "A Nose That Looks Like a Hand and Acts Like an Eye: The Unusual Mechanosensory System of the Star-Nosed Mole," *Journal of Comparative Physiology, Series A: Neuroethology, Sensory, Neural, and Behavioral Physiology* 185, no. 4 (1999): 367–72; and Kenneth C. Catania and Fiona E. Remple, "Tactile Foveation in the Star-Nosed Mole," *Brain, Behavior and Evolution* 63, no. 1 (2004): 1–12.

150. Kenneth C. Catania and Fiona E. Remple, "Asymptotic Prey Profitability Drives Star-Nosed Moles to the Foraging Speed Limit," *Nature* 433, no. 7025 (2005): 519–22.

151. J. M. Loomis, "An Investigation of Tactile Hyperacuity," *Sensory Processes* 3, no. 4 (1979): 289–302.

152. A. W. Mills, "On the Minimum Audible Angle," *Journal of the Acoustical Society of America* 30 (1958): 237–46.

153. Stefan Schöneich and Berthold Hedwig, "Hyperacute Directional Hearing and Phonotactic Steering in the Cricket (*Gryllus bimaculatus*

de Geer)," *PLoS One* 5, no. 12 (2010), e15141, https://doi.org/10.1371/journal.pone.0015141.

154. Dennis McFadden, "Sex Differences in the Auditory System," *Developmental Neuropsychology* 14, no. 2–3 (1998): 261–98.
155. Jerry V. Tobias, "Consistency of Sex Differences in Binaural-Beat Perception," *International Audiology* 4, no. 2 (1965): 179–82.
156. Jerry V. Tobias, "Curious Binaural Phenomena," in *Foundations of Modern Auditory Theory*, vol. 2, ed. Jerry V. Tobias (New York: Academic Press, 1972), 465–86.
157. Tobias, "Curious Binaural Phenomena," 474.
158. Hannah M. Moir, Joseph C. Jackson, and James F. C. Windmill, "Extremely High Frequency Sensitivity in a 'Simple' Ear," *Biology Letters* 9, no. 4 (2013), 20130241, https://doi.org/10.1098/rsbl.2013.0241.
159. Winfried Denk and Watt W. Webb, "Thermal-Noise-Limited Transduction Observed in Mechanosensory Receptors of the Inner Ear," *Physical Review Letters* 63, no. 2 (1989): 207–10.
160. See, for example, Martin Braun, "Tuned Hair-Cells for Hearing, but Tuned Basilar-Membrane for Overload Protection: Evidence from Dolphins, Bats, and Desert Rodents," *Hearing Research* 78, no. 1 (1994): 98–114.
161. See the overview by A. J. Hudspeth, "How the Ear's Works Work," *Nature* 341, no. 6241 (1989): 397–404.
162. Martin C. Göpfert and Daniel Robert, "Nanometre-Range Acoustic Sensitivity in Male and Female Mosquitoes," *Proceedings of the Royal Society of London, Series B: Biological Sciences* 267, no. 1442 (2000): 453–57.
163. Friedrich G. Barth, "Spider Mechanoreceptors," *Current Opinion in Neurobiology* 14, no.4 (2004): 415–22.
164. Tateo Shimozawa, Jun Murakami, and Tsuneko Kumagai, "Cricket Wind Receptors: Thermal Noise for the Highest Sensitivity Known," in *Sensors and Sensing in Biology and Engineering*, ed. Friedrich G. Barth, Joseph A. C. Humphrey, and Timothy W. Secomb (New York: Springer-Verlag: Wien, 2003), 145–57.

165. Marie Dacke, Emily Baird, Marcus Byrne, Clarke H. Scholtz, and Eric J. Warrant, "Dung Beetles Use the Milky Way for Orientation," *Current Biology* 23, no. 4 (2013): 298–300; commentary in James L. Gould, "Animal Navigation: A Galaxy of Cues," *Current Biology* 23, no. 4 (2013): R149–50.
166. Simon Conway Morris, "Predicting What Extra-terrestrials Will Be Like: And Preparing for the Worst," *Philosophical Transactions of the Royal Society of London, Series A: Mathematical, Physical, and Engineering Sciences* 369, no. 1936 (2011): 555–71.
167. For a succinct review, see William G. Eberhard and William T. Wcislo, "Plenty of Room at the Bottom?" *American Scientist* 100 (May–June 2012): 226–33.
168. See, for example, Stefan Fischer, Zhiyuan Lu, and Ian A. Meinertzhagen, "From Two to Three Dimensions: The Importance of the Third Dimension for Evaluating the Limits to Neuronal Miniaturization in Insects," *Journal of Comparative Neurobiology* 526, no. 4 (2018): 653–62.
169. Alexey A. Polilov, "The Smallest Insects Evolve Anucleate Neurons," *Arthropod Structure and Development* 41 (2012): 29–34; also for evidence for associative learning in a 0.5 mm beetle with a brain of approximately 10,000 cells, see Alexey A. Polilov, Anastasia A. Makarova, and Uliana K. Kolesnikova, "Cognitive Abilities with a Tiny Brain: Neuronal Structures and Associative Learning in the Minute *Nephanes titan* (Coleoptera: Ptiliidae)," *Arthropod Structure and Development* 48 (2019): 98–102.
170. Michel A. Hofman, "Brain Evolution in Hominids: Are We at the End of the Road?" in *Evolutionary Anatomy of the Primate Cerebral Cortex*, ed. Dean Falk and Kathleen R. Gibson (Cambridge: Cambridge University Press, 2001), 113–27.
171. For a robustly skeptical view of artificial intelligence, see J. Mark Bishop, "Artificial Intelligence Is Stupid and Causal Reasoning Will Not Fix It," *Frontiers in Psychology* 11 (2021), 513474, https://doi.org/10.3389/fpsyg.2020.513474; crucially he identifies a lack of causal reasoning that may find an echo in animal mentality.

CHAPTER 2

1. For a useful corrective, B. Davis Barnes, Judith A. Sclafani, and Andrew Zaffos, "Dead Clades Walking Are a Pervasive Macroevolutionary Pattern," *Proceedings of the National Academy of Sciences of the United States of America* 118, no. 15 (2021), e2019208118.
2. Douglas H. Erwin and Mary L. Droser, "Elvis Taxa," *Palaios* 8, no. 6 (1993): 623–24.
3. The gastropods associated with the end-Permian extinctions (PTE) provide one such example. Here we see the place of the Permian pseudozygopleurids being taken by the Triassic "genuine" zygopleurids; see Alexander Nützel, "Recovery of Gastropods in the Early Triassic," *Comptes Rendus Palevol* 4, no. 6–7 (2005): 433–47.
4. Thomas E. Yancey and Donald W. Boyd, "Revision of the Alatoconchidae: A Remarkable Family of Permian Bivalves," *Palaeontology* 26, no. 3 (1983): 497–520.
5. See, for example, Yukio Isozaki and Dunja Aljinović, "End-Guadalupian Extinction of the Permian Gigantic Bivalve Alatoconchidae: End of Gigantism in Tropical Seas by Cooling," *Palaeogeography, Palaeoclimatology, Palaeoecology* 284, no. 1–2 (2009): 11–21.
6. This view is contested by K. Asato et al., "Morphology, Systematics and Paleoecology of *Shikamaia*, Aberrant Permian Bivalves (Alatoconchidae: Ambonychioidea) from Japan," *Paleontological Research* 21, no. 4 (2017): 358–79, at least so far as it pertains to *Shikamaia*.
7. See, for example, Fayao Chen, Wuqiang Xue, Jiaxin Yan, Paul B. Wignall, Qi Meng, Jinxiong Luo, and Qinglai Feng, "Alatoconchids: Giant Permian Bivalves from South China," *Earth-Science Reviews* 179 (2018): 147–67.
8. Yukio Isozaki, "Guadalupian (Middle Permian) Giant Bivalve Alatoconchidae from a Mid-Panthalassan Paleo-Atoll Complex in Kyushu, Japan: A Unique Community Associated with Tethyan Fusulines and Corals," *Proceedings of the Japanese Academy, B* 82, no. 1 (2006): 25–32.
9. The end-Guadalupian event; see, for example, Fayao Chen, Wuqiang Xue, Jiaxin Yan, and Qi Meng, "The Implications of the Giant Bivalve

Family Alatoconchidae for the End-Guadalupian (Middle Permian) Extinction Event," *Geological Journal* (2021), https://doi.org/10.1002/gj.4151; also Isozaki and Aljinović, "End-Guadalupian Extinction."

10. T. A. de Freitas, F. Brunton, and T. Bernecker, "Silurian Megalodont Bivalves of the Canadian Arctic and Australia: Paleoecology and Evolutionary Significance," *Palaios* 8, no. 5 (1993): 450–64.
11. See, for example, Thomas E. Yancey, George D. Stanley Jr., Werner E. Piller, and Mark A. Woods, "Biogeography of the Late Triassic Wallowaconchid Megalodontoid Bivalves," *Lethaia* 38 (2005): 351–65.
12. George D. Stanley Jr., Thomas E. Yancey, and Hannah M. E. Shepherd, "Giant Upper Triassic Bivalves of Wrangellia, Vancouver Island, Canada," *Canadian Journal of Earth Sciences* 50 (2013): 142–47.
13. Valentin Fischer, Nathalie Bardet, Myette Guiomar, and Pascal Godefroit, "High Diversity in Cretaceous Ichthyosaurs from Europe prior to Their Extinction," *PLoS One* 9, no. 1 (2014), e84709.
14. Nicholas R. Longrich, Nathalie Bardet, Anne S. Schulp, and Nour-Eddine Jalil, "*Xenodens calminechari* gen. et sp. nov., a Bizarre Mosasaurid (Mosasauridae, Squamata) with Shark-Like Cutting Teeth from the Upper Maastrichtian of Morocco, North Africa," *Cretaceous Research* 123 (2021), 104764, as well as a variety of other feeding types, including shell crushing (durophagy).
15. Johan Lindgren, Michael W. Caldwell, and John W. M. Jagt, "New Data on the Postcranial Anatomy of the California Mosasaur *Plotosaurus bennisoni* (Camp, 1942) (Upper Cretaceous: Maastrichtian), and the Taxonomic Status of *P. tuckeri* (Camp, 1942)," *Journal of Vertebrate Paleontology* 28, no. 4 (2008): 1043–54.
16. See p. 159 of Johan Lindgren, John W. M. Jagt, and Michael W. Caldwell, "A Fishy Mosasaur: The Axial Skeleton of *Plotosaurus* (Reptilia, Squamata) Reassessed," *Lethaia* 40 (2007): 153–60.
17. Johan Lindgren, Michael W. Caldwell, Takuya Konishi, and Luis M. Chiappe, "Convergent Evolution in Aquatic Tetrapods: Insights from an Exceptional Fossil Mosasaur," *PLoS One* 5, no. 8 (2010), e11998.
18. For an excellent overview, see Z.-X. Luo, "Transformation and Diversification in Early Mammal Evolution," *Nature* 450, no. 7172 (2007): 1011–19; also Roger A. Close, Matt Friedman, Graeme T. Lloyd, and Roger B. J.

Benson, "Evidence for a Mid-Jurassic Adaptive Radiation in Mammals," *Current Biology* 25, no. 16 (2015): 2137–42.

19. Concerning the mammaliaforms known as docodontans, see Zhe-Xi Luo, Qing-Jin Meng, Qiang Ji, Di Liu, Yu-Guang Zhang, and April I. Neander, "Mammalian Evolution. Evolutionary Development in Basal Mammaliaforms as Revealed by a Docodontan," *Science* 347, no. 6223 (2015): 760–64; and Thomas Martin, "Postcranial Anatomy of *Haldanodon exspectatus* (Mammalia, Docodonta) from the Late Jurassic (Kimmeridgian) of Portugal and Its Bearing for Mammalian Evolution," *Zoological Journal of the Linnean Society* 145, no. 2 (2005): 219–48. With respect to fossoriality in the eutriconodontans and tritylodontids, see Fangyuan Mao, Chi Zhang, Cunyu Liu, and Jin Meng, "Fossoriality and Evolutionary Development in Two Cretaceous Mammaliamorphs," *Nature* 592, no. 7855 (2021): 577–82.
20. Zhe Xi Luo and John R. Wible, "A Late Jurassic Digging Mammal and Early Mammalian Diversification," *Science* 308, no. 5718 (2005): 103–7.
21. Qiang Ji, Zhe-Xi Luo, Chong-Xi Yuan, and Alan R. Tabrum, "A Swimming Mammaliaform from the Middle Jurassic and Ecomorphological Diversification of Early Mammals," *Science* 311, no. 5764 (2006): 1123–27; with commentary by Thomas Martin, "Paleontology. Early Mammalian Evolutionary Experiments," *Science* 311, no. 5764 (2006): 1109–10.
22. See, for example, Pamela G. Gill et al., "Dietary Specializations and Diversity in Feeding Ecology of the Earliest Stem Mammals," *Nature* 512, no. 7514 (2014): 303–5; Zhe Xi Luo et al., "New Evidence for Mammaliaform Ear Evolution and Feeding Adaptation in a Jurassic Ecosystem," *Nature* 548, no. 7667 (2017): 326–29.
23. Meng Chen and Gregory P. Wilson, "A Multivariate Approach to Infer Locomotor Modes in Mesozoic Mammals," *Paleobiology* 41, no. 2 (2015): 280–312.
24. See, for example, Qing-Jin Meng, Qiang Ji, Yu-Guang Zhang, Di Liu, David M. Grossnickle, and Zhe-Xi Luo, "An Arboreal Docodont from the Jurassic and Mammaliaform Ecological Diversification," *Science* 347, no. 6223 (2015): 764–68. Their suggestion, however, that this animal was

a sap or gum feeder seems less likely; see John R. Wible and Anne M. Burrows, "Does the Jurassic *Agilodocodon* (Mammaliaformes, Docodonta) Have Any Exudativorous Dental Features?" *Palaeontologia Polonica* 67 (2016): 289–99.

25. Jin Meng, Yaoming Hu, Yuanqing Wang, Xiaolin Wang, and Chuankui Li, "A Mesozoic Gliding Mammal from Northeastern China," *Nature* 444, no. 7121 (2006): 889–93 [also correction of a specific name in *Nature* 446, no. 7131 (2007): 102]; and parallel reports by Qing-Jin Meng et al., "New Gliding Mammaliaforms from the Jurassic," *Nature* 548, no. 7667 (2017): 291–96; Zhe-Xi Luo et al., "New Evidence for Mammaliaform Ear Evolution and Feeding Adaptation in a Jurassic Ecosystem," *Nature* 548, no. 7667 (2017): 326–29; and Gang Han, Fangyuan Mao, Shundong Bi, Yuanqing Wang, and Jin Meng, "A Jurassic Gliding Euharamiyidan Mammal with an Ear of Five Auditory Bones," *Nature* 551, no. 7681 (2017): 451–56.
26. Zhe-Xi Luo, Irina Ruf, and Thomas Martin, "The Petrosal and Inner Ear of the Late Jurassic Cladotherian Mammal *Dryolestes leiriensis* and Implications for Ear Evolution in Therian Mammals," *Zoological Journal of the Linnean Society* 166, no. 2 (2012): 433–63.
27. Thomas Martin, Jesús Marugán-Lobón, Romain Vullo, Hugo Martín-Abad, Zhe-Xi Luo, and Angela D. Buscalioni. "A Cretaceous Eutriconodont and Integument Evolution in Early Mammals," *Nature* 526, no. 7573 (2015): 380–84.
28. Erika J. Edwards, "The Inevitability of C_4 Photosynthesis," *eLife* 3 (2014), e03702.
29. See, for example, Szabolcs Lengyel, Aaron D. Gove, Andrew M. Latimer, Jonathan D. Majer, and Robert R. Dunn, "Convergent Evolution of Seed Dispersal by Ants, and Phylogeny and Biogeography in Flowering Plants: A Global Survey," *Perspectives in Plant Ecology, Evolution and Systematics* 12, no. 1 (2010): 43–55.
30. See Christopher D. K. Cook, *Aquatic Plant Book* (Amsterdam: SPB Academic, 1990).
31. As interest in convergence has grown, so there has been a flurry of publications. Although tangential to my present purposes, contributions such as those by Jacob D. Washburn, Kevin A. Bird, Gavin C. Conant,

and J. Chris Pires, "Convergent Evolution and the Origin of Complex Phenotypes in the Age of Systems Biology," *International Journal of Plant Sciences* 177, no. 4 (2016): 305–18; and D. Luke Mahler, Marjorie G. Weber, Catherine E. Wagner, and Travis Ingram, "Pattern and Process in the Comparative Study of Convergent Evolution," *American Naturalist* 190 (2017): S13–S28, are useful markers.

32. So too this conceit has led to various responses. See, for example, Derek D. Turner, "Gould's Replay Revisited," *Biology and Philosophy* 26, no. 1 (2011): 65–79; Alexander E. Lobkovsky and Eugene V. Koonin, "Replaying the Tape of Life: Quantification of the Predictability of Evolution," *Frontiers in Genetics* 3 (2012), 246; Trevor Pearce, "Convergence and Parallelism in Evolution: A Neo-Gouldian Account," *British Journal for the Philosophy of Science* 63, no. 2 (2011): 429–48; and Russell Powell and Carlos Mariscal, "Convergent Evolution as Natural Experiment: The Tape of Life Reconsidered," *Interface Focus* 5, no. 6 (2015), 20150040.
33. John Beatty, "Replaying Life's Tape," *Journal of Philosophy* 103, no. 7 (2006): 336–62.
34. See also critiques by Russell Powell, "Contingency and Convergence in Macroevolution: A Reply to John Beatty," *Journal of Philosophy* 106, no. 7 (2009): 390–403; and Russell Powell, "Convergent Evolution and the Limits of Natural Selection," *European Journal of the Philosophy of Science* 2, no. 3 (2012): 355–73.
35. See, for example, Arnaud Martin and Virginie Orgogozo, "The Loci of Repeated Evolution: A Catalog of Genetic Hotspots of Phenotypic Variation," *Evolution* 67, no. 5 (2013): 1235–50.
36. J. A. Russell, J. G. Sanders, and C. S. Moreau, "Hotspots for Symbiosis: Function, Evolution, and Specificity of Ant-Microbe Associations from Trunk to Tips of the Ant Phylogeny (Hymenoptera: Formicidae)," *Myrmecological News* 24 (2017): 43–69.
37. Nicolas A. Blouin and Christopher E. Lane, "Red Algal Parasites: Models for a Life History Evolution That Leaves Photosynthesis behind Again and Again," *BioEssays* 34, no. 3 (2012): 226–35; also Jillian M. Freese and Christopher E. Lane, "Parasitism Finds Many Solutions to the Same Problems in Red Algae (Florideophyceae, Rhodophyta)," *Molecular and Biochemical Parasitology* 214 (2017): 105–11.

38. Daniel L. Nickrent, "Parasitic Angiosperms: How Often and How Many?" *Taxon* 69, no. 1 (2020): 5–27.
39. Christopher J. Still, Joseph A. Berry, G. James Collatz, Ruth S. DeFries, "Global Distribution of C_3 and C_4 Vegetation: Carbon Cycle Implications," *Global Biogeochemical Cycles* 17, no. 1 (2003), 1006.
40. See, however, Defeng Shen et al., "A Homeotic Mutation Changes Legume Nodule Ontogeny into Actinorhizal-Type Ontogeny," *Plant Cell* 32, no. 6 (2020): 1868–85, who argue for a common ancestor, possibly actinorhizal.
41. For an overview, see Carole Santi, Didier Bogusz, and Claudine Franche, "Biological Nitrogen Fixation in Non-legume Plants," *Annals of Botany* 111, no. 5 (2013): 743–67.
42. See, for example, Jeff J. Doyle, "Phylogenetic Perspectives on the Origins of Nodulation," *Molecular Plant-Microbe Interactions* 24, no. 11 (2011): 1289–95.
43. See, for example, Sergio Svistoonoff et al., "The Independent Acquisition of Plant Root Nitrogen-Fixing Symbiosis in Fabids Recruited the Same Genetic Pathway for Nodule Organogenesis," *PLoS One* 8, no. 5 (2013), e64515.
44. Reported from the famous Rhynie Chert by T. N. Taylor, W. Remy, H. Hass, and H. Kerp, "Fossil Arbuscular Mycorrhizae from the Early Devonian," *Mycologia* 87, no. 4 (1995): 560–73; see also Carla J. Harper, Christopher Walker, Andrew B. Schwendemann, Hans Kerp, and Michael Krings, "*Archaeosporites rhyniensis* gen. et sp. nov. (Glomeromycota, Archaeosporaceae) from the Lower Devonian Rhynie Chert: A Fungal Lineage Morphologically Unchanged for More Than 400 Million Years," *Annals of Botany* 126, no. 5 (2020): 915–28.
45. See, for example, the overview by Martin Parniske, "Arbuscular mycorrhiza: The Mother of Plant Root Endosymbioses," *Nature Reviews: Microbiology* 6, no. 10 (2008): 763–75.
46. Hong-Lei Li et al., "Large-Scale Phylogenetic Analyses Reveal Multiple Gains of Actinorhizal Nitrogen-Fixing Symbioses in Angiosperms Associated with Climate Change," *Science Reports* 5 (2015), 14023. They suggest elevated temperatures and CO_2 facilitated the process in plants growing in low-nutrient soils.

47. Gijsbert D. A. Werner, William K. Cornwell, Janet I. Sprent, Jens Kattge, and E. Toby Kiers, “A Single Evolutionary Innovation Drives the Deep Evolution of Symbiotic N_2-Fixation in Angiosperms,” *Nature Communications* 5 (2014): 4087.
48. For a more skeptical view of such a precursor, at least so far as the rhizobial symbiosis is concerned, see Robin van Velzen et al., “Comparative Genomics of the Nonlegume *Parasponia* Reveals Insights into Evolution of Nitrogen-Fixing Rhizobium Symbioses,” *Proceedings of the National Academy of Sciences of the United States of America* 115, no. 20 (2018): E4700–E4709; and Robin van Velzen, Jeff J. Doyle, and Rene Geurts, “A Resurrected Scenario: Single Gain and Massive Loss of Nitrogen-Fixing Nodulation,” *Trends in Plant Science* 24, no. 1 (2019): 49–57.
49. Jeff Doyle, “Chasing Unicorns: Nodulation Origins and the Paradox of Novelty,” *American Journal of Botany* 103, no. 11 (2016): 1865–68.
50. Mark Pagani, James C. Zachos, Katherine H. Freeman, Brett Tipple, and Stephen Bohaty, “Marked Decline in Atmospheric Carbon Dioxide Concentrations during the Paleogene,” *Science* 309, no. 5734 (2005): 600–603; and with specific reference to C_4 grasses, see P.-A. Christin et al., “Oligocene CO_2 Decline Promoted C_4 Photosynthesis in Grasses,” *Current Biology* 18, no. 1 (2008): 37–43.
51. See, for example, Richard M. Sharpe and Sascha Offermann, “One Decade after the Discovery of Single-Cell C_4 Species in Terrestrial Plants: What Did We Learn about the Minimal Requirements of C_4 Photosynthesis?” *Photosynthesis Research* 119, no. 1–2 (2014): 169–80.
52. For an excellent overview, see Rowan F. Sage, “A Portrait of the C_4 Photosynthetic Family on the 50th Anniversary of Its Discovery: Species Number, Evolutionary Lineages, and Hall of Fame,” *Journal of Experimental Botany* 68, no. 2 (2017): e11–e28.
53. See, for example, Pascal-Antoine Christin and Colin P. Osborne, “The Evolutionary Ecology of C_4 Plants,” *New Phytologist* 204, no. 4 (2014): 765–81.
54. See Ben P. Williams, Iain G. Johnston, Sarah Covshoff, and Julian M. Hibberd, “Phenotypic Landscape Inference Reveals Multiple Evolutionary Paths to C_4 Photosynthesis,” *eLife* 2 (2013), e00961.

55. For an overview, see Ben P. Williams, Sylvain Aubry, and Julian M. Hibberd, "Molecular Evolution of Genes Recruited into C_4 Photosynthesis," *Trends in Plant Sciences* 17, no. 4 (2012): 213–20.
56. See, for example, Pascal-Antoine Christin, Blaise Petitpierre, Nicolas Salamin, Lucie Büchi, and Guillaume Besnard, "Evolution of C_4 Phosphoenolpyruvate Carboxykinase in Grasses, from Genotype to Phenotype," *Molecular Biology and Evolution* 26, no. 2 (2009): 357–65.
57. See the comparison of amino acid substitutions in Josh J. Rosnow, Marc A. Evans, Maxim V. Kapralov, Asaph B. Cousins, Gerald E. Edwards, and Eric H. Roalson, "Kranz and Single-Cell Forms of C_4 Plants in the Subfamily Suaedoideae Show Kinetic C_4 Convergence for PEPC and Rubisco with Divergent Amino Acid Substitutions," *Journal of Experimental Botany* 66, no. 22 (2015): 7347–58.
58. Erika J. Edwards, "Inevitability of C_4 Photosynthesis," *eLife* (2014): e03702.
59. Described as a "Mount Fuji fitness landscape" by David Heckmann et al., "Predicting C_4 Photosynthesis Evolution: Modular, Individually Adaptive Steps on a Mount Fuji Fitness Landscape," *Cell* 153, no. 7 (2013): 1579–88; with commentary by Karlyn D. Beer, Mónica V. Orellana, and Nitin S. Baliga, "Modeling the Evolution of C_4 Photosynthesis," *Cell* 153, no. 7 (2013): 1427–29.
60. Still et al., "Global Distribution of C_3 and C_4 Vegetation," 1006.
61. The acronym for six subfamilies of the poacean grasses: Panicoideae, Arundinoideae, Chloridoideae, Micrairoideae, Aristidoideae, Danthonioideae; see Robert J. Soreng et al., "A Worldwide Phylogenetic Classification of the Poaceae (Gramineae) II: An Update and a Comparison of Two 2015 Classifications," *Journal of Systematics and Evolution* 55, no. 4 (2017): 259–90.
62. Erika J. Edwards and R. Matthew Ogburn, "Angiosperm Responses to a Low-CO_2 World: CAM and C_4 Photosynthesis as Parallel Evolutionary Trajectories," *International Journal of Plant Sciences* 173, no. 6 (2012): 724–33.
63. See the overview by O. H. Sayed, "Crassulacean Acid Metabolism 1975–2000, a Check List," *Photosynthetica* 39, no. 3 (2001): 339–52.
64. See, for example, Sylvain Aubry, Naomi J. Brown, and Julian M. Hibberd, "The Role of Proteins in C_3 Plants Prior to Their Recruitment into

the C_4 Pathway," *Journal of Experimental Botany* 62, no. 9 (2011): 3049–59; and Naomi J. Brown,, Christine A. Newell, Susan Stanley, Jit E. Chen, Abigail J. Perrin, Kaisa Kajala, and Julian M. Hibberd, "Independent and Parallel Recruitment of Preexisting Mechanisms Underlying C_4 Photosynthesis," *Science* 331, no. 6023 (2011): 1436–39.

65. See, for example, Ivan Reyna-Llorens and Julian M. Hibberd, "Recruitment of Pre-existing Networks during the Evolution of C_4 Photosynthesis," *Philosophical Transactions of the Royal Society of London, Series B: Biological Sciences* 372, no. 1730 (2017): 20160386.
66. See, for example, Pascal-Antoine Christin, Susanna F. Boxall, Richard Gregory, Erika J. Edwards, James Hartwell, and Colin P. Osborne, "Parallel Recruitment of Multiple Genes into C_4 Photosynthesis," *Genome Biology and Evolution* 5, no. 11 (2013): 2174–87.
67. See, for example, Pascal-Antoine Christin, Mónica Arakaki, Colin P. Osborne, and Erika J. Edwards, "Genetic Enablers Underlying the Clustered Evolutionary Origins of C_4 Photosynthesis in Angiosperms," *Molecular Biology and Evolution* 32, no. 4 (2015): 846–58; and Jose J. Moreno-Villena, Luke T. Dunning, Colin P. Osborne, and Pascal-Antoine Christin, "Highly Expressed Genes Are Preferentially Co-opted for C_4 Photosynthesis," *Molecular Biology and Evolution* 35, no. 1 (2018): 94–106.
68. See, for example, Yimin Tao, Ming-Ju Amy Lyu, and Xin-Guang Zhu, "Transcriptome Comparisons Shed Light on the Pre-condition and Potential Barrier for C_4 Photosynthesis Evolution in Eudicots," *Plant Molecular Biology* 91, no. 1–2 (2016): 193–209.
69. See, for example, Pascal-Antoine Christin et al., "Anatomical Enablers and the Evolution of C_4 Photosynthesis in Grasses," *Proceedings of the National Academy of Sciences of the United States of America* 110, no. 4 (2013): 1381–86; and Howard Griffiths, George Weller, Lydia F. M. Toy, and Ross J. Dennis, "You're So Vein: Bundle Sheath Physiology, Phylogeny and Evolution in C_3 and C_4 Plants," *Plant Cell and Environment* 36, no. 2 (2013): 249–61.
70. See, for example, Pascal-Antoine Christin and Colin P. Osborne, "The Recurrent Assembly of C_4 Photosynthesis, an Evolutionary Tale," *Photosynthesis Research* 117 (2013): 163–75.

71. The most notable exception is *Euphorbia*; see the overview by Sophie N. R. Young, Lawren Sack, Margaret J. Sporck-Koehler, and Marjorie R. Lundgren, "Why Is C_4 Photosynthesis So Rare in Trees?" *Journal of Experimental Botany* 71, no. 16 (2020): 4629–38.
72. Rowan F. Sage and Stefanie Sultmanis, "Why Are There No C_4 Forests?" *Journal of Plant Physiology* 203 (2016): 55–68.
73. Marc Jamon, Sabine Renous, Jean Pierre Gasc, Vincent Bels, and John Davenport, "Evidence of Force Exchanges during the Six-Legged Walking of the Bottom-Dwelling Fish, *Chelidonichthys lucerna*," *Journal of Experimental Zoology A. Ecology, Genetics and Physiology* 307, no. 9 (2007): 542–47.
74. Geerat Vermeij, "Forbidden Phenotypes and the Limits of Evolution," *Interface Focus* 5, no. 6 (2015): 2015.0028, https://doi.org/10.1098/rsfs.2015.0028.
75. See the excellent biography by Peter Pringle, *The Murder of Nikolai Vavilov: The Story of Stalin's Persecution of One of the Twentieth Century's Greatest Scientists* (London: JR Books, 2009).
76. N. I. Vavilov, "The Law of Homologous Series in Variation," *Journal of Genetics* 12 (1922): 47–89.
77. Exploring occupation of biological hypervolumes by lizards, in Eric R. Pianka, Laurie J. Vitt, Nicolás Pelegrin, Daniel B. Fitzgerald, and Kirk O. Winemiller, "Toward a Periodic Table of Niches, or Exploring the Lizard Niche Hypervolume," *American Naturalist* 190, no. 5 (2017): 601–16; see also the overview by Kirk O. Winemiller, Daniel B. Fitzgerald, Luke M. Bower, and Eric R. Pianka, "Functional Traits, Convergent Evolution, and Periodic Tables of Niches," *Ecology Letters* 18, no. 8 (2017): 737–51.
78. But not entirely, because some fungi are effective degraders but this capacity is convergent; see Iván Ayuso-Fernández, Francisco J. Ruiz-Dueñas, and Angel T. Martínez, "Evolutionary Convergence in Lignin-Degrading Enzymes," *Proceedings of the National Academy of Sciences of the United States of America* 115, no. 25 (2018): 6428–33.
79. See, for example, Jing-Ke Weng and Clint Chapple, "The Origin and Evolution of Lignin Biosynthesis," *New Phytologist* 187, no. 2 (2010): 273–85; and J. M. Espiñeira, E. Novo Uzal, L. V. Gómez Ros, J. S. Carrión, F. Merino, A. Ros Barceló, and F. Pomar, "Distribution of Lignin Monomers

and the Evolution of Lignification among Lower Plants," *Plant Biology* 13, no. 1 (2011): 59–68.

80. See, for example, Leen Labeeuw, Patrick T. Martone, Yan Boucher, and Rebecca J. Case, "Ancient Origin of the Biosynthesis of Lignin Precursors," *Biology Direct* 10 (2015): 23, https://doi.org/10.1186/s13062-015-0052-y; also Hugues Renault et al., "A Phenol-Enriched Cuticle Is Ancestral to Lignin Evolution in Land Plants," *Nature Communications* 8 (2017): 14713.
81. Patrick T. Martone et al., "Discovery of Lignin in Seaweed Reveals Convergent Evolution of Cell-Wall Architecture," *Current Biology* 19, no. 2 (2009): 169–75; so too the geniculation of the algal stems that employ lignin also is convergent when it comes to bending methods; see Kyra G. Janot and Patrick T. Martone, "Bending Strategies of Convergently Evolved, Articulated Coralline Algae," *Journal of Phycology* 54, no. 3 (2018): 305–16.
82. See, for example, Jing-Ke Weng, Xu Li, Jake Stout, and Clint Chapple, "Independent Origins of Syringyl Lignin in Vascular Plants," *Proceedings of the National Academy of Sciences of the United States of America* 105, no. 22 (2008): 7887–92.
83. See, for example, Benjamin Backfisch et al., "Stable Transgenesis in the Marine Annelid *Platynereis dumerilii* Sheds New Light on Photoreceptor Evolution," *Proceedings of the National Academy of Sciences of the United States of America* 110, no. 1 (2013): 193–98.
84. See, for example, Martine Manuel and David J. Price, "Role of *Pax6* in Forebrain Regionalization," *Brain Research Bulletin* 66, no. 4–6 (2005): 387–93.
85. T. Ian Simpson and David J. Price, "Pax6; a Pleiotropic Player in Development," *BioEssays* 24, no. 11 (2002): 1041–51.
86. Qing Luan, Qing Chen, and Markus Friedrich, "The *Pax6* Genes *Eyeless* and *Twin of Eyeless* Are Required for Global Patterning of the Ocular Segment in the *Tribolium* Embryo," *Developmental Biology* 394, no. 2 (2014): 367–81.
87. See, for example, T. A. Heanue et al., "Synergistic Regulation of Vertebrate Muscle Development by *Dach2, Eya2,* and *Six1,* Homologs of Genes

Required for *Drosophila* Eye Formation," *Genes and Development* 13, no. 24 (1999): 3231–43.

88. Ryuichi Kimura, Kaichi Yoshizaki, and Noriko Osumi, "Dynamic Expression Patterns of *Pax6* during Spermatogenesis in the Mouse," *Journal of Anatomy* 227, no. 1 (2015): 1–9.
89. See, for example, Koji Sato, Maurizio Pellegrino, Takao Nakagawa, Tatsuro Nakagawa, Leslie B. Vosshall, and Kazushige Touhara, "Insect Olfactory Receptors Are Heteromeric Ligand-Gated Ion Channels," *Nature* 452, no. 7190 (2008): 1002–7.
90. See, for example, Murray Badger, "The Roles of Carbonic Anhydrases in Photosynthetic CO_2 Concentrating Mechanisms," *Photosynthesis Research* 77, no. 22 (2003): 83–94.
91. A. Liljas and M. Laurberg, "A Wheel Invented Three Times: The Molecular Structures of the Three Carbonic Anhydrases," *EMBO Reports* 1, no. 1 (2000): 16–17.
92. β-carbonic anhydrases, for example, are now known to occur in many animal groups, probably arriving by horizontal gene transfer; Leo Syrjänen et al., "Characterization of the First Beta-Class Carbonic Anhydrase from an Arthropod (*Drosophila melanogaster*) and Phylogenetic Analysis of Beta-Class Carbonic Anhydrases in Invertebrates," *BMC Biochemistry* 11 (2010): 28. Among the fungi, this carbonic anhydrase is found in basidiomycetes and yeasts, whereas α-carbonic anhydrase occurs in filamentous ascomycetes; Skander Elleuche and Stefanie Pöggeler, "Evolution of Carbonic Anhydrases in Fungi," *Current Genetics* 55, no. 2 (2009): 211–22.
93. See, for example, J. V. Moroney, S. G. Bartlett, and G. Samuelsson, "Carbonic Anhydrases in Plants and Algae," *Plant, Cell and Environment* 24, no. 2 (2001): 141–53.
94. Kerry S. Smith, Cheryl Ingram-Smith, and James G. Ferry, "Roles of the Conserved Aspartate and Arginine in the Catalytic Mechanism of an Archael Beta-Class Carbonic Anhydrase," *Journal of Bacteriology* 184, no. 15 (2002): 4240–45.
95. James G. Ferry, "The γ Class of Carbonic Anhydrases," *Biochimica et Biophysica Acta—Proteins and Proteomics* 1804 (2010): 374–81.

96. Gustavo Parisi et al., "Gamma Carbonic Anhydrases in Plant Mitochondria," *Plant Molecular Biology* 55 (2004): 193–207.
97. For example, the γ-variety belongs to a much larger group of proteins (γ CASRPs) and most likely acquired its zinc before it began to function as a carbonic anhydrase. Xiang Fu, Long-Jiang Yu, Li Mao-Teng, Li Wei, Chen Wu, and Ma Yun-Feng, "Evolution of Structure in γ-Class Carbonic Anhydrase and Structurally Related Proteins," *Molecular Phylogenetics and Evolution* 47, no. 1 (2008): 211–20.
98. For overviews, see Anthony K.-C. So and George S. Espie, "Cyanobacterial Carbonic Anhydrases," *Canadian Journal of Botany* 83 (2005): 721–34; and Robert J. DiMario, Marylou C. Machingura, Grover L. Waldrop, and James V. Moroney, "The Many Types of Carbonic Anhydrases in Photosynthetic Organisms," *Plant Science* 268 (2018): 11–17.
99. Michael R. Sawaya, Gordon C. Cannon, Sabine Heinhorst, Shiho Tanaka, Eric B. Williams, Todd O. Yeates, and Cheryl A. Kerfeld, "The Structure of Beta-Carbonic Anhydrase from the Carboxysomal Shell Reveals a Distinct Subclass with One Active Site for the Price of Two," *Journal of Biological Chemistry* 281, no. 11 (2006): 7546–55.
100. M. S. Kimber and E. F. Pai, "The Active Site Architecture of *Pisum sativum* Beta-Carbonic Anhydrase Is a Mirror Image of That of Alpha-Carbonic Anhydrase," *EMBO Journal* 19, no. 7 (2000): 1407–18.
101. Regarding θ-carbonic anhydrase, see Sae Kikutani, Kensuke Nakajima, Chikako Nagasato, Yoshinori Tsuji, Ai Miyatake, and Yusuke Matsuda, "Thylakoid Luminal θ-Carbonic Anhydrase Critical for Growth and Photosynthesis in the Marine Diatom *Phaeodactylum tricornutum*," *Proceedings of the National Academy of Sciences of the United States of America* 113, no. 35 (2016): 9828–33.
102. For example, Renee Bee Yong Lee, J. Andrew C. Smith, and Rosalind E. M. Rickaby, "Cloning, Expression and Characterization of the δ-Carbonic Anhydrase of *Thalassiosira weissflogii* (Bacillariophyceae)," *Journal of Phycology* 49, no. 1 (2013): 170–77.
103. Todd W. Lane, Mak A. Saito, Graham N. George, Ingrid J. Pickering, Roger C. Prince, and François M. M. Morel, "A Cadmium Enzyme from a Marine Diatom," *Nature* 435, no. 7038 (2005): 42.

104. Yan Xu, Liang Feng, Philip D. Jeffrey, Yigong Shi, and François M. M. Morel, "Structure and Metal Exchange in the Cadmium Carbonic Anhydrase of Marine Diatoms," *Nature* 452, no. 7183 (2008): 56–61.
105. See, for example, Haewon Park, Bongkeun Song, and François M. M. Morel, "Diversity of the Cadmium-Containing Carbonic Anhydrase in Marine Diatoms and Natural Waters," *Environmental Microbiology* 9, no. 2 (2007): 403–13.
106. Espen Granum, John A. Raven, and Richard C. Leegood, "How Do Marine Diatoms Fix 10 Billion Tonnes of Inorganic Carbon Every Year?" *Canadian Journal of Botany* 83 (2005): 898–908.
107. Erik L. Jensen, Romain Clement, Artemis Kosta, Stephen C. Maberly, and Brigitte Gontero, "A New Widespread Subclass of Carbonic Anhydrase in Marine Phytoplankton," *ISME Journal* 13 (2019): 2094–106; also Sonia Del Prete, Alessio Nocentini, Claudiu T. Supuran, and Clemente Capasso, "Bacterial ι-Carbonic Anhydrase: A New Active Class of Carbonic Anhydrase Identified in the Genome of the Gram-Negative Bacterium *Burkholderia territorii*," *Journal of Enzyme Inhibition and Medical Chemistry* 35, no. 1 (2020): 1060–68.
108. James G. Ferry, "Carbonic Anhydrase of Anaerobic Microbes," *Bioorganic and Medicinal Chemistry* 21 (2013): 1392–95.
109. See, for example, J. L. Pierre and M. Fontencave, "Iron and Activated Oxygen Species in Biology: The Basic Chemistry," *BioMetals* 12, no. 3 (1999): 195–99.
110. See, for example, Ronald Eugene Stenkamp, "Dioxygen and Hemerythrin," *Chemical Reviews* 94, no. 3 (1994): 715–26.
111. See, for example, Alina Roman, Iulia Lupan, Donald M. Kurtz Jr., and Radu Silaghi-Dumitrescu, "Towards the Development of Hemerythrin-Based Blood Substitutes," *Protein Journal* 29, no. 6 (2010): 387–93.
112. See, for example, R. C. Long, J. H. Zhang, D. M. Kurtz Jr, A. Negri, G. Tedeschi, and F. Bonomi, "Myohemerythrin from the Sipunculid, *Phascolopsis gouldii*: Purification, Properties and Amino Acid Sequence," *Biochimica et Biophysica Acta. Protein Structure and Molecular Enzymology* 1122, no. 2 (1992): 136–42.
113. See, for example, M. Brunori and B. Vallone, "A Globin for the Brain," *FASEB Journal* 20, no. 13 (2006): 2192–97.

114. See, for example, Paolo Ascenzi et al., "Neuroglobin: From Structure to Function in Health and Disease," *Molecular Aspects of Medicine* 52 (2016): 1–48.
115. David Vergote et al., "Up-regulation of Neurohemerythrin Expression in the Central Nervous System of the Medicinal Leech, *Hirudo medicinalis,* Following Septic Injury," *Journal of Biological Chemistry* 279, no. 42 (2004): 43828–37.
116. See, for example, Thorsten Burmester, Bettina Ebner, Bettina Weich, and Thomas Hankeln, "Cytoglobin: A Novel Globin Type Ubiquitously Expressed in Vertebrate Tissue," *Molecular Biology and Evolution* 19, no. 4 (2002): 416–21.
117. David Hoogewijs et al., "Androglobin: A Chimeric Globin in Metazoans That Is Preferentially Expressed in Mammalian Testes," *Molecular Biology and Evolution* 29, no. 4 (2012): 1105–14.
118. J.-L. Baert, M. Britel, P. Sautiere, and J. Malecha, "Ovohemerythrin, a Major 14-kDa Yolk Protein Distinct from Vitellogenin in Leech," *European Journal of Biochemistry* 209, no. 2 (1992): 563–69.
119. Elisa M. Costa-Paiva, Nathan V. Whelan, Damien S. Waits, Scott R. Santos, Carlos G. Schrago, and Kenneth M. Halanych, "Discovery and Evolution of Novel Hemerythrin Genes in Annelid Worms," *BMC Evolutionary Biology* 17, no. 1 (2017), 85.
120. A. Negri, G. Tedeschi, F. Bonomi, J. H. Zhang, and D. M. Kurtz Jr., "Amino-Acid Sequences of the Alpha- and Beta-Subunits of Hemerythrin from *Lingula reevii,*" *Biochemica et Biophysica Acta. Protein Structure and Molecular Enzymology* 1208, no. 2 (1994): 277–85.
121. R. E. Weber, R. Binge, and K. K. Rasmussen, "Respiratory Significance of Priapulid Hemerythrin," *Marine Biology Letters* 1 (1979): 87–97.
122. Reported by Xavier Bailly, Stefano Vanin, Christine Chabasse, Kenji Mizuguchi, and Serge N. Vinogradov, "A Phylogenomic Profile of Hemerythrins, the Nonheme Diiron Binding Respiratory Proteins," *BMC Evolutionary Biology* 8 (2008), 244.
123. Here it is associated with an E3 ligase and is probably involved with the regulation of intracellular iron; see Ameen A. Salahudeen et al., "An E3 Ligase Possessing an Iron-Responsive Hemerythrin Domain Is a Regulator of Iron Homeostasis," *Science* 326, no. 5953 (2009): 722–26;

Ajay A. Vashisht et al., "Control of Iron Homeostasis by an Iron-Regulated Ubiquitin Ligase," *Science* 326, no. 5953 (2009): 718–21; and commentary by Tracy A. Roualt, "Cell Biology. An Ancient Gauge for Iron," *Science* 326, no. 5953 (2009): 676–77.

124. Claudia Alvarez-Carreño, Arturo Becerra, and Antonio Lazcano, "Molecular Evolution of the Oxygen-Binding Hemerythrin Domain," *PLoS One* 11, no. 6 (2016): e0157904.
125. Christopher E. French, Jennifer M. L. Bell, and F. Bruce Ward, "Diversity and Distribution of Hemerythrin-Like Proteins in Prokaryotes," *FEMS Microbiology Letters* 279, no. 2 (2008): 131–45.
126. Bailly et al., "Phylogenomic Profile," (2008), 7.
127. Donald M. Kurtz Jr., "Oxygen-Carrying Proteins: Three Solutions to a Common Problem," *Essays in Biochemistry* 34 (1999): 85–100.
128. Kurtz, "Oxygen-Carrying Proteins," 97.
129. Edwin Blalock, "The Syntax of Immune-Neuroendocrine Communication," *Immunology Today* 15, no. 11 (1994): 504–11, p. 504.
130. Antonio Benítez-Burraco and Juan Uriagereka, "The Immune Syntax Revisited: Opening New Windows on Language Evolution," *Frontiers in Molecular Neuroscience* 8 (2016), 84.
131. See the overview by L. M. Boulanger, G. S. Huh, and C. J. Shatz, "Neuronal Plasticity and Cellular Immunity: Shared Molecular Mechanisms," *Current Opinion in Neurobiology* 11, no. 5 (2001): 568–78.
132. L. M. Boulanger, "Immune Proteins in Brain Development and Synaptic Plasticity," *Neuron* 64, no. 1 (2009): 93–109.
133. N. Hilschmann, H. U. Barnikol, S. Barnikol-Watanabe, H. Götz, H. Kratzin, and F. P. Thinnes, "The Immunoglobulin-Like Genetic Predetermination of the Brain: The Protocadherins, Blueprint of the Neuronal Network," *Naturwissenschaften* 88, no. 1 (2001): 2–12.
134. See, for example, B. E. Chen, "The Molecular Diversity of *Dscam* Is Functionally Required for Neuronal Wiring Specificity in *Drosophila*," *Cell* 125, no. 3 (2006): 607–20; and Daisuke Hattori, Ebru Demir, Ho Won Kim, Erika Viragh, S. Lawrence Zipursky, and Barry J. Dickson, "Dscam Diversity Is Essential for Neuronal Wiring and Self-recognition," *Nature* 449, no. 7159 (2007): 223–27; see also Yongfeng Jin and Hao Li, "Revisiting Dscam Diversity: Lessons from Clustered

Protocadherins," *Cellular and Molecular Life Sciences* 76, no. 4 (2019): 667–80.

CHAPTER 3

1. Specifically the Late Ordovician (ca. 445 Ma), Late Devonian (ca. 365 Ma), end Permian (ca. 252 Ma), very Late Triassic (ca. 210 Ma), and end Cretaceous (ca. 66 Ma).
2. More precisely, and the resolution is impressive, the end-Permian extinction begins at 251.941 ± 0.037 Ma; see Seth D. Burgess, Samuel Bowring, and Shu-zhong Shen, "High Precision Timeline for Earth's Most Severe Extinction," *Proceedings of the National Academy of Sciences of the United States of America* 111, no. 9 (2014): 3316–21, also p. 5060. The impact event that contributed to the end-Cretaceous extinctions clocks in 66.021 ± 0.024/0.039 Ma; see William C. Clyde, Jahandar Ramezani, Kirk R. Johnson, Samuel A. Bowring, and Matthew M. Jones, "Direct High-Precision U-Pb Geochronology of the End-Cretaceous Extinction and Calibration of Paleocene Astronomical Timescales," *Earth and Planetary Science Letters* 452 (2016): 272–80.
3. See Charles Darwin, *On the Origin of Species by Means of Natural Selection, or the Preservation of Favoured Races in the Struggle for Life* (London: John Murray, 1859), chapter 9.
4. John Phillips, *Life on the Earth; Its Origin and Succession* (London: Macmillan, 1860).
5. See, for example, Peter J. Wagner, Matthew A. Kosnik, and Scott Lidgard, "Abundance Distributions Imply Elevated Complexity of Post-Paleozoic Marine Ecosystems," *Science* 314, no. 5803 (2006): 1289–92.
6. But not for all, not least the plants where by no stretch of imagination do the words mass extinction apply; see, for example, Hendrik Nowak, Elke Schneebeli-Hermann, and Evelyn Kustatscher, "No Mass Extinction for Land Plants at the Permian-Triassic Transition," *Nature Communications* 10, no. 1 (2019), 384. Much the same may apply to the insects; see Matteo Montagna, K. Jun Tong, Giulia Magoga, Laura Strada, Andrea Tintori, Simon Y. W. Ho, and Nathan Lo, "Recalibration of the Insect Evolutionary Time Scale Using Monte San Giorgio Fossils Sug-

gests Survival of Key Lineages through the End-Permian Extinction," *Proceedings of the Royal Society of London, Series B: Biological Sciences* 286, no. 1912 (2019): 20191854.

7. David M. Raup, "Size of the Permian/Triassic Bottleneck and Its Evolutionary Implications," *Science* 206, no. 4415 (1979): 217–18; but see the reconsideration in Steven M. Stanley, "Estimates of the Magnitudes of Major Marine Mass Extinctions in Earth History," *Proceedings of the National Academy of Sciences of the United States of America* 113, no. 42 (2016): E6325–34.
8. Michael J. Benton and Richard J. Twitchett, "How to Kill (Almost) All Life: The End-Permian Extinction Event," *Trends in Ecology and Evolution* 18, no. 7 (2003): 358–65.
9. A convenient overview is provided by Zhong-Qiang Chen and Michael J. Benton, "The Timing and Pattern of Biotic Recovery following the End-Permian Mass Extinction," *Nature Geoscience* 5 (2012): 375–83.
10. Raup, "Size of the Permian/Triassic Bottleneck," abstract.
11. Aaron W. Hunter and Kenneth J. McNamara, "Prolonged Co-existence of 'Archaic' and 'Modern' Palaeozoic Ophiuroids—Evidence from the Early Permian, Southern Carnarvon Basin, Western Australia," *Journal of Systematic Palaeontology* 16, no. 11 (2018): 891–907.
12. With respect to the ophiocistioids, identifiable on account of their distinctive teeth, see Mike Reich, Tanja R. Stegemann, Imelda M. Hausmann, Vanessa J. Roden, and Alexander Nützel, "The Youngest Ophiocistioid: A First Palaeozoic-Type Echinoderm Group Representative from the Mesozoic," *Palaeontology* 61, no. 6 (2018): 803–11. So too Triassic blastoids are identified in a Chinese section (Huangzhishan); see Z. Q. Chen et al., "Environmental and Biotic Turnover across the Permian–Triassic Boundary on a Shallow Carbonate Platform in Western Zhejiang, South China," *Australian Journal of Earth Sciences* 56, no. 6 (2009): 775–97.
13. For an overview, see Ben Thuy, Hans Hagdorn, and Andy S. Gale, "Paleozoic Echinoderm Hangovers: Waking Up in the Triassic," *Geology* 45, no. 6 (2017): 531–34, who address the sea urchins, brittle stars, and starfish; their study generated debate with critiques by Daniel B. Blake,

Geology 45, no 7 (2017): e417, with authors' response (e418); Mariusz A. Salamon and Przemysław Gorzelak, *Geology* 45, no. 7 (2017): e419, with authors' response (e420); and Aaron W. Hunter and Kenneth J. McNamara, *Geology* 45, no. 11: e431, with authors' response (e432).

14. Jeffrey R. Thompson et al., "A New Stem Group Echinoid from the Triassic of China Leads to a Revised Macroevolutionary History of Echinoids during the End-Permian Mass Extinction," *Royal Society Open Science* 5, no. 1 (2018), 171548; Jeffrey R. Thompson, Renato Posenato, David J. Bottjer, and Elizabeth Petsios, "Echinoids from the Tesero Member (Werfen Formation) of the Dolomites (Italy): Implications for Extinction and Survival of Echinoids in the Aftermath of the End-Permian Mass Extinction," *PeerJ* 7 (2019), e7361; Hans Hagdorn, "Slipped through the Bottleneck: *Lazarechinus mirabeti* gen. et sp. nov., a Paleozoic-like Echinoid from the Triassic Muschelkalk (Late Anisian) of East France," *Paläontologische Zeitschrift* 92, no. 2 (2018): 267–82; and Carlie Pietsch, Kathleen A. Ritterbush, Jeffrey R. Thompson, Elizabeth Petsios, and David J. Bottjer, "Evolutionary Models in the Early Triassic Marine Realm," *Palaeogeography, Palaeoclimatology, Palaeoecology* 513 (2019): 65–85.
15. Tatsuo Oji and Richard J. Twitchett, "The Oldest Post-Palaeozoic Crinoid and Permian-Triassic Origins of the Articulata (Echinodermata)," *Zoological Science* 32, no. 2 (2015): 211–15; also Mariusz A. Salamon, Przemysław Gorzelak, Nils M. Hanken, Henrik E. Riise, and Bruno Ferré, "Crinoids from Svalbard in the Aftermath of the End-Permian Mass Extinction," *Polish Polar Research* 36, no. 3 (2015): 225–38.
16. See, for example, Marco Balini, Spencer G. Lucas, James F. Jenks, and Justin A. Spielmann, "Triassic Ammonoid Biostratigraphy: An Overview," *Special Publications of the Geological Society of London* 334 (2010): 221–62.
17. Alistair J. McGowan and Andrew B. Smith, "Ammonoids across the Permian/Triassic Boundary: A Cladistic Perspective," *Palaeontology* 50, no. 3 (2007): 573–90; also Arnaud Brayard et al., "*Proharpoceras* Chao: A New Ammonoid Lineage Surviving the End-Permian Mass Extinction," *Lethaia* 40, no. 2 (2007): 175–81.

18. Alistair J. McGowan, "Ammonoid Recovery from the Late Permian Mass Extinction Event," *Comptes Rendus Palevol* 4, no. 6–7 (2005): 517–30.
19. Morgane Brosse, Arnaud Brayard, Emmanuel Fara, and Pascal Neige, "Ammonoid Recovery after the Permian-Triassic Mass Extinction: A Re-exploration of Morphological and Phylogenetic Diversity Patterns," *Journal of the Geological Society of London* 170 (2013): 225–36; and Steven M. Stanley, "Evidence from Ammonoids and Conodonts for Multiple Early Triassic Extinctions," *Proceedings of the National Academy of Sciences of the United States of America* 106, no. 36 (2009): 15264–67.
20. Such might apply to the so-called archosauriforms, the reptilian group that encompasses the birds/dinosaurs as well as the so-called pseudosuchians; see, for example, Massimo Bernardi, Hendrik Klein, Fabio Massimo Petti, and Martín D. Ezcurra, "The Origin and Early Radiation of Archosauriforms: Integrating the Skeletal and Footprint Record," *PLoS One* 10, no. 6 (2015), e0128449; see also Lorenzo Marchetti, Sebastian Voigt, and Hendrik Klein, "Revision of Late Permian Tetrapod Tracks from the Dolomites (Trentino-Alto Adige, Italy)," *Historical Biology* 31 (2019): 748–83. So too the plesiosaurs may have been assembling their game-plan in the Permian; see R. L. Carroll, "Plesiosaur Ancestors from the Upper Permian of Madagascar," *Philosophical Transactions of the Royal Society of London, Series B: Biological Sciences* 293, no. 1066 (1981): 315–83; and R. L. Carroll and P. Gaskill, "The Nothosaur *Pachypleurosaurus* and the Origin of Plesiosaurs," 309, no. 1139 (1985): 343–93.
21. See Ryosuke Motani, Da-Yong Jiang, Andrea Tintori, Cheng Ji, and Jian-Dong Huang, "Pre- versus Post-Mass Extinction Divergence of Mesozoic Marine Reptiles Dictated by Time-Scale Dependence of Evolutionary Rates," *Proceedings of the Royal Society of London, Series B: Biological Sciences* 284, no. 1854 (2017): 2017.0241; also Da-Yong Jiang et al., "A Large Aberrant Stem Ichthyosauriform Indicating Early Rise and Demise of Ichthyosauromorphs in the Wake of the End-Permian Extinction," *Scientific Reports* 6 (2016), 26232.
22. Thus, Benjamin C. Moon and Thomas L. Stubbs, "Early High Rates and Disparity in the Evolution of Ichthyosaurs," *Communications Biology* 3,

no. 1 (2020), 68. They note that although the first ichthyosaurs are Triassic, the group appeared to have originated in the Late Permian.

23. Heinz W. Kozur and Robert E. Weems, "Detailed Correlation and Age of Late Continental Changhsingian and Earliest Triassic Beds: Implications for the Role of the Siberian Trap in the Permian–Triassic Biotic Crisis," *Palaeogeography, Palaeoclimatology, Palaeoecology* 308, no. 1–2 (2011): 22–40; and J. B. Waterhouse and G. R. Shi, "Evolution in a Cold Climate," *Palaeogeography, Palaeoclimatology, Palaeoecology* 298, no. 1–2 (2010): 17–30.
24. Jörg Fröbisch, Kenneth D. Angielczyk, and Christian A. Sidor, "The Triassic Dicynodont *Kombuisia* (Synapsida, Anomodonta) from Antarctica, a Refuge from the Terrestrial Permian–Triassic Mass Extinction," *Naturwissenschaften* 97, no. 2 (2010): 187–96.
25. Tyler W. Beatty, J.-P. Zonneveld, and Charles M. Henderson, "Anomalously Diverse Early Triassic Ichnofossil Assemblages in Northwest Pangea: A Case for a Shallow-Marine Habitable Zone," *Geology* 36, no. 10 (2008): 771–74; and John-Paul Zonneveld, Murray K. Gingras, and Tyler W. Beatty, "Diverse Ichnofossil Assemblages following the P-T Mass Extinction, Lower Triassic, Alberta and British Columbia, Canada: Evidence for Shallow Marine Refugia on the Northwestern Coast of Pangaea," *Palaios* 25, no. 6 (2010): 368–92.
26. Regarding vertebrate recoveries, see, for example, J. Botha and R. M. H. Smith, "Rapid Vertebrate Recuperation in the Karoo Basin of South Africa following the End-Permian Extinction," *Journal of African Earth Sciences* 45 (2006): 502–14; and David A. Tarailo and David E. Fastovsky, "Post-Permo-Triassic Terrestrial Vertebrate Recovery: Southwestern United States," *Paleobiology* 38, no. 4 (2012): 644–63. For similar stories among microfossils, see Haijun Song et al., "Recovery Tempo and Pattern of Marine Ecosystems after the End-Permian Mass Extinction," *Geology* 39, no. 8 (2011): 739–742.
27. See, for example, Maria Ovtcharova, Hugo Bucher, Urs Schaltegger, Thomas Galfetti, Arnaud Brayard, and Jean Guex, "New Early to Middle U-Pb Ages from South China: Calibration with Ammonoid Biochronozones and Implications for the Timing of the Triassic Biotic Recovery," *Earth and Planetary Science Letters* 243, no. 3–4 (2006): 463–75;

and Ian Metcalfe and Yukio Isozaki, "Current Perspectives on the Permian-Triassic Boundary and End-Permian Mass Extinctions: Preface," *Journal of Asian Earth Sciences* 36, no. 6 (2009): 407–12.

28. See, for example, Arnaud Brayard, Gilles Escarguela, and Hugo Bucher, "The Biogeography of Early Triassic Ammonoid Faunas: Clusters, Gradients, and Networks," *Geobios* 40, no. 6 (2007): 749–65.
29. See, for example, Michael J. Orchard, "Conodont Diversity and Evolution through the Latest Permian and Early Triassic Upheavals," *Palaeogeography, Palaeoclimatology, Palaeoecology* 252, no. 1–2 (2007): 93–117.
30. Haishui Jiang, Richard J. Aldridge, Xulong Lai, Chunbo Yan, and Yadong Sun, "Phylogeny of the Conodont Genera *Hindeodus* and *Isarcicella* across the Permian-Triassic Boundary," *Lethaia* 44, no. 4 (2011): 374–82.
31. Fiann M. Smithwick and Thomas L. Stubbs, "Phanerozoic Survivors: Actinopterygian Evolution through the Permo-Triassic and Triassic-Jurassic Mass Extinction Events," *Evolution* 72, no. 2 (2018): 348–62.
32. See, for example, Arnaud Brayard et al., "Unexpected Early Triassic Marine Ecosystem and the Rise of the Modern Evolutionary Fauna," *Science Advances* 3, no. 2 (2017), e1602159; Dirk Knaust, "The End-Permian Mass Extinction and Its Aftermath on an Equatorial Carbonate Platform: Insights from Ichnology," *Terra Nova* 22, no. 3 (2010): 195–202; James R. Wheeley and Richard J. Twitchett, "Palaeoecological Significance of a New Griesbachian (Early Triassic) Gastropod Assemblage from Oman," *Lethaia* 38, no. 1 (2005): 37–45; and Richard J. Twitchett, L. Krystyn, A. Baud, James R. Wheeley, and S. Richoz, "Rapid Marine Recovery after the End-Permian Mass-Extinction Event in the Absence of Marine Anoxia," *Geology* 32, no. 9 (2004): 805–8.
33. For bivalves, see, for example, Michael Hautmann, Hugo Bucher Thomas Brühwiler, Nicolas Goudemand, Andrzej Kaim, and Alexander Nützel, "An Unusually Diverse Mollusc Fauna from the Earliest Triassic of South China and Its Implications for Benthic Recovery after the End-Permian Biotic Crisis," *Geobios* 44, no. 1 (2011): 71–85. For gastropods, see Xin Sun, Xincheng Qiu, Erik Tihelka, Hao Yang, Dongying Sun, Jinnan Tong, and Li Tian, "A Diverse Gastropod Fauna from the Shallow Marine Carbonate Platform of the Yangou Section (South China) in the Immediate Aftermath of the Permian–Triassic Mass Extinction," *Geological*

Journal (2021), https://doi.org/10.1002/gj.4083. In reference to Utah, see Michael Hautmann, Andrew B. Smith, Alistair J. McGowan, and Hugo Bucher, "Bivalves from the Olenekian (Early Triassic) of South-western Utah: Systematics and Evolutionary Significance," *Journal of Systematic Palaeontology* 11, no. 3 (2013): 263–93.

34. In addition to the examples given below, in the context of Permian floras, see, for example, Patrick Blomenkemper et al., "A Hidden Cradle of Plant Evolution in Permian Tropical Lowlands," *Science* 362, no. 6421 (2018): 1414–16; and commentary Elizabeth Pennisi, "Fossils Push Back Origin of Key Plant Groups Millions of Years," *Science* 362, no. 6421 (2018): 1340; and Patrick Blomenkemper, Robert Bäumer, Malte Backer, Abdalla Abu Hamad, Jun Wang, Hans Kerp, and Benjamin Bomfleur, "Bennettitalean Leaves from the Permian of Equatorial Pangea—The Early Radiation of an Iconic Mesozoic Gymnosperm Group," *Frontiers in Earth Science* 9 (2021), 652699.
35. David M. Raup, "Extinction: Bad Genes or Bad Luck?" *Acta Geologica Hispanica* 16, no. 1–2 (1981): 25–33.
36. Robert M. Owens, "The Stratigraphical Distribution and Extinctions of Permian Trilobites," *Special Papers in Palaeontology* 70 (2003): 377–97; and chapter by Rudy Lerosey-Aubril and Raimund Feist, "Quantitative Approach to Diversity and Decline in Late Palaeozoic Trilobites," in *Earth and Life: Global Diversity, Extinction Intervals and Biogeographic Perturbations through Time*, ed. John A. Talent (Dordrecht: Springer, 2012), 535–55.
37. David K. Brezinski, "The Rise and Fall of Late Paleozoic Trilobites of the United States," *Journal of Paleontology* 73, no. 2 (1999): 164–75.
38. Robert M. Owens, "A Review of Permian Trilobite Genera," *Special Papers in Paleontology* 30 (1983): 15–41.
39. Danita S. Brandt, "Ecdysial Efficiency and Evolutionary Efficacy among Marine Arthropods: Implications for Trilobite Survivorship," *Alcheringa* 26, no. 3 (2002): 399–421.
40. See, for example, Rudy Lerosey-Aubril and Lucia Angiolini, "Permian Trilobites from Antalya Province, Turkey and Enrollment in Late Palaeozoic Trilobites," *Turkish Journal of Earth Sciences* 18, (2009): 427–48.

41. Juan José Rustán, Diego Balseiro, Beatriz G. Waisfeld, Rodolfo D. Foglia, and N. Emilio Vaccari, "Infaunal Molting in Trilobita and Escalatory Responses against Predation," *Geology* 39, no. 5 (2011): 495–98.
42. Stephen Jay Gould and C. Bradford Calloway, "Clams and Brachiopods: Ships That Pass in the Night," *Paleobiology* 6, no. 4 (1980): 383–96.
43. J. Chen, Z. Q. Chen, and J. N. Tong, "Environmental Determinants and Ecological Selectivity of Benthic Faunas from Nearshore to Bathyal Zones in the End-Permian Mass Extinction: Brachiopod Evidence from South China," *Palaeogeography, Palaeoclimatology, Palaeoecology* 308 (2011): 84–97.
44. Dimitri A. Ruban, "The Permian/Triassic Mass Extinction among Brachiopods in the Northern Caucasus (Northern Palaeo-Tethys)," *GeoBios* 43, no. 3 (2010): 355–63.
45. David A. T. Harper and Anders Drachen, "The Orthida: The Rise and Fall of a Great Palaeozoic Clade," *Special Papers in Palaeontology* 84 (2010): 107–17.
46. See, for example, Lee Hsiang Liow, Trond Reitan, and Paul G. Harnik, "Ecological Interactions on Macroevolutionary Time Scales: Clams and Brachiopods Are More Than Ships That Pass in the Night," *Ecology Letters* 18, no. 10 (2015): 1030–39.
47. See, for example, Charles W. Thayer, "Are Brachiopods Better Than Bivalves? Mechanisms of Turbidity Tolerance and Their Interaction with Feeding in Articulates," *Paleobiology* 12, no. 2 (1986): 161–74; and Stephen K. Donovan and Andrew S. Gale, "Predatory Asteroids and the Decline of the Articulate Brachiopods," *Lethaia* 23, no. 1 (1990): 77–86.
48. J. John Sepkoski, "Competition in Macroevolution: The Double Wedge Revisited," in *Evolutionary Paleobiology: In Honor of James W. Valentine*, ed. David Jablonski, Douglas H. Erwin, and Jere H. Lipps (Chicago: University of Chicago Press, 1996), 211–55.
49. Charles N. Ciampaglio, "Measuring Changes in Articulate Brachiopod Morphology before and after the Permian Mass Extinction Event: Do Developmental Constraints Limit Morphological Innovation?" *Evolution and Development* 6, no. 4 (2004): 260–74.

50. With respect to the spiriferinids, see, for example, Zhen Guo, Zhong-Qiang Chen, and David A. T. Harper, "Phylogenetic and Ecomorphologic Diversifications of Spiriferinid Brachiopods after the End-Permian Extinction," *Paleobiology* 46, no. 4 (2020): 495–510.
51. S. E. Greene, D. J. Bottjer, H. Hagdorn, and J. P. Zonneveld, "The Mesozoic Return of Paleozoic Faunal Constituents: A Decoupling of Taxonomic and Ecological Dominance during the Recovery from the End-Permian Mass Extinction," *Palaeogeography, Palaeoclimatology, Palaeoecology* 308 (2011): 224–32.
52. D. Jablonski, J. J. Sepkoski Jr., D. J. Bottjer, and P. M. Sheehan, "Onshore-Offshore Patterns in the Evolution of Phanerozic Shelf Communities," *Science* 222, no. 4628 (1983): 1123–25.
53. Sepkoski, "Competition in Macroevolution," 243 (my emphasis).
54. Jonathan L. Payne, Noel A. Heim, Matthew L. Knope, and Craig R. McClain, "Metabolic Dominance of Bivalves Predates Brachiopod Diversity Decline by More Than 150 Million Years," *Proceedings of the Royal Society of London, Series B: Biological Sciences* 281, no. 1783 (2014): 20133122; this study focuses on an exceptionally preserved Carboniferous assemblage. Also see Shannon Hsieh, Andrew M. Bush, and J. Bret Bennington, "Were Bivalves Ecologically Dominant over Brachiopods in the Late Paleozoic? A Test Using Exceptionally Preserved Fossil Assemblages," *Paleobiology* 45, no. 2 (2019): 265–79, who found more of an equal allocation of resources between brachiopods and bivalves.
55. Arnold L. Miller, "Spatio-Temporal Transitions in Paleozoic Bivalvia: An Analysis of North American Fossil Assemblages," *Historical Biology* 1, no. 3 (1988): 251–73. Moreover, this ecological category rapidly recovered after the PTE; see Toshifumi Komatsu, Dang T. Huyen, and Nguyen D. Huu, "Radiation of Middle Triassic Bivalve: Bivalve Assemblages Characterized by Infaunal and Semi-infaunal Burrowers in a Storm- and Wave-dominated Shelf, An Chau Basin, North Vietnam," *Palaeogeography, Palaeoclimatology, Palaeoecology* 291, no. 3–4 (2010): 190–204.
56. Lindsey R. Leighton, Amelinda E. Webb, and Jennifer A. Sawyer, "Ecological Effects of the Paleozoic-Modern Faunal Transition: Comparing Predation on Paleozoic Brachiopods and Molluscs," *Geology* 41, no. 2 (2013): 275–78.

57. William J. Foster and Richard J. Twitchett, "Functional Diversity of Marine Ecosystems after the Late Permian Mass Extinction Event," *Nature Geoscience* 7 (2014): 233–38; also commentary Martin Aberhan, "Ecological Diversity Maintained," *Nature Geoscience* 7, no. 3 (2014): 171–72.
58. Michael J. Vendrasco, Richard D. Hoare, and Gorden L. Bell Jr., "The Youngest Rostroconch Mollusc from North America, *Minycardita capitanesis*," *Zootaxa* 2603, no. 1 (2010): 61–64.
59. Evidence for near-synchronicity of marine and terrestrial extinctions is reviewed by Jennifer Botha, Adam K. Huttenlocker, Roger M. H. Smith, Rose Prevec, Pia Viglietti, and Sean P. Modesto, "New Geochemical and Palaeontological Data from the Permian-Triassic Boundary in the South African Karoo Basin Test the Synchronicity of Terrestrial and Marine Extinctions," *Palaeogeography, Palaeoclimatology, Palaeoecology* 540 (2020), 109467; and Michael R. Rampino, Yoram Eshet-Alkalai, Athanasios Koutavas, and Sedelia Rodriguez, "End-Permian Stratigraphic Timeline Applied to the Timing of Marine and Non-marine Extinctions," *Palaeoworld* 29, no. 3 (2020): 577–89. For an argument that the marine extinctions were about 350,000 years later, see Robert A. Gastaldo, Sandra L. Kamo, Johann Neveling, John W. Geissman, Cindy V. Looy, and Anna M. Martini, "The Base of the *Lystrosaurus* Assemblage Zone, Karoo Basin, Predates the End-Permian Marine Extinction," *Nature Communications* 11 (2020), 1428.
60. See the works listed in notes 20 and 34, respectively.
61. See, for example, Robert R. Reisz and Jörg Fröbisch, "The Oldest Caseid Synapsid from the Late Pennsylvanian of Kansas, and the Evolution of Herbivory in Terrestrial Vertebrates," *PLoS One* 9, no. 4 (2014), e94518.
62. T. S. Kemp, "The Concept of Correlated Progression as the Basis of a Model for the Evolutionary Origin of Major New Taxa," *Proceedings of the Royal Society of London, Series B: Biological Sciences* 274, no. 1618 (2007): 1667–73.
63. T. S. Kemp, "The Origin of Higher Taxa: Macroevolutionary Processes, and the Case of the Mammals," *Acta Zoologica* 88, no. 1 (2007): 3–22.
64. David P. Ford and Roger B. J. Benson, "The Phylogeny of Early Amniotes and the Affinities of Parareptilia and Varanopidae," *Nature Ecology and Evolution* 4 (2020): 57–65. They argue that the varanopids are

not synapsids but fall within the reptiles. This is an important development but tangential to the arguments given here.

65. See, for example, Sean P. Modesto, Roger M. H. Smith, Nicolás E. Campione, and Robert R. Reisz, "The Last 'Pelycosaur': A Varanopid Synapsid from the *Pristerognathus* Assemblage Zone, Middle Permian of South Africa," *Naturwissenschaften* 98, no. 12 (2011): 1027–34; and Jun Liu, Bruce Rubidge, and Jinling Li, "New Basal Synapsid Supports Laurasian Origin for Therapsids," *Acta Palaeontologica Polonica* 54, no. 3 (2009): 393–400.
66. T. S. Kemp, "The Origin and Early Radiation of the Therapsid Mammal-like Reptiles: A Palaeobiological Hypothesis," *Journal of Evolutionary Biology* 19, no. 4 (2006): 1231–47; also Jacqueline K. Lungmus and Kenneth D. Angielczyk, "Antiquity of Forelimb Ecomorphological Diversity in the Mammalian Stem Lineage (Synapsida)," *Proceedings of the National Academy of Sciences of the United States of America* 116, no. 14 (2019): 6903–7, which explains how the roots of forelimb versatility go deep into the Permian.
67. Michael Laaß and Anders Kaestner, "Evidence for Convergent Evolution of a Neocortex-like Structure in a Late Permian Therapsid," *Journal of Morphology* 278, no. 8 (2017): 1033–57.
68. See, for example, Ricardo N. Martinez, Cathleen L. May, and Catherine A. Forster, "A New Carnivorous Cynodont from the Ischigualasto Formation (Late Triassic, Argentina) with Comments on Eucynodont Phylogeny," *Journal of Vertebrate Paleontology* 16, no. 2 (1996): 271–84.
69. Kenneth D. Angielczyk, "Phylogenetic Evidence for and Implications of a Dual Origin of Propaliny in Anomodont Therapsids (Synapsida)," *Paleobiology* 30, no. 2 (2004): 268–96.
70. Juan Carlos Cisneros, Fernando Abdala, Bruce S. Rubidge, Paula Camboim Dentzien-Dias, and Ana de Oliveira Bueno, "Dental Occlusion in a 260-Million-Year-Old Therapsid with Saber Canines from the Permian of Brazil," *Science* 331, no. 6024 (2011): 1603–5; with commentary by Jörg Fröbisch, "On Dental Occlusion and Saber Teeth," *Science* 331, no. 6024 (2011): 1525–28. For multicusped postcanine teeth from the Lower Triassic, see Leandro C. Gaetano, Helke Mocke, Fernando Abdala, and P. John Hancox, "Complex Multicusped Postcanine Teeth from the

Lower Triassic of South Africa," *Journal of Vertebrate Paleontology* 32, no. 6 (2012): 1411–20.

71. T. S. Kemp, "Stance and Gait in the Hindlimb of a Therocephalian Mammal-like Reptile," *Journal of Zoology, London* 186, no. 2 (1978): 143–61.
72. James A. Hopson, "Patterns of Evolution in the Manus and Pes of Non-mammalian Therapsids," *Journal of Vertebrate Paleontology* 15, no. 3 (1995): 615–39.
73. Jörg Fröbisch and Robert R. Reisz, "The Late Permian Herbivore *Suminia* and the Early Evolution of Arboreality in Terrestrial Ecosystems," *Proceedings of the Royal Society of London, Series B: Biological Sciences* 276, no. 1673 (2009): 3611–18.
74. Tobias Nasterlack, Aurore Canoville, and Anusuya Chinsamy, "New Insights into the Biology of the Permian Genus *Cistecephalus* (Therapsida, Dicynodonta)," *Journal of Vertebrate Paleontology* 32, no. 6 (2012): 1396–410.
75. Christian A. Sidor, "Simplification as a Trend in Synapsid Cranial Evolution," *Evolution* 55, no. 7 (2001): 1419–42.
76. Christian A. Sidor, "The Naris and Palate of *Lycaenodon longiceps* (Therapsida: Biarmosuchia), with Comments on Their Early Evolution in Therapsida," *Journal of Paleontology* 77 (2003): 977–84.
77. See, for example, Michael Laaß, "The Origins of the Cochlea and Impedance Matching Hearing in Synapsids," *Acta Palaeontological Polonica* 61, no. 2 (2016): 267–80.
78. Kévin Rey et al., "Oxygen Isotopes Suggest Elevated Thermometabolism within Multiple Permo-Triassic Therapsid Clades," *eLife* 6 (2017), e28589.
79. See, for example, Qiang Ji, Zhe-Xi Luo, Chong-Xi Yuan, and Alan R. Tabrum, "A Swimming Mammaliaform from the Middle Jurassic and Ecomorphological Diversification of Early Mammals," *Science* 311, no. 5764 (2006): 1123–27.
80. Piotr Bajdek, Martin Qvarnström, Krzysztof Owocki, Tomasz Sulej, Andrey G. Sennikov, Valeriy K. Golubev, and Grzegorz Niedźwiedzki, "Microbiota and Food Residues Including Possible Evidence of Pre-mammalian Hair in Upper Permian Coprolites from Russia," *Lethaia* 49, no. 4 (2016): 455–77.

81. See the linkage between bone structure and inferred size of red blood cells (linked to metabolic rate) by Adam K. Huttenlocker and C. G. Farmer, "Bone Microvasculature Tracks Red Blood Cell Size Diminution in Triassic Mammal and Dinosaur Forerunners," *Current Biology* 27, no. 1 (2017): 48–54; also Mathieu G. Faure-Brac and Jorge Cubo, "Were the Synapsids Primitively Endotherms? A Palaeohistological Approach Using Phylogenetic Eigenvector Maps," *Philosophical Transactions of the Royal Society of London, Series B: Biological Sciences* 375, no. 1793 (2020), 20190138.
82. Willem J. Hillenius, "The Evolution of Nasal Turbinates and Mammalian Endothermy," *Paleobiology* 18, no. 1 (1992): 17–29.
83. Willem J. Hillenius, "Turbinates in Therapsids: Evidence for Late Permian Origins of Mammalian Endothermy," *Evolution* 48, no. 2 (1994): 207–29.
84. J. Benoit, P. R. Manger, and B. S. Rubidge, "Palaeoneurological Clues to the Evolution of Defining Mammalian Soft Tissue Traits," *Scientific Reports* 6 (2016), e25604. Concerning the related theme of parental care, see, for example, Sandra C. Jasinoski and Fernando Abdala, "Aggregations and Parental Care in the Early Triassic Basal Cynodonts *Galesaurus planiceps* and *Thrinaxodon liorhinus*," *PeerJ* 5 (2017), e2875.
85. Marcello Ruta, Jennifer Botha-Brink, Stephen A. Mitchell, and Michael J. Benton, "The Radiation of Cynodonts and the Groundplan of Mammalian Morphological Diversity," *Proceedings of the Royal Society of London, Series B: Biological Sciences* 280, no. 1769 (2013), 20131865.
86. Exemplified by the anomodont *Lystrosaurus,* until you read the cautionary article by Sean P. Modesto, "The Disaster Taxon *Lystrosaurus:* A Paleontological Myth," *Frontiers in Earth Science* 8 (2020), 610463.
87. Spencer G. Lucas, "Timing and Magnitude of Tetrapod Extinctions across the Permo-Triassic Boundary," *Journal of Asian Earth Sciences* 36, no. 6 (2009): 491–502.
88. Roger Smith and Jennifer Botha, "The Recovery of Terrestrial Vertebrate Diversity in the South African Karoo Basin after the End-Permian Event," *Compte Rendus Palevol* 4, no. 6–7 (2005): 623–36.
89. Randall B. Irmis and Jessica H. Whiteside, "Delayed Recovery of Non-marine Tetrapods after the End-Permian Mass Extinction Tracks Global

Carbon Cycle," *Proceedings of the Royal Society of London, Series B: Biological Sciences* 279, no. 1732 (2012): 1310–18.

90. Randall B. Irmis, "Evolutionary Hypotheses for the Early Diversification of Dinosaurs," *Transactions of the Royal Society of Edinburgh: Earth and Environmental Science* 101, no. 3–4 (2010): 397–426.
91. Tomasz Sulej and Grzegorz Niedźwiedzki, "An Elephant-Sized Late Triassic Synapsid with Erect Limbs," *Science* 363, no. 6422 (2019): 78–80. However, Marco Romano and Fabio Manucci, "Resizing *Lisowicia bojani:* Volumetric Body Mass Estimate and 3D Reconstruction of the Giant Late Triassic Dicynodont," *Historical Biology* 33, no. 4 (2021): 474–79, suggest a somewhat smaller size.
92. A. W. Crompton, "The Enigma of the Evolution of Mammals," *Optima* 18, no. 3 (1968): 137–51, p. 140.
93. Concerning this miniaturization, see José F. Bonaparte, "Miniaturisation and the Origin of Mammals," *Historical Biology* 24, no. 1 (2012): 43–48; and Stephan Lautenschlager, Pamela G. Gill, Zhe-Xi Luo, Michael J. Fagan, and Emily J. Rayfield, "The Role of Miniaturization in the Evolution of the Mammalian Jaw and Middle Ear," *Nature* 561, no. 7724 (2018): 533–37.
94. Timothy B. Rowe, Thomas E. Macrini, and Zhe-Xi Luo, "Fossil Evidence on Origin of the Mammalian Brain," *Science* 332, no. 6032 (2011): 955–57; with commentary by R. Glenn Northcutt, "Evolving Large and Complex Brains," *Science* 332, no. 6032 (2011): 926–27.
95. Zhe-Xi Luo, Alfred W. Crompton, and Ai-Lin Sun, "A New Mammaliaform from the Early Jurassic and Evolution of Mammalian Characteristics," *Science* 292, no. 5521 (2001): 1535–40.
96. Yaoming Hu, Jin Meng, Yuanqing Wang, and Chuankui Li, "Large Mesozoic Mammals Fed on Young Dinosaurs," *Nature* 433 (2005): 149–52; also commentary by Anne Weil, "Living Large in the Cretaceous," *Nature* 433, no. 7022 (2005): 116–17.
97. Brian M. Davis, "Evolution of the Tribosphenic Molar Pattern in Early Mammals, with Comments on the 'Dual-Origin' Hypothesis," *Journal of Mammalian Evolution* 18 (2011): 227–44.
98. Chang-Fu Zhou, Shaoyuan Wu, Thomas Martin, and Zhe-Xi Luo, "A Jurassic Mammaliaform and the Earliest Mammalian Evolutionary

Adaptations," *Nature* 500, no. 7461 (2013): 163–67, with corrections in 505 (2014): 436; also commentary by Richard L. Cifelli and Brian M. Davis, "Jurassic Fossils and Mammalian Antiquity," *Nature* 500, no. 7461 (2013): 160–61. It now transpires that hypsodonty per se evolved even earlier, as documented in a Triassic traversodontid (a group of stem mammals); see Tomaz P. Melo, Ana Maria Ribeiro, Agustín G. Martinelli, and Marina Bento Soares, "Early Evidence of Molariform Hypsodonty in a Triassic Stem-Mammal," *Nature Communications* 10, no. 1 (2019), 2841.

99. On the hyoid bone of the diminutive docodontan *Microdocodon*, see Chang-Fu Zhou, Bhart-Anjan S. Bhullar, April I. Neander, Thomas Martin, and Zhe-Xi Luo, "New Jurassic Mammaliaform Sheds Light on Early Evolution of Mammal-Like Hyoid Bones," *Science* 365, no. 6450 (2019): 276–79; with commentary by Simone Hoffmann and David W. Krause, "Tongues Untied," *Science* 365, no. 6450 (2019): 222–23.

100. Zhe-Xi Luo, Irina Ruf, and Thomas Martin, "The Petrosal and Inner Ear of the Late Jurassic Cladotherian Mammal *Dryolestes leiriensis* and Implications for Ear Evolution in Therian Mammals," *Zoological Journal of the Linnean Society* 166, no. 2 (2012): 433–63.

101. Irina Ruf, Zhe-Xi Luo, and Thomas Martin, "Reinvestigation of the Basicranium of *Haldanodon exspectatus* (Mammaliaformes, Docodonta)," *Journal of Vertebrate Paleontology* 33, no. 2 (2013): 382–400.

102. Zhe-Xi Luo, Chong-Xi Yuan, Qing-Jin Meng, and Qiang Ji, "A Jurassic Eutherian Mammal and Divergence of Marsupials and Placentals," *Nature* 476, no. 7361 (2011): 442–45.

103. See, for example, Steven C. Sweetman, Grant Smith, and David M. Martill, "Highly Derived Eutherian Mammals from the Earliest Cretaceous of Southern Britain," *Acta Palaeontologica Polonica* 62, no. 4 (2017): 657–65; and Shundong Bi, Xiaoting Zheng, Xiaoli Wang, Natalie E. Cignetti, Shiling Yang, and John R. Wible, "An Early Cretaceous Eutherian and the Placental-Marsupial Dichotomy," *Nature* 558, no. 7710 (2018): 390–95.

104. Meng Chen and Zhe-Xi Luo, "Postcranial Skeleton of the Cretaceous Mammal *Akidolestes cifellii* and Its Locomotor Adaptations," *Journal of Mammalian Evolution* 20, no. 3 (2013): 159–89.

105. James E. Martin, Judd A. Case, John W. M. Jagt, Anne S. Schulp, and Eric W. A. Mulder, "A New European Marsupial Indicates a Late Cretaceous High-Latitude Transatlantic Dispersal Route," *Journal of Mammalian Evolution* 12 (2005): 495–511.
106. David Archibald, *Extinction and Radiation: How the Fall of the Dinosaurs Led to the Rise of Mammals* (Baltimore: Johns Hopkins University Press, 2011), 67.
107. See, for example, P. Raia et al., "Rapid Action in the Palaeogene, the Relationship between Phenotypic and Taxonomic Diversification in Coenozoic Mammals," *Proceedings of the Royal Society of London, Series B: Biological Sciences* 280, no. 1750 (2013), 20122244; and Thomas J. D. Halliday, Paul Upchurch, and Anjali Goswami, "Resolving the Relationships of Paleocene Placental Mammals," *Biological Reviews* 92, no. 1 (2016): 521–50.
108. Felisa Smith et al., "The Evolution of Maximum Body Size of Terrestrial Mammals," *Science* 330, no. 6008 (2010): 1216–19, p. 1216 [abstract].
109. Matthew J. Phillips and Carmelo Fruciano, "The Soft Explosive Model of Placental Mammal Evolution," *BMC Evolutionary Biology* 18, no. 1 (2018), 104.
110. Liang Liu et al., "Genomic Evidence Reveals a Radiation of Placental Mammals Uninterrupted by the KPg Boundary," *Proceedings of the National Academy of Sciences of the United States of America* 114, no. 35 (2017): E7282–90; also commentary by John Gatesby and Mark S. Springer, "Phylogenomic Red Flags: Homology Errors and Zombie Lineages in the Evolutionary Diversification of Placental Mammals," *Proceedings of the National Academy of Sciences of the United States of America* 114, no. 45 (2017): E9431–32, with authors' response (E9433–34). For more general overviews, see David M. Grossnickle and Ellis Newham, "Therian Mammals Experience an Ecomorphological Radiation during the Late Cretaceous and Selective Extinction at the K–Pg Boundary," *Proceedings of the Royal Society of London, Series B: Biological Sciences* 283, no. 1832 (2016), 20160256; and Mathias M. Pires, Brian D. Rankin, Daniele Silvestro, and Tiago B. Quental, "Diversification Dynamics of Mammalian

Clades during the K-Pg Mass Extinction," *Biology Letters* 14, no. 9 (2018), 2018.0458.

111. Specifically, the ratio of synonymous to nonsynonymous substitutions (in the latter case, a change in the DNA results in the coding for a different amino acid) and the nature of the third codon in the amino acid triplets (for example, the amino acid histidine can be coded as either the triplet CA*U* or CAC).
112. J. Romiguier, V. Ranwez, E. J. P. Douzery, and N. Galtier, "Genomic Evidence for Large, Long-Lived Ancestors to Placental Mammals," *Molecular Biology and Evolution* 30, no. 1 (2013): 5–13.
113. Cynthia L. Gordon, "A First Look at Estimating Body Size in Dentally Conservative Marsupials," *Journal of Mammalian Evolution* 10, no. 1–2 (2003): 1–21.
114. Gregory P. Wilson, Eric G. Ekdale, John W. Hoganson, Jonathan J. Calede, and Abby Vander Linden, "A Large Carnivorous Mammal from the Late Cretaceous and the North American Origin of Marsupials," *Nature Communications* 7 (2016), 13734.
115. Octávio Mateus et al., "Angolan Ichnosite in a Diamond Mine Shows the Presence of a Large Terrestrial Mammaliamorph, a Crocodylomorph, and Sauropod Dinosaurs in the Early Cretaceous of Africa," *Palaeogeography, Palaeoclimatology, Palaeoecology* 471 (2017): 220–32.
116. Giuseppe Leonardi and Ismar de Souza Carvalho, "Review of the Early Mammal *Brasilichnium* and *Brasilichnium*-like Tracks from the Lower Cretaceous of South America," *Journal of South American Earth Sciences* 106 (2021), 102940, where they review the occurrences of fossil footprints across the Paraná basin, including on the pavement slabs in various towns.
117. See Thomas J. D. Halliday, Mario dos Reis, Asif U. Tamuri, Henry Ferguson-Gow, Ziheng Yang, and Anjali Goswami, "Rapid Morphological Evolution in Placental Mammals Post-dates the Origin of the Crown Group," *Proceedings of the Royal Society of London, Series B: Biological Sciences* 286, no. 1898 (2019), 20182418.
118. To this roster we can add such groups as the extinct taeniodonts that, although identified as a Tertiary group, appear in the Cretaceous; see

Richard C. Fox, "The Status of *Schowalteria clemensi,* the Late Cretaceous Taeniodont (Mammalia)," *Journal of Vertebrate Paleontology* 36, no. 6 (2016), e1211666.

119. See, for example, Robert W. Meredith et al., "Impacts of the Cretaceous Terrestrial Revolution and KPg Extinction on Mammal Diversification," *Science* 334, no. 6055 (2011): 521–24; and commentary by Kristopher M. Helgen, "The Mammal Family Tree," *Science* 334, no. 6055 (2011): 458–59. See also the critique (in favor of deeper originations) by Olaf R. P. Bininda-Emonds and Andy Purvis, *Science* 337, no. 6090 (2012), 34, and the authors' response: William J. Murphy et al., *Science* 337, no. 6090 (2012). For the argument that eulipotyphlan diversification occurred shortly after the KTE, see Jun J. Sato et al., "Post K-Pg Diversification of the Mammalian Order Eulipotyphla as Suggested by Phylogenomic Analyses of Ultra-conserved Elements," *Molecular Phylogenetics and Evolution* 141 (2019), 106605.
120. See J. D. Archibald, A. O. Averianov, and E. G. Ekdale, "Late Cretaceous Relatives of Rabbits, Rodents, and Other Extant Eutherian Mammals," *Nature* 414, no. 6859 (2001): 62–65.
121. See, for example, Robert D. Martin, Christophe Soligo, and Simon Tavaré, "Primate Origins: Implications of a Cretaceous Ancestry," *Folia Primatologica* 78, no. 5–6 (2007): 277–96; and Richard D. Wilkinson, Michael E. Steiper, Christophe Soligo, Robert D. Martin, Ziheng Yang, and Simon Tavaré, "Dating Primate Divergences through an Integrated Analysis of Palaeontological and Molecular Data," *Systematic Biology* 60, no. 1 (2011): 16–31.
122. Shaoyuan Wu et al., "Molecular and Paleontological Evidence for a Post-Cretaceous Origin of Rodents," *PLoS One* 7, no. 10 (2012), e46445.
123. See A. Weil and D. W. Krause, "Multituberculata," in *Evolution of Tertiary Mammals in North America,* vol. 2, ed. Christine M. Janis, Gregg F. Gunnell, and Mark D. Uhen (Cambridge: Cambridge University Press, 2008), 19–38.
124. Lazzari and colleagues point out that this rodent-like dentition extends to the tritylodont mammals and must be an important component in the success of all three groups; see Vincent Lazzari, Julia A. Schultz, Paul Tafforeau, and Thomas Martin, "Occlusal Pattern in

Paulchoffatiid Multituberculates and the Evolution of Cusp Morphology in Mammaliamorphs with Rodent-like Dentitions," *Journal of Mammalian Evolution* 17 (2010): 177–92.

125. Gregory P. Wilson, Alistair R. Evans, Ian J. Corfe, Peter D. Smits, Mikael Fortelius, and Jukka Jernvall, "Adaptive Radiation of Multituberculate Mammals before the Extinction of Dinosaurs," *Nature* 483 (2012): 457–60; see also Neil F. Adams, Emily J. Rayfield, Philip G. Cox, Samuel N. Cobb, and Ian J. Corfe, "Functional Tests of the Competitive Exclusion Hypothesis for Multituberculate Extinction," *Royal Society Open Science* 6, no. 3 (2019), e181536.
126. This includes evidence for sociality; see Lucas N. Weaver, David J. Varricchio, Eric J. Sargis, Meng Chen, William J. Freimuth, and Gregory P. Wilson Mantilla, "Early Mammalian Social Behaviour Revealed by Multituberculates from a Dinosaur Nesting Site," *Nature Ecology and Evolution* 5, no. 1 (2021): 32–37, who also note this is most likely convergent with the therian mammals.
127. R. W. Wilson, "Late Cretaceous (Fox Hills) Multituberculates from the Red Owl Local Fauna of Western South Dakota," *Dakoterra* 3 (1987): 118–22. The generic name refers to owl (*Bubo*) and teeth (*dens*).
128. See, for example, U. Heimhofer, P. A. Hochuli, S. Burla, J. M. L. Dinis, and H. Weissert, "Timing of Early Cretaceous Angiosperm Diversification and Possible Links to Major Paleoenvironmental Change," *Geology* 33, no. 2 (2005): 141–44.
129. See, for example, Zhong-Jian Liu and Xin Wang, "*Yuhania:* A Unique Angiosperm from the Middle Jurassic of Inner Mongolia, China," *Historical Biology* 29, no. 4 (2017): 431–41. For a more skeptical overview, see Mario Coiro, James A. Doyle, and Jason Hilton, "How Deep Is the Conflict between Molecular and Fossil Evidence on the Age of Angiosperms?" *New Phytologist* 223, no. 1 (2019): 83–99. Similar disagreements accompany *Nanjinganthus,* interpreted as an angiosperm by Qiang Fu et al., "An Unexpected Noncarpellate Epigynous Flower from the Jurassic of China," *eLife* 7 (2018), e38827; but disputed by Dmitry D. Sokoloff, Margarita V. Remizowa, Elena S. El, Paula J. Rudall, and Richard M. Bateman, "Supposed Jurassic Angiosperms Lack Pentamery, an Important Angiosperm-Specific Feature," *New*

Phytologist 228, no. 2 (2020): 420–26. The debate will rumble on, but overall the evidence for Jurassic angiosperms seems to grow; see, for example, D. Silvestro et al., "Fossil Data Support a Pre-Cretaceous Origin of Flowering Plants," *Nature Ecology and Evolution* 5 (2021): 449–57.

130. See, for example, Peter A. Hochuli and Susanne Feist-Burkhardt, "A Boreal Early Cradle of Angiosperms? Angiosperm-like Pollen from the Middle Triassic of the Barents Sea (Norway)," *Journal of Micropaleontology* 23 (2004): 97–104.
131. See, for example, Jin-Hua Ran, Ting-Ting Shen, Ming-Ming Wang, and Xiao-Quan Wang, "Phylogenomics Resolves the Deep Phylogeny of Seed Plants and Indicates Partial Convergent or Homoplastic Evolution between Gnetales and Angiosperms," *Proceedings of the Royal Society of London, Series B: Biological Sciences* 285, no. 1881 (2018), 20181012.
132. Taylor S. Feild, Nan Crystal Arens, James A. Doyle, Todd E. Dawson, and Michael J. Donoghue, "Dark and Disturbed: A New Image of Early Angiosperm Ecology," *Paleobiology* 30, no. 1 (2004): 82–107.
133. See, for example, C. Coiffard and B. Gomez, "Influence of Latitude and Climate on Spread, Radiation and Rise to Dominance of Early Angiosperms during the Cretaceous in the Northern Hemisphere," *Geologica Acta* 10, no. 2 (2012): 181–88.
134. Clément Coiffard, Bernard Gomez, Véronique Daviero-Gomez, and David L. Dilcher, "Rise to Dominance of Angiosperm Pioneers in European Cretaceous Environments," *Proceedings of the National Academy of Sciences of the United States of America* 109, no. 51 (2012): 20955–59.
135. Marcela Martínez-Millán, "Fossil Record and Age of the Asteridae," *Botanical Review* 76 (2010): 83–135.
136. Wang Heng-chang et al., "Rosid Radiation and the Rapid Rise of Angiosperm-Dominated Forests," *Proceedings of the National Academy of Sciences of the United States of America* 106, no. 10 (2009): 3853–58.
137. M. A. Gandolfo, K. C. Nixon, and W. L. Crepet, "Cretaceous Flowers of Nymphaeaceae and Implications for Complex Insect Entrapment

Pollination Mechanisms in Early Angiosperms," *Proceedings of the National Academy of Sciences of the United States of America* 101, no. 21 (2004): 8056–60. Interestingly, their flower structure indicates the insects in question were beetles.

138. See, for example, Steven R. Manchester, Thomas M. Lehman, and Elisabeth A. Wheeler, "Fossil Palms (Arecaceae, Coryphoideae) Associated with Juvenile Herbivorous Dinosaurs in the Upper Cretaceous Aguja Formation, Big Bend National Park, Texas," *International Journal of Plant Sciences* 171, no. 6 (2010): 679–89.
139. V. Prasad et al., "Dinosaur Coprolites and the Early Evolution of Grasses and Grazers," *Science* 310, no. 5751 (2007): 1177–80; also the commentary by Dolores R. Piperno and Hans-Dieter Sues, "Dinosaurs Dined on Grass," *Science* 310, no. 5751 (2007): 1126–28. And not only grasses, but also representatives of the Oryzeae, the rice group; see V. Prasad et al., "Late Cretaceous Origin of the Rice Tribe Provides Evidence for Early Diversification in Poaceae," *Nature Communications* 2 (2011), 480.
140. Steven R. Manchester, Dashrath K. Kapgate, and Jun Wen, "Oldest Fruits of the Grape Family (Vitaceae) from the Late Cretaceous Deccan Cherts of India," *American Journal of Botany* 100, no. 9 (2013): 1849–59.
141. Santiago R. Ramírez, Barbara Gravendeel, Rodrigo B. Singer, Charles R. Marshall, and Naomi E. Pierce, "Dating the Origin of Orchidaceae from a Fossil Orchid with Its Pollinator," *Nature* 448, no. 7157 (2007): 1042–45; also Thomas J. Givnish et al., "Orchid Phylogenomics and Multiple Drivers of Their Extraordinary Diversification," *Proceedings of the Royal Society of London, Series B: Biological Sciences* 282, no. 1814 (2015), 20151553.
142. Julia Naumann et al., "Single-Copy Nuclear Genes Place Haustorial Hydnoraceae within Piperales and Reveal a Cretaceous Origin of Multiple Parasitic Angiosperm Lineages," *PLoS One* 8, no. 11 (2013), e79204.
143. See, for example, Taylor S. Feild and Sandrine Isnard, "Climbing Habit and Ecophysiology of *Schisandra glabra* (Schisandraceae): Implications for the Early Evolution of Angiosperm Lianescence," *Interna-*

tional Journal of Plant Sciences 174, no. 8 (2013): 1121–33; and Selena Y. Smith, Stefan A. Little, Ranessa L. Cooper, Robyn J. Burnham, and Ruth A. Stockey, "A Ranunculalean Liana Stem from the Cretaceous of British Columbia, Canada: *Atli morinii* gen. et sp. nov.," *International Journal of Plant Sciences* 174, no. 5 (2013): 818–31.

144. Sarah Mathews et al., "Adaptive Evolution in the Photosensory Domain of Phytochrome A in Early Angiosperms," *Molecular Biology and Evolution* 20, no. 7 (2003): 1087–97.
145. Taylor S. Feild and Jonathan P. Wilson, "Evolutionary Voyage of Angiosperm Vessel Structure-Function and Its Significance for Early Angiosperm Success," *International Journal of Plant Sciences* 173, no. 6 (2012): 596–609.
146. Taylor S. Feild et al., "Fossil Evidence for Cretaceous Escalation in Angiosperm Leaf Vein Evolution," *Proceedings of the National Academy of Sciences of the United States of America* 108, no. 20 (2011): 8363–66; and Hugo Jan de Boer, Maarten B. Eppinga, Martin J. Wassen, and Stefan C. Dekker, "A Critical Transition in Leaf Evolution Facilitated the Cretaceous Angiosperm Revolution," *Nature Communications* 3 (2012), 1221.
147. C. Kevin Boyce and Andrew B. Leslie, "The Paleontological Context of Angiosperm Vegetative Evolution," *International Journal of Plant Sciences* 173, no. 6 (2012): 561–68.
148. See, for example, Tim J. Brodribb and Taylor S. Feild, "Leaf Hydraulic Evolution Led a Surge in Leaf Photosynthetic Capacity during Early Angiosperm Diversification," *Ecology Letters* 13, no. 2 (2010): 175–83.
149. C. Kevin Boyce, Jung-Eun Lee, Taylor S. Feild, Tim J. Brodribb, and Maciej A. Zwieniecki, "Angiosperms Helped Put the Rain in the Rainforests: The Impact of Plant Physiological Evolution on Tropical Biodiversity," *Annals of the Missouri Botanical Garden* 97, no. 4 (2010): 527–40.
150. On, respectively, the malpighialeans (a very diverse group but important in rainforests), palms, menispermaceans (including the tropical plants from which curare is obtained), and meliaceans (mahogany): Charles C. Davis, Campbell O. Webb, Kenneth J. Wurdack, Carlos A. Jaramillo, and Michael J. Donoghue, "Explosive Radiation of Malpighiales Supports a

Mid-Cretaceous Origin of Modern Tropical Rain Forests," *American Naturalist* 165, no. 3 (2005): E36–65; Thomas L. P. Couvreur, Félix Forest, and William J. Baker, "Origin and Global Diversification Patterns of Tropical Rain Forests: Inferences from a Complete Genus-Level Phylogeny of Palms," *BMC Biology* 9 (2011), 44; Wei Wang et al., "Menispermaceae and the Diversification of Tropical Rainforests near the Cretaceous-Paleogene Boundary," *New Phytologist* 195, no. 2 (2012): 470–78; and Brian A. Atkinson, "Fossil Evidence for a Cretaceous Rise of the Mahogany Family," *American Journal of Botany* 107, no. 1 (2020): 139–47.

151. See, for example, Andrew C. Scott and Ian J. Glasspool, "The Diversification of Paleozoic Fire Systems and Fluctuations in Atmospheric Oxygen Concentration," *Proceedings of the National Academy of Sciences of the United States of America* 103, no. 29 (2006): 10861–65.
152. William J. Bond and Jeremy J. Midgley, "Fire and Angiosperm Revolutions," *International Journal of Plant Sciences* 173, no. 6 (2012): 569–83. The rise of angiosperms and their occasional incineration also may have played a role in regulating the amount of atmospheric oxygen and among other things favored the rainforests; see Claire M. Belcher, Benjamin J. W. Mills, Rayanne Vitali, Sarah J. Baker, Timothy M. Lenton, and Andrew J. Watson, "The Rise of Angiosperms Strengthened Fire Feedbacks and Improved the Regulation of Atmospheric Oxygen," *Nature Communications* 12, no. 1 (2021), 503.
153. Frank Berendse and Marten Scheffer, "The Angiosperm Radiation Revisited, an Ecological Explanation for Darwin's 'Abominable Mystery,'" *Ecology Letters* 12, no. 9 (2009): 865–72.
154. See, for example, Emiliano Peralta-Medina and Howard J. Falcon-Lang, "Cretaceous Forest Composition and Productivity Inferred from a Global Fossil Wood Database," *Geology* 40, no. 3 (2012): 219–22; also Fabien L. Condamine, Daniele Silvestro, Eva B. Koppelhus, and Alexandre Antonelli, "The Rise of Angiosperms Pushed Conifers to Decline during Global Cooling," *Proceedings of the National Academy of Sciences of the United States of America* 117, no. 46 (2020): 28867–75, with commentary by H. John B. Birks, "Angiosperms versus Gymnosperms in the Cretaceous," *Proceedings of the National Acad-*

emy of Sciences of the United States of America 117, no. 49 (2020): 30879–81.

155. See, for example, Howard J. Falcon-Lang, Viola Mages, and Margaret Collinson, "The Oldest *Pinus* and Its Preservation by Fire," *Geology* 44, no. 4 (2016): 303–6; also commentary by Jason Hilton, James B. Riding, and G. W. Rothwell, "Age and Identity of the Oldest Pine Fossils: Comment," *Geology* 44, no. 8 (2016): e400–e401.
156. See Xin-Yu Du et al., "Simultaneous Diversification of Polypodiales and Angiosperms in the Mesozoic," *Cladistics* (2021), https://doi.org/10.1111/cla.12457.
157. Harald Schneider, Eric Schuettpelz, Kathleen M. Pryer, Raymond Cranfill, Susana Magallón, and Richard Lupia, "Ferns Diversified in the Shadow of Angiosperms," *Nature* 428, no. 6982 (2004): 553–57; see also Eric Schuettpelz and Kathleen M. Pryer, "Evidence for a Cenozoic Radiation of Ferns in an Angiosperm-Dominated Canopy," *Proceedings of the National Academy of Sciences of the United States of America* 106, no. 27 (2009): 11200–205.
158. J. E. Watkins Jr., and Catherine L. Cardelús, "Ferns in an Angiosperm World: Cretaceous Radiation into the Epiphytic Niche and Diversification on the Forest Floor," *International Journal of Plant Sciences* 173, no. 6 (2012): 695–710.
159. David M. Grossnickle and P. David Polly, "Mammal Disparity Decreases during the Cretaceous Angiosperm Radiation," *Proceedings of the Royal Society of London, Series B: Biological Sciences* 280, no. 1771 (2013), 20132110; and David M. Grossnickle and Elis Newham, "Therian Mammals Experience an Ecomorphological Radiation during the Late Cretaceous and Selective Extinction at the K–Pg Boundary," *Proceedings of the Royal Society of London, Series B: Biological Sciences* 283, no. 1832 (2016), 20160256.
160. Hallie J. Sims, "Paleolatitudinal Gradients in Seed Size during the Cretaceous-Tertiary Radiation of Angiosperms," *International Journal of Plant Sciences* 171, no. 2 (2010): 216–20.
161. Ove Eriksson, Else Marie Friis, and Per Löfgren, "Seed Size, Fruit Size, and Dispersal Systems in Angiosperms from the Early Cretaceous to the Late Tertiary," *American Naturalist* 156, no. 1 (2000): 47–58.

162. See, for example, David Peris, Ricardo Pérez-de la Fuente, Enrique Peñalver, Xavier Delclòs, Eduardo Barrón, and Conrad C. Labandeira, "False Blister Beetles and the Expansion of Gymnosperm-Insect Pollination Modes before Angiosperm Dominance," *Current Biology* 27, no. 6 (2017): 897–904; and Chenyang Cai, Hermes E. Escalona, Liqin Li, Ziwei Yin, Diying Huang, and Michael S. Engel, "Beetle Pollination of Cycads in the Mesozoic," *Current Biology* 28, no. 17 (2018): 2806–12.e1.
163. Enrique Peñalver et al., "Thrips Pollination of Mesozoic Gymnosperms," *Proceedings of the National Academy of Sciences of the United States of America* 109, no. 22 (2012): 8623–28.
164. David Peris, Conrad C. Labandeira, Eduardo Barrón, Xavier Delclòs, Jes Rust, and Bo Wang, "Generalist Pollen-Feeding Beetles during the Mid-Cretaceous," *iScience* 23, no. 3 (2020), 100913.
165. Niklas Wahlberg, Christopher W. Wheat, and Carlos Peña, "Timing and Patterns in the Taxonomic Diversification of Lepidoptera (Butterflies and Moths)," *PLoS One* 8, no. 11 (2013), e80875; also Akiko Y. Kawahara et al., "Phylogenomics Reveals the Evolutionary Timing and Pattern of Butterflies and Moths," *Proceedings of the National Academy of Sciences of the United States of America* 116, no. 45 (2019): 22657–63.
166. Sophie Cardinal and Bryan N. Danforth, "Bees Diversified in the Age of Eudicots," *Proceedings of the Royal Society of London, Series B: Biological Sciences* 280, no. 1755 (2013), 2012.2686; also commentary by Simone C. Cappellari, Hanno Schaefer, and Charles C. Davis, "Evolution: Pollen or Pollinators—Which Came First?" *Current Biology* 23, no. 8 (2013): R316–R318; also Jorge Fernando Genise et al., "100 Ma Sweat Bee Nests: Early and Rapid Codiversification of Crown Bees and Flowering Plants," *PLoS One* 15, no. 1 (2020), e0227789.
167. One needs, however, to be careful. For example, see Santiago R. Ramírez, Thomas Eltz, Mikiko K. Fujiwara, Günter Gerlach, Benjamin Goldman-Huertas, Neil D. Tsutsui, and Naomi E. Pierce, "Asynchronous Diversification in a Specialized Plant-Pollinator Mechanism," *Science* 333, no. 6050 (2011): 1742–46. They show that, notwithstanding a close link between euglossine bees and orchids, and one that depends on collec-

tion of fragrances and reciprocal pollination, the diversifications of either group were asynchronous with the bees some 12 Ma ahead.

168. See, for example, William L. Crepet and Kevin C. Nixon, "Fossil Clusiaceae from the Late Cretaceous (Turonian) of New Jersey and Implications Regarding the History of Bee Pollination," *American Journal of Botany* 85, no. 8 (1998): 1122–33.
169. Shusheng Hu, David L. Dilcher, David M. Jarzen, and David Winship Taylor, "Early Steps of Angiosperm-Pollinator Coevolution," *Proceedings of the National Academy of Sciences of the United States of America* 105, no. 1 (2008): 240–45.
170. Bo Wang, Haichun Zhang, and Edmund A. Jarzembowski, "Early Cretaceous Angiosperms and Beetle Evolution," *Frontiers in Plant Science* 4 (2013), 360; also Tong Bao, Bo Wang, Jianguo Li, and David Dilcher, "Pollination of Cretaceous Flowers," *Proceedings of the National Academy of Sciences of the United States of America* 116, no. 49 (2019): 24707–11; and E. Tihelka et al., *Nature Plants* 7, no. 49 (2021): 445–51.
171. See the overview by Qingqing Zhang and Bo Wang, "Evolution of Lower Brachyceran Flies (Diptera) and Their Adaptive Radiation with Angiosperms," *Frontiers in Plant Science* 8 (2017), 631; also, for example, with respect to the bee flies, see Xuankun Li et al., "Phylogenomics Reveals Accelerated Late Cretaceous Diversification of Bee Flies (Diptera: Bombyliidae)," *Cladistics* 37, no. 3 (2021): 276–97.
172. Harald Schneider, "The Ghost of the Cretaceous Terrestrial Revolution in the Evolution of Fern-Sawfly Associations," *Journal of Systematics and Evolution* 54, no. 2 (2016): 93–103.
173. See, for example, Luis M. Chiappe and Gareth J. Dyke, "The Mesozoic Radiation of Birds," *Annual Review of Ecology and Systematics* 33 (2002): 91–124.
174. See, for example, Francisco J. Serrano, Paul Palmqvist, Luis M. Chiappe, and José L. Sanz, "Inferring Flight Parameters of Mesozoic Avians through Multivariate Analyses of Forelimb Elements in Their Living Relatives," *Paleobiology* 43, no. 1 (2017): 144–69.
175. See, for example, Darren Naish, Gareth Dyke, Andrea Cau, François Escuillié, and Pascal Godefroit, "A Gigantic Bird from the Upper

Cretaceous of Central Asia," *Biology Letters* 8, no. 1 (2012): 97–100. This bird is known from its enormous mandibles; it is not clear if it was somewhat albatross-like or flightless.

176. See Jessie Atterholt, J. Howard Hutchison, and Jingmai K. O'Connor, "The Most Complete Enantiornithine from North America and a Phylogenetic Analysis of the Avisauridae," *PeerJ* 6 (2018), e5910; also Di Liu, Luis M. Chiappe, Francisco Serrano, Michael Habib, Yuguang Zhang, and Qinjing Meng, "Flight Aerodynamics in Enantiornithines: Information from a New Chinese Early Cretaceous Bird," *PLoS One* 12, no. 10 (2017), e0184637.
177. For overviews in the context of the Cretaceous, see Gareth J. Dyke and Marcel van Tuinen, "The Evolutionary Radiation of Modern Birds (Neornithes): Reconciling Molecules, Morphology and the Fossil Record," *Zoological Journal of the Linnean Society* 141, no. 2 (2004): 153–77; and Per G. P. Ericson et al., "Diversification of Neoaves: Integration of Molecular Sequence Data and Fossils," *Biology Letters* 2, no. 4 (2006): 543–47.
178. See, for example, Takahiro Yonezawa et al., "Phylogenomics and Morphology of Extinct Paleognaths Reveal the Origin and Evolution of Ratites," *Current Biology* 27, no. 1 (2017): 68–77.
179. See, for example, R. Will Stein, Joseph W. Brown, and Arne Ø. Mooers, "A Molecular Genetic Time Scale Demonstrates Cretaceous Origins and Multiple Diversification Rate Shifts within the Order Galliformes (Aves)," *Molecular Phylogenetics and Evolution* 92 (2015): 155–64; see also a corresponding employment of morphological clocks in a paper by Michael S. Y. Lee, Andrea Cau, Darren Naish, and Gareth J. Dyke, "Morphological Clocks in Paleontology, and a Mid-Cretaceous Origin of Crown Aves," *Systematic Biology* 63, no. 3 (2014): 442–49.
180. See, for example, Allan J. Baker, Sérgio L. Pereira, and Tara A. Paton, "Phylogenetic Relationships and Divergence Times of Charadriiformes Genera: Multigene Evidence for the Cretaceous Origin of at Least 14 Clades of Shorebirds," *Biology Letters* 3, no. 2 (2007): 205–9; and M. Andreína Pacheco, Fabia U. Battistuzzi, Miguel Lentino, Roberto F. Aguilar, Sudhir Kumar, and Ananias A. Escalante, "Evolu-

tion of Modern Birds Revealed by Mitogenomics: Timing the Radiation and Origin of Major Orders," *Molecular Biology and Evolution* 28, no. 6 (2011): 1927–42.

181. See, for example, Gerald Mayr, Vanesa L. De Pietri, and R. Paul Scofield, "A New Fossil from the Mid-Paleocene of New Zealand Reveals an Unexpected Diversity of World's Oldest Penguins," *Science of Nature* 104, no. 3–4 (2017), e9.
182. Neil Brocklehurst, Paul Upchurch, Philip D. Mannion, and Jingmai O'Connor, "The Completeness of the Fossil Record of Mesozoic Birds: Implications for Early Avian Evolution," *PLoS One* 7, no. 6 (2012), e39056.
183. See, for example, Gerald Mayr, Vanesa Lopez De Pietri, R. Scofield, and Trevor H. Worthy, "On the Taxonomic Composition and Phylogenetic Affinities of the Recently Proposed Clade Vegaviidae Agnolín et al., 2017—Neornithine Birds from the Upper Cretaceous of the Southern Hemisphere," *Cretaceous Research* 86 (2018): 178–85.
184. See, for example, Alan Feduccia, "Avian Extinction at the End of the Cretaceous: Assessing the Magnitude and Subsequent Explosive Radiation," *Cretaceous Research* 50 (2014): 1–15.
185. Andrzej Elzanowski and Thomas A. Stidham, "A Galloanserine Quadrate from the Late Cretaceous Lance Formation of Wyoming," *Auk* 128, no. 1 (2011): 138–45.
186. Carolina Acosta Hospitaleche and Trevor W. Worthy, "New Data on the *Vegavis iaai* Holotype from the Maastrichtian of Antarctica," *Cretaceous Research* 124 (2021), 104818; Julia A. Clarke, Claudia P. Tambussi, Jorge I. Noriega, Gregory M. Erickson, and Richard A. Ketcham, "Definitive Fossil Evidence for the Extant Avian Radiation in the Cretaceous," *Nature* 433, no. 7023 (2005): 305–8; and Federico L. Agnolín, Federico Brissón Egli, Sankar Chatterjee, Jordi Alexis Garcia Marsà, and Fernando E. Novas, "Vegaviidae, a New Clade of Southern Diving Birds That Survived the K/T Boundary," *Naturwissenschaften* 104, no. 11–12 (2017), 87.
187. Daniel J. Field, Juan Benito, Albert Chen, John W. M. Jagt, and Daniel T. Ksepka, "Late Cretaceous Neornithine from Europe Illuminates the Origins of Crown Birds," *Nature* 579, no. 7799 (2020): 393–401;

with commentary by Kevin Padian, "Poultry through Time," *Nature* 579, no. 7799 (2020): 351–52.

188. Julia A. Clarke et al., "Fossil Evidence of the Avian Vocal Organ from the Mesozoic," *Nature* 538, no. 7626 (2016): 502–505; also commentary by Patrick M. O'Connor, "Ancient Avian Aria from Antarctica," *Nature* 538, no. 7626 (2016): 468–69.
189. Storrs L. Olson, "*Neogaeornis wetzeli* Lambrecht, a Cretaceous Loon from Chile (Aves: Gaviidae)," *Journal of Vertebrate Paleontology* 12, no. 1 (1992): 122–24.
190. Carolina Acosta Hospitaleche and Javier N. Gelfo, "New Antarctic Findings of Upper Cretaceous and Lower Eocene Loons (Aves: Gaviiformes)," *Annales de Paleontologie* 101, no. 4 (2015): 315–24.
191. Per G. P. Ericson, Seraina Klopfstein, Martin Irestedt, Jacqueline M. T. Nguyen, and Johan A. A. Nylander, "Dating the Diversification of the Major Lineages of Passeriformes (Aves)," *BMC Evolutionary Biology* 14 (2014), 8.
192. See, for example, Timothy F. Wright et al., "A Multilocus Molecular Phylogeny of the Parrots (Psittaciformes): Support for a Gondwanan Origin during the Cretaceous," *Molecular Biology and Evolution* 25, no. 10 (2008): 2141–56.
193. Thomas A. Stidham, "The Lower Jaw from a Cretaceous Parrot," *Nature* 396 (1998): 29–30.
194. See the critique by Gareth J. Dyke and Gerald Mayr, "Did Parrots Exist in the Cretaceous Period?" *Nature* 399 (1999): 317–18; and the reply by Stidham (318).
195. Michela Contessi and Federico Fanti, "First Record of Bird Tracks in the Late Cretaceous (Cenomanian) of Tunisia," *Palaios* 27, no. 7 (2012): 455–64.
196. Martin G. Lockley, Rihui Li, Jerald D. Harris, Masaki Matsukawa, and Mingwei Liu, "Earliest Zygodactyl Bird Feet: Evidence from Early Cretaceous Roadrunner-Like Tracks," *Naturwissenschaften* 94, no. 8 (2007): 657–65.
197. See, for example, A. McAnena et al., "Atlantic Cooling Associated with a Marine Biotic Crisis during the Mid-Cretaceous Period," *Nature Geoscience* 6 (2013): 558–61.

198. Jack A. Wolfe, "Palaeobotanical Evidence for a June 'Impact Winter' at the Cretaceous/Tertiary Boundary," *Nature* 352 (1991): 420–25; followed by critiques: Douglas J. Nichols, "Plants at the K/T Boundary," *Nature* 356 (1992): 295; and Leo J. Hickey and Lucinda J. McWeeney, "Plants at the K/T Boundary," *Nature* 356 (1992): 295–96, and the response from Wolfe (296).
199. Joanna Morgan et al., "Analysis of Shocked Quartz at the Global K-P Boundary Indicate an Origin from a Single, High-Angle, Oblique Impact at Chicxulub," *Earth and Planetary Science Letters* 251, no. 3–4 (2006): 264–79; G. S. Collins, N. Patel, T. M. Davison, A. S. P. Rae, J. V. Morgan, S. P. S. Gulick, the IODP-ICDP Expedition 364 Science Party; and Third-Party Scientists, "A Steeply-Inclined Trajectory for the Chicxulub Impact," *Nature Communications* 11, no. 1 (2020), 1480. The latter supports the high angle but argues that the asteroid came in from the northeast.
200. Mark A. Richards et al., "Triggering of the Largest Deccan Eruptions by the Chicxulub Impact," *Geological Society of America Bulletin* 127, no. 11–12 (2015): 1507–20; also the comment by Gauri Dole, Shilpa Patil Pillai, Devdutt Upasani, and Vivek S. Kale, "Triggering of the Largest Deccan Eruptions by the Chicxulub Impact: Comment," *Geological Society of America Bulletin* 129, no. 1–2 (2017): 253–55, and the authors' response (256).
201. David Penney, C. Philip Wheater, and Paul A. Selden, "Resistance of Spiders to Cretaceous-Tertiary Extinction Events," *Evolution* 57, no. 11 (2003): 2599–607.
202. Vincent S. Smith, Tom Ford, Kevin P. Johnson, Paul C. D. Johnson, Kazunori Yoshizawa, and Jessica E. Light, "Multiple Lineages of Lice Pass through the K-Pg Boundary," *Biology Letters* 7, no. 5 (2011): 782–85.
203. Elizabeth C. Sibert, Pincelli M. Hull, and Richard D. Norris, "Resilience of Pacific Pelagic Fish across the Cretaceous/Palaeogene Mass Extinction," *Nature Geoscience* 7, no. 9 (2014): 667–70; and Elizabeth C. Sibert, Matt Friedman, Pincelli Hull, Gene Hunt, and Richard Norris, "Two Pulses of Morphological Diversification in Pacific Pelagic Fishes following the Cretaceous-Palaeogene Mass Extinction,"

Proceedings of the Royal Society of London, Series B: Biological Sciences 285, no. 1888 (2018), 20181194.

204. See, for example, Andrew J. Ross, A. Jarzembowski, and Stephen J. Brooks, "The Cretaceous and Cenozoic Record of Insects (Hexapoda) with Regard to Global Change," in *Biotic Response to Global Change: The Last 145 Million Years,* ed. Steven J. Culver and Peter F. Rawson (Cambridge: Cambridge University Press, 2000), 288–302.
205. See, for example, Shigun Jiang, Timothy J. Bralower, Mark E. Patzkowsky, Lee R. Kump, and Jonathan D. Schueth, "Geographic Controls on Nannoplankton Extinction across the Cretaceous/Palaeogene Boundary," *Nature Geoscience* 3 (2010): 280–85; also commentary by Paul B. Wignall, "Safer in the South," *Nature Geoscience* 3 (2010): 228–29. This even applies to the impact site itself; see Christopher M. Lowery et al., "Rapid Recovery of Life at Ground Zero of the End-Cretaceous Mass Extinction," *Nature* 558, no. 7709 (2018): 288–91.
206. Viviana D. Barreda, Nestor R. Cúneo, Peter Wilf, Ellen D. Currano, Roberto A. Scasso, and Henk Brinkhuis, "Cretaceous/Paleogene Floral Turnover in Patagonia: Drop in Diversity, Low Extinction, and a *Classopolis* Spike," *PLoS One* 7, no. 12 (2012), e52455; also see Elena Stiles, Peter Wilf, Ari Iglesias, María A. Gandolfo, and N. Rubén Cúneo, "Cretaceous–Paleogene Plant Extinction and Recovery in Patagonia," *Paleobiology* 46, no. 4 (2020): 445–69, who identify a more severe extinction, but less severe than in the Northern Hemisphere and also at higher taxonomic levels.
207. W. A. Green and L. J. Hickey, "Leaf Architectural Profiles of Angiosperm Floras across the Cretaceous/Tertiary Boundary," *American Journal of Science* 305, no. 10 (2005): 983–1013.
208. Kirk R. Johnson and Beth Ellis, "A Tropical Rainforest in Colorado 1.4 Million Years after the Cretaceous-Tertiary Boundary," *Science* 296, no. 5577 (2002): 2379–83; and Beth Ellis, Kirk R. Johnson, and Regan E. Dunn, "Evidence for an in Situ Early Paleocene Rainforest from Castle Rock, Colorado," *Rocky Mountain Geology* 38, no. 1 (2003): 73–100.
209. Ari Iglesias, Peter Wilf, Kirk R. Johnson, Alba B. Zamuner, N. Rubén Cúneo, Sergio D. Matheos, and Bradley S. Singer, "A Paleocene Lowland Macroflora from Patagonia Reveals Significantly Greater Rich-

ness than North American Analogs," *Geology* 35, no. 10 (2007): 947–50; and William C. Clyde et al., "New Age Constraints for the Salamanca Formation and Lower Río Chico Group in the Western San Jorge Basin, Patagonia, Argentina: Implications for Cretaceous-Paleogene Extinction Recovery and Land Mammal Age Correlations," *Geological Society of America Bulletin* 126, no. 3–4 (2014): 289–306. Clyde and colleagues provide radiometric data to suggest that the age of these floras was up to 3 Ma earlier than Iglesias and colleagues had thought, again pointing to a rapid recovery.

210. For the nymphalid butterflies, see Niklas Wahlberg, Julien Leneveu, Ullasa Kodandaramaiah, Carlos Peña, Sören Nylin, André V. L. Freitas, and Andrew V. Z. Brower, "Nymphalid Butterflies Diversify following Near Demise at the Cretaceous/Tertiary Boundary," *Proceedings of the Royal Society of London, Series B: Biological Sciences* 276, no. 1677 (2009): 4295–302. For long-tongued bees (xylocopines), see Sandra M. Rehan, Remko Leys, and Michael P. Schwarz, "First Evidence for a Massive Extinction Event Affecting Bees Close to the K-T Boundary," *PLoS One* 8, no. 10 (2013), e76683.

211. For the mammals, see N. R. Longrich, J. Scriberas, and M. A. Wills, "Severe Extinction and Rapid Recovery of Mammals across the Cretaceous-Palaeogene Boundary, and the Effects of Rarity on Patterns of Extinction and Recovery," *Journal of Evolutionary Biology* 29, no. 8 (2016): 1495–512. For birds, see Daniel T. Ksepka, Thomas A. Stidham, and Thomas E. Williamson, "Early Paleocene Landbird Supports Rapid Phylogenetic and Morphological Diversification of Crown Birds after the K-Pg Mass Extinction," *Proceedings of the National Academy of Sciences of the United States of America* 114, no. 30 (2017): 8047–52.

212. See, for example, James E. Fassett, Larry M. Heaman, and Antonio Simonetti, "Direct U-Pb Dating of Cretaceous and Paleocene Dinosaur Bones, San Juan Basin, New Mexico," *Geology* 39, no. 2 (2011): 159–62. Not surprisingly, these results have attracted criticism; see Kenneth R. Ludwig, *Geology* 40, no. 4 (2011): e258, with authors' response (e260); Paul R. Renne and Mark B. Goodwin, *Geology* 40, no. 4 (2012), e259, with authors' response (e261); and Alan E. Koenig et al., *Geology* 40, no. 4 (2012), e262, with authors' response (e264).

213. See, for example, Marcin Machalski and Claus Heinberg, "Evidence for Ammonite Survival into the Danian (Paleogene) from the Cerithium Limestone at Stevns Klint, Denmark," *Bulletin of the Geological Society of Denmark* 52, no. 2 (2005): 97–111; and Marcin Machalski, "Late Maastrichtian and Earliest Danian scaphitid Ammonites from Central Europe: Taxonomy, Evolution, and Extinction," *Acta Palaeontologica Polonica* 50, no. 4 (2005): 653–96.
214. Joseph A. Bonsor, Paul M. Barrett, Thomas J. Raven, and Natalie Cooper, "Dinosaur Diversification Rates Were Not in Decline prior to the K-Pg Boundary," *Royal Society Open Science* 7, no. 11 (2020), 1195, suggest this was not the case. See also Klara K. Nordén, Thomas L. Stubbs, Albert Prieto-Márquez, and Michael J. Benton, "Multifaceted Disparity Approach Reveals Dinosaur Herbivory Flourished before the End-Cretaceous Mass Extinction," *Paleobiology* 44, no. 4 (2018): 620–37.
215. See, for example, Stephen L. Brusatte, Carlos R. A. Candeiro, and Felipe M. Simbras, "The Last Dinosaurs of Brazil: The Bauru Group and Its Implications for the End-Cretaceous Mass Extinction," *Anais da Academia Brasileira de Ciências* 89, no. 3 (2017): 1465–85.
216. See, for example, Nicolás E. Campione and David C. Evans, "Cranial Growth and Variation in Edmontosaurs (Dinosauria: Hadrosauridae): Implications for Latest Cretaceous Megaherbivore Diversity in North America," *PLoS One* 6, no. 9 (2011), e25186; and Nieves López-Martínez et al., "New Dinosaur Sites Correlated with Upper Maastrichtian Pelagic Deposits in the Spanish Pyrenees: Implications for the Dinosaur Extinction," *Cretaceous Research* 22, no. 1 (2001): 41–61.
217. See, for example, Stephen L. Brusatte, Richard J. Butler, Albert Prieto-Márquez, and Mark A. Norell, "Dinosaur Morphological Diversity and the End-Cretaceous Extinction," *Nature Communications* 3 (2012), 804.
218. Jonathan S. Mitchell, Peter D. Roopnarine, and Kenneth D. Angielczyk, "Late Cretaceous Restructuring of Terrestrial Communities Facilitated the End-Cretaceous Mass Extinction in North America," *Proceedings of the National Academy of Sciences of the United States of America* 109, no. 46 (2012): 18857–61.

219. See, for example, William B. Gallagher et al., "On the Last Mosasaurs: Late Maastrichtian Mosasaurs and the Cretaceous-Paleogene Boundary in New Jersey," *Bulletin de la Société Géologique de France* 183, no. 2 (2012): 145–50.
220. See, for example, Fabio M. Dalla Vecchia, Violeta Riera, Josep Oriol Oms, Jaume Dinarès-Turell, Rodrigo Gaete, and Angel Galobart, "The Last Pterosaurs: First Record from the Uppermost Maastrichtian of the Tremp Syncline (Northern Spain)," *Acta Geologica Sinica (English Edition)* 87, no. 5 (2013): 1198–227.
221. László Makádi, Michael W. Caldwell, and Attila Ősi, "The First Freshwater Mosasauroid (Upper Cretaceous, Hungary) and a New Clade of Basal Mosasauroids," *PLoS One* 7, no. 12 (2012), e51781.
222. Robert Holmes, Michael W. Caldwell, and Stephen L. Cumbaa, "A New Specimen of *Plioplatecarpus* (Mosasauridae) from the Lower Maastrichtian of Alberta: Comments on Allometry, Functional Morphology, and Paleoecology," *Canadian Journal of Earth Sciences* 36, no. 3 (1999): 363–69. Oxygen isotopes also indicate frequent visits to estuarine habitats, perhaps to obtain fresh water; see Leah T. Taylor et al., "Oxygen Isotopes from the Teeth of Cretaceous Marine Lizards Reveal Their Migration and Consumption of Freshwater in the Western Interior Seaway, North America," *Palaeogeography, Palaeoclimatology, Palaeoecology* 573 (2021), 110406.
223. See, for example, Douglas S. Robertson, William M. Lewis, Peter M. Sheehan, and Owen B. Toon, "K-Pg Extinction Patterns in Marine and Freshwater Environments: The Impact Winter Model," *Journal of Geophysical Research: Biogeosciences* 118, no. 3 (2013): 1006–14.
224. See, for example, A. J. McGowan and G. J. Dyke, "A Morphospace-based Test for Competitive Exclusion among Flying Vertebrates: Did Birds, Bats and Pterosaurs Get in Each Other's Space?" *Journal of Evolutionary Biology* 20, no. 3 (2007): 1230–36; and Richard J. Butler, Paul M. Barrett, Stephen Nowbath, and Paul Upchurch, "Estimating the Effects of Sampling Biases on Pterosaur Diversity Patterns: Implications for Hypotheses of Bird/Pterosaur Competitive Replacement," *Paleobiology* 35, no. 3 (2009): 432–46.

225. Mark P. Witton and Michael B. Habib, "On the Size and Flight Diversity of Giant Pterosaurs, the Use of Birds as Pterosaur Analogues and Comments on Pterosaur Flightlessness," *PLoS One* 5, no. 11 (2010), e13982.
226. Roger B. J. Benson, Rachel A. Frigot, Anjali Goswami, Brian Andres, and Richard J. Butler, "Competition and Constraint Drove Cope's Rule in the Evolution of Giant Flying Reptiles," *Nature Communications* 5 (2014), 3567. Not that expiration was necessarily total—a cat-sized azhdarchoid from the late Cretaceous that appears not to be a juvenile has been reported; see Elizabeth Martin-Silverstone, Mark P. Witton, Victoria M. Arbour, and Philip J. Currie, "A Small Azhdarchoid Pterosaur from the Latest Cretaceous, the Age of Flying Giants," *Royal Society Open Science* 3, no. 8 (2016), 160333.
227. Leon P. A. M. Claessens, Patrick M. O'Connor, and David M. Unwin, "Respiratory Evolution Facilitated the Origin of Pterosaur Flight and Aerial Gigantism," *PLoS One* 4, no. 2 (2009), e4497.
228. Current evidence points to a mass extinction of the pterosaurs with little sign of an earlier decline; see Nicholas R. Longrich, David M. Martill, and Brian Andres, "Late Maastrichtian Pterosaurs from North Africa and Mass Extinction of Pterosauria at the Cretaceous-Paleogene Boundary," *PLoS Biology* 16, no. 3 (2018), e2001663.
229. Nicholas R. Longrich, Bhart-Anjan S. Bhullar, and Jacques A. Gauthier, "Mass Extinction of Lizards and Snakes at the Cretaceous-Paleogene Boundary," *Proceedings of the National Academy of Sciences of the United States of America* 109, no. 52 (2012): 21396–401; with correction in 110, no. 16 (2013): 6608.
230. See Mateusz Tałanda, "An Exceptionally Preserved Jurassic Skink Suggests Lizard Diversification Preceded Fragmentation of Pangaea," *Palaeontology* 61, no. 5 (2018): 659–77.
231. See Jorge A. Herrera-Flores, Thomas L. Stubbs, and Michael J. Benton, "Ecomorphological Diversification of Squamates in the Cretaceous," *Royal Society Open Science* 8, no. 3 (2021), 201961.
232. See Hang-Jae Lee, Yuong-Nam Lee, Anthony R. Fiorillo, and Junchang Lü, "Lizards Ran Bipedally 110 Million Years Ago," *Scientific Reports* 8, no. 1 (2018), 2617; and Damián Villaseñor-Amador, Nut Xanat

Suárez, and J. Alberto Cruz, "Bipedalism in Mexican Albian Lizard (Squamata) and the Locomotion Type in Other Cretaceous Lizards," *Journal of South American Earth Sciences* 109 (2021), 103299.

233. Allison Y. Hsiang, Daniel J. Field, Timothy H. Webster, Adam D. B. Behlke, Matthew B. Davis, Rachel A. Racicot, and Jacques A. Gauthier, "The Origin of Snakes: Revealing the Ecology, Behavior, and Evolutionary History of Early Snakes Using Genomics, Phenomics, and the Fossil Record," *BMC Evolutionary Biology* 15 (2015), 87; and Michael W. Caldwell, Randall L. Nydam, Alessandro Palci, and Sebastián Apesteguía, "The Oldest Known Snakes from the Middle Jurassic-Lower Cretaceous Provide Insights on Snake Evolution," *Nature Communications* 6 (2015), 5996. More recent finds, such as Cretaceous blind snakes, confirm this viewpoint; see Thiago Schineider Fachini, Silvio Onary, Alessandro Palci, Michael S. Y. Lee, Mario Bronzati, and Annie Schmaltz Hsiou, "Cretaceous Blind Snake from Brazil Fills Major Gap in Snake Evolution," *iScience* 23, no. 12 (2020), e101834.
234. See, for example, Randall L. Nydam, "Squamates from the Jurassic and Cretaceous of North America," *Palaeobiodiversity and Palaeoenvironments* 93 (2013): 535–65; and Adriana María Albino and Santiago Brizuela, "An Overview of the South American Fossil Squamates," *Anatomical Record* 297, no. 3 (2014): 349–68.
235. Juan D. Daza, Aaron M. Bauer, and Eric D. Snively, "On the Fossil Record of the Gekkota," *Anatomical Record* 297, no. 3 (2014): 433–62.
236. E. Nicholas Arnold and George Poinar, "A 100 Million Year Old Gecko with Sophisticated Adhesive Toe Pads Preserved in Amber from Myanmar," *Zootaxa* 1847, no. 1 (2008): 62–68; also Gabriela Fontanarrosa, Juan D. Daza, and Virginia Abdala, "Cretaceous Fossil Gecko Hand Reveals a Strikingly Modern Scansorial Morphology: Qualitative and Biometric Analysis of an Amber-Preserved Lizard Hand," *Cretaceous Research* 84 (2018): 120–33.
237. Jason J. Head, Gregg F. Gunnell, Patricia A. Holroyd, J. Howard Hutchison, and Russell L. Ciochon, "Giant Lizards Occupied Herbivorous Mammalian Ecospace during the Paleogene Greenhouse in Southeast Asia," *Proceedings of the Royal Society of London, Series B: Biological Sciences* 280, no. 1763 (2013): 20130665.

238. See Gilbert J. Price et al., "Temporal Overlap of Humans and Giant Lizards (Varanidae; Squamata) in Pleistocene Australia," *Quaternary Science Reviews* 125 (2015): 98–105.
239. Gregory M. Erickson, Armand De Ricqles, Vivian De Buffrénil, Ralph E. Molnar, and Mark K. Bayless, "Vermiform Bones and the Evolution of Gigantism of *Megalania*—How a Reptilian Fox Became a Lion," *Journal of Vertebrate Paleontology* 23, no. 4 (2003): 966–70.
240. See, for example, Aradhna Tripati and Dennis Darby, "Evidence for Ephemeral Middle Eocene to Early Oligocene Greenland Glacial Ice and Pan-Arctic Sea Ice," *Nature Communications* 9, no. 1 (2018), 1038.
241. See, for example, Anusuya Chinsamy, Daniel B. Thomas, Allison R. Tumarkin-Deratzian, and Anthony R. Fiorillo, "Hadrosaurs Were Perennial Polar Residents," *Anatomical Record* 295, no. 4 (2012): 610–14.
242. See, for example, Meike Köhler, Nekane Marín-Moratalla, Xavier Jordana, and Ronny Aanes, "Seasonal Bone Growth and Physiology in Endotherms Shed Light on Dinosaur Physiology," *Nature* 487, no. 7407 (2012): 358–61; also commentary by Kevin Padian, "Evolutionary Physiology: A Bone for All Seasons," *Nature* 487, no. 7407 (2012): 310–11.
243. Robert A. Eagle et al., "Dinosaur Body Temperatures Determined from Isotopic (^{13}C-^{18}O) Ordering in Fossil Biominerals," *Science* 333, no. 6041 (2011): 443–45; also Robin R. Dawson, Daniel J. Field, Pincelli M. Hull, Darla K. Zelenitsky, François Therrien, and Hagit P. Affek, "Eggshell Geochemistry Reveals Ancestral Metabolic Thermoregulation in Dinosauria," *Science Advances* 6, no. 7 (2020), eaax9361.
244. Andrew Clarke, "Dinosaur Energetics: Setting Bounds on Feasible Physiologies and Ecologies," *American Naturalist* 182, no. 3 (2013): 283–97; and John M. Grady, Brian J. Enquist, Eva Dettweiler-Robinson, Natalie A. Wright, and Felisa A. Smith, "Evidence for Mesothermy in Dinosaurs," *Science* 344, no. 6189 (2014): 1268–72; and commentary by Michael Balter, "Dinosaur Metabolism Neither Hot Nor Cold, but Just Right," *Science* 344, no. 6189 (2014): 2216–17. See also critiques by M. D. D'Emic, *Science* 348, no. 6238 (2015): 982-b, and N. P. Myhrvold, *Science* 348, no. 6238 (2015): 982-c, and the authors' response to both (982-d). Finally, Jan Werner and Eva Maria Griebeler, "Allometries of Maximum Growth Rate versus Body Mass at Maximum Growth Indi-

cate that Non-avian Dinosaurs Had Growth Rates Typical of Fast Growing Ectothermic Sauropsids," *PLoS One* 9, no. 2 (2014), e88834, suggest body temperatures more akin to ectotherms; see also yet more critiques by Nathan P. Myhrvold, "Response to Formal Comment on Myhrvold (2016) submitted by Griebeler and Werner (2017)," *PLoS One* 13, no. 2 (2018), e0192912 and the authors' response: Eva Maria Griebeler and Jan Werner, "Formal Comment on: Myhrvold (2016) Dinosaur Metabolism and the Allometry of Maximum Growth Rate. *PLoS One*; 11(11): e0163205," *PLoS One* 13, no. 2 (2018), e0184756.

245. Aurélien Bernard et al., "Regulation of Body Temperature by Some Mesozoic Reptiles," *Science* 328, no. 5984 (2010): 1379–82; also commentary by Ryosuke Motani, "Warm-Blooded 'Sea Dragons'?" *Science* 328, no. 5984 (2010): 1361–62.
246. Mark P. Witton, "New Insights into the Skull of *Istiodactylus latidens* (Ornithocheiroidea, Pterodactyloidea)," *PLoS One* 7, no. 3 (2012), e33170.

CHAPTER 4

1. Jingxia Zhao, Yunyun Zhao, Chungkun Shih, Dong Ren, and Yongjie Wang, "Transitional Fossil Earwigs—A Missing Link in Dermaptera Evolution," *BMC Evolutionary Biology* 10 (2010), 344.
2. For a dated but classic study, see J. L. B. Smith, *Old Fourlegs: The Story of the Coelacanth* (London: Longman, Green, 1956).
3. See, for example, Naoko Takezaki and Hidenori Nishihara, "Support for Lungfish as the Closest Relative of Tetrapods by Using Slowly Evolving Ray-Finned Fish as the Outgroup," *Genome Biology and Evolution* 9, no. 1 (2017): 93–101.
4. See, for example, M. I. Coates, "The Devonian Tetrapod *Acanthostega gunnari* Jarvik: Postcranial Anatomy, Basal Tetrapod Interrelationships and Patterns of Skeletal Evolution," *Transactions of the Royal Society of Edinburgh: Earth Sciences* 87, no. 3 (1996): 363–421.
5. See, for example, Edward B. Daeschler, Neil H. Shubin, and Farish A. Jenkins Jr., "A Devonian Tetrapod-Like Fish and the Evolution of the Tetrapod Body Plan," *Nature* 440, no. 7085 (2006): 757–63; and commentary by Per E. Ahlberg and Jennifer A. Clark, "Palaeontology: A Firm Step from Water to Land," *Nature* 440, no. 7085 (2006): 747–49.

6. For coelacanths, see Chris T. Amemiya et al., "The African Coelacanth Genome Provides Insights into Tetrapod Evolution," *Nature* 496, no. 7445 (2013): 311–16; for lungfish, see Axel Meyer et al., "Giant Lungfish Genome Elucidates the Conquest of Land by Vertebrates," *Nature* 590, no. 7845 (2021): 284–89.
7. Timothy Holland, "The Endocranial Anatomy of *Gogonasus andrewsae* Long, 1985 Revealed through Micro CT-Scanning," *Transactions of the Royal Society of Edinburgh: Earth Sciences* 105, no. 1 (2014): 9–34.
8. See, for example, Sophie Sanchez, Paul Tafforeau, Jennifer A. Clack, and Per E. Ahlberg, "Life History of the Stem Tetrapod *Acanthostega* Revealed by Synchroton Microtomography," *Nature* 537, no. 7620 (2016): 408–11; and commentary by Nadia B. Fröbisch, "Teenage Tetrapods," *Nature* 537, no. 7620 (2016): 311–12.
9. M. I. Coates and J. A. Clack, "Fish-like Gills and Breathing in the Earliest Known Tetrapod [letter]," *Nature* 352 (1991): 234–36.
10. See, for example, James L. Edwards, "Two Perspectives on the Evolution of the Tetrapod Limb," *American Zoologist* 29 (1989): 235–54.
11. Brooke E. Flammang, Aphinun Suvarnaraksha, Julie Markiewicz, and Daphne Soares, "Tetrapod-like Pelvic Girdle in a Walking Cavefish," *Scientific Reports* 6 (2016), 23711.
12. Theodore H. Eaton Jr., "The Aquatic Origin of Tetrapods," *Transactions of the Kansas Academy of Sciences* 63, no. 3 (1960): 115–20.
13. Jenny Clack, *Gaining Ground: The Origin and Evolution of Tetrapods*, 2nd ed. (Bloomington: Indiana University Press, 2012).
14. Neil Shubin, *Your Inner Fish: A Journey into the 3.5 Billion-Year History of the Human Body* (London: Allen Lane, 2008).
15. Spencer G. Lucas, "*Thinopus* and a Critical Review of Devonian Tetrapod Footprints," *Ichnos* 22, no. 3–4 (2015): 136–54.
16. See, for example, Per E. Ahlberg, *Transactions of the Royal Society of Edinburgh: Earth and Environmental Science* 109, no. 1–2 (2018): 115–37.
17. See, for example, John A. Long, Gavin C. Young, Tim Holland, Tim J. Senden, and Erich M. G. Fitzgerald, "An Exceptional Devonian Fish from Australia Sheds Light on Tetrapod Origins," *Nature* 444, no. 7116 (2006): 199–202; and Catherine A. Boisvert, Elga Mark-Kurik, and Per E. Ahlberg, "The Pectoral Fin of *Panderichthys* and the Origin of Digits,"

Nature 456, no. 7222 (2008): 636–38; also an earlier contribution by Catherine A. Boisvert, "The Pelvic Fin and Girdle of *Panderichthys* and the Origin of Tetrapod Locomotion," *Nature* 438, no. 7071 (2005): 1145–47.

18. Jennifer A. Clack, Per E. Ahlberg, Henning Blom, and Sarah M. Finney, "A New Genus of Devonian Tetrapod from North-east Greenland, with New Information on the Lower Jaw of *Ichthyostega*," *Palaeontology* 55, no. 1 (2012): 73–86.
19. See, for example, Min Zhu, Per E. Ahlberg, Wen-Jin Zhao, and Lian-Tao Jia, "A Devonian Tetrapod-like Fish Reveals Substantial Parallelism in Stem Tetrapod Evolution," *Nature Ecology and Evolution* 1, no. 10 (2017): 1470–76; also, with an emphasis on mosaicism and parallelisms, note the prescient remarks by E. I. Vorobyeva, "A New Approach to the Problem of Tetrapod Origin," *Paleontological Journal* 37 (2003): 449–60.
20. See Thomas A. Stewart, Justin B. Lemberg, Natalia K. Taft, Ihna Yoo, Edward B. Daeschler, and Neil H. Shubin, "Fin Ray Patterns at the Fin-to-Limb Transition," *Proceedings of the National Academy of Sciences of the United States of America* 117, no. 3 (2020): 1612–20. These authors document the changes in these fin rays in *Sauripterus, Eusthenopteron,* and *Tiktaalik*, including a curious asymmetry between dorsal and ventral sets, perhaps related to increasing interactions with the substrate (and convergent with benthic bony fish).
21. Oddly in one rhizodontid the usual 1 + 2 formula is replaced by 1 + 3 arrangement in the pelvic fin, but this example of less constraint does not find parallels in other sarcopterygians; see Jonathan E. Jeffery, Glenn W. Storrs, Timothy Holland, Clifford J. Tabin, and Per E. Ahlberg, "Unique Pelvic Fin in a Tetrapod-like Fossil Fish, and the Evolution of Limb Patterning," *Proceedings of the National Academy of Sciences of the United States of America* 115, no. 47 (2018): 12005–10.
22. Edward Daeschler and Neil Shubin, "Fish with Fingers?" *Nature* 391 (1998), 133.
23. Marcus C. Davis, Neil Shubin, and Edward B. Daeschler, "Immature Rhizodontids from the Devonian of North America," *Bulletin of the Museum of Comparative Zoology* 156, no. 1 (2001): 171–87.

24. Marcus C. Davis, Neil Shubin, and Edward B. Daeschler, "A New Specimen of *Sauripterus taylori* (Sarcopterygii, Osteichthyes) from the Fammenian Catskill Formation of North America," *Journal of Vertebrate Paleontology* 24, no. 1 (2004): 26–40.
25. Jillian M. Garvey, Zerina Johanson, and Anne Warren, "Redescription of the Pectoral Fin and Vertebral Column of the Rhizodontid Fish *Barameda decipiens* from the Lower Carboniferous of Australia," *Journal of Vertebrate Paleontology* 25, no. 1 (2005): 8–18.
26. Martin D. Brazeau, "A New Genus of Rhizodontid (Sarcopterygii, Tetrapodomorpha) from the Lower Carboniferous Horton Bluff Formation of Nova Scotia, and the Evolution of the Lower Jaw in this Group," *Canadian Journal of Earth Sciences* 42, no. 8 (2005): 1481–99.
27. See, for example, Gaël Clément, "Large Tristichopteridae (Sarcopterygii, Tetrapodomorpha) from the Late Famennian of Belgium," *Palaeontology* 45, no. 3 (2002): 577–93; and Daniel Snitting, "A Redescription of the Anatomy of the Late Devonian *Spodichthys buetleri* Jarvik, 1985 (Sarcopterygii, Tetrapodomorpha) from East Greenland," *Journal of Vertebrate Paleontology* 28, no. 3 (2008): 637–55.
28. Zerina Johanson, Per Ahlberg, and Alex Ritchie, "The Braincase and Palate of the Tetrapodomorph Sarcopterygian *Mandageria fairfaxi*: Morphological Variability near the Fish-Tetrapod Transition," *Palaeontology* 46, no. 2 (2003): 271–93.
29. François J. Meunier and Michel Laurin, "A Microanatomical and Histological Study of the Fin Long Bones of the Devonian Sarcopterygian *Eusthenopteron foordi*," *Acta Zoologica* 93, no. 1 (2012): 88–97; and S. Sanchez, P. Tafforeau, and Per E. Ahlberg, "The Humerus of *Eusthenopteron*: A Puzzling Organization Presaging the Establishment of Tetrapod Limb Bone Marrow," *Proceedings of the Royal Society of London, Series B: Biological Sciences* 281, no. 1782 (2014), 2014.0299.
30. Brian Swartz, "A Marine Stem-Tetrapod from the Devonian of Western North America," *PLoS One* 7, no. 3 (2012), e33683.
31. See, for example, Min Zhu and Xiabo Zhu, "Stem Sarcopterygians Have Primitive Polybasal Fin Articulation," *Biology Letters* 5, no. 3 (2009): 372–75; Min Zhu, Wenjin Zhao, Liantao Jia, Jing Lu, Tuo Qiao, and Qingming Qu, "The Oldest Articulated Osteichthyan Reveals Mosaic Gna-

thostome Characters," *Nature* 458, no. 7237 (2009): 469–74; and commentary by Michael I. Coates, "Beyond the Age of Fishes," *Nature* 458, no. 7237 (2009): 413–14.

32. Norifumi Tatsumi, Ritsuko Kobayashi, Tohru Yano, Masatsugu Noda, Koji Fujimura, Norihiro Okada, and Masataka Okabe, "Molecular Developmental Mechanisms in Polypterid Fish Provides Insights into the Origin of Vertebrate Lungs," *Scientific Reports* 6 (2016), 30580.
33. Jeffery B. Graham et al., "Spiracular Air Breathing in Polypterid Fishes and Its Implications for Aerial Respiration in Stem Tetrapods," *Nature Communications* 5 (2014), 3022.
34. Benjamin C. Wilhelm, Trina Y. Du, Emily M. Standen, and Hans C. E. Larsson, "*Polypterus* and the Evolution of Fish Pectoral Muscle," *Journal of Anatomy* 226, no. 6 (2015): 511–22; for details of the pelvic equivalent, see Julia L. Molnar, Peter S. Johnston, Borja Esteve-Altava, and Rui Diogo, "Musculoskeletal Anatomy of the Pelvic Fin of *Polypterus*: Implications for Phylogenetic Distribution and Homology of Pre- and Postaxial Pelvic Appendicular Muscles," *Journal of Anatomy* 230, no. 4 (2017): 532–41.
35. Emily M. Standen, Trina Y. Du, and Hans C. E. Larsson, "Developmental Plasticity and the Origins of Tetrapods," *Nature* 513, no. 7516 (2014): 54–58; with commentary by John Hutchinson, "Dynasty of the Plastic Fish," *Nature* 513, no. 7516 (2014): 37–38; and subsequently Emily M. Standen, Trina Y. Du, Philippe Laroche, and Hans C. E. Larsson, "Locomotor Flexibility of *Polypterus senegalus* across Various Aquatic and Terrestrial Substrates," *Zoology* 119, no. 5 (2016): 447–54.
36. Sam Giles, Guang-Hui Xu, Thomas J. Near, and Matt Friedman, "Early Members of 'Living Fossil' Lineage Imply Later Origin of Modern Ray-Finned Fishes," *Nature* 549, no. 7671 (2017): 265–68; with commentary by Michael Coates, "Plenty of Fish in the Sea," *Nature* 549, no. 7671 (2017): 167–69.
37. Rui Diogo, Peter Johnston, Julia L. Molnar, and Borja Esteve-Altava, "Characteristic Tetrapod Musculoskeletal Limb Phenotype Emerged More Than 400 MYA in Basal Lobe-Finned Fishes," *Scientific Reports* 6 (2016), 37592.
38. That, of course, is not the end of the story; otherwise, fins would be limbs. For the transition, see, for example, Julia L. Molnar, Rui Diogo, John R.

Hutchinson, and Stephanie E. Pierce, "Reconstructing Pectoral Appendicular Muscle Anatomy in Fossil Fish and Tetrapods over the Fins-to-Limbs Transition," *Biological Reviews of the Cambridge Philosophical Society* 93, no. 2 (2017): 1077–107; and Borja Esteve-Altava, Julia L. Molnar, Peter Johnston, John R. Hutchinson, and Rui Diogo, "Anatomical Network Analysis of the Musculoskeletal System Reveals Integration Loss and Parcellation Boost during the Fins-to-Limbs Transition," *Evolution* 72, no. 3 (2018): 601–18.

39. Notably these are *HoxA* and *HoxD*, along with a battery of associated genes. See, for example, Andrew R. Gehrke et al., "Deep Conservation of Wrist and Digit Enhancers in Fish," *Proceedings of the National Academy of Sciences of the United States of America* 112, no. 3 (2015): 803–8; Tetsuya Nakamura, Andrew R. Gehrke, Justin Lemberg, Julie Szymaszek, and Neil H. Shubin, "Digits and Fin Rays Share Common Developmental Histories," *Nature* 537, no. 7619 (2016): 225–28; and commentary by Aditya Saxena and Kimberly L. Cooper, "Fin to Limb within Our Grasp," *Nature* 537, no. 7619 (2016): 176–77.
40. See, for example, Mikiko Tanaka, "Fins into Limbs: Autopod Acquisition and Anterior Elements Reduction by Modifying Gene Networks Involving *5′Hox*, *Gli3*, and *Shh*," *Developmental Biology* 413, no. 1 (2016): 1–7.
41. Günter P. Wagner and Alexander O. Vargas, "On the Nature of Thumbs," *Genome Biology* 9, no. 3 (2008), 213.
42. See, for example, Frank J. Tulenko et al., "Fin-fold Development in Paddlefish and Catshark and Implications for the Evolution of the Autopod," *Proceedings of the Royal Society of London, Series B: Biological Sciences* 284, no. 1855 (2017), 2016.2780.
43. Robert E. Gill Jr., Theunis Piersma, Gary Hufford, Rene Servranckx, and Adrian Riegen, "Crossing the Ultimate Ecological Barrier: Evidence for an 11 000-km-Long Nonstop Flight from Alaska to New Zealand and Eastern Australia by Bar-Tailed Godwits," *Condor* 107, no. 1 (2005): 1–20.
44. See the overview by Sabine L. Laguë, "High-Altitude Champions: Birds That Live and Migrate at Altitude," *Journal of Applied Physiology* 123, no. 4 (2017): 942–50.

45. See, for example, Benjamin Ponitz, Anke Schmitz, Dominik Fischer, Horst Bleckmann, and Christoph Brücker, "Diving-Flight Aerodynamics of a Peregrine Falcon (*Falco peregrinus*)," *PLoS One* 9, no. 2 (2014), e86506.
46. Danielle Dhouailly, "A New Scenario for the Evolutionary Origin of Hair, Feathers and Avian Scales," *Journal of Anatomy* 214, no. 4 (2009): 587–606.
47. Notably cysteines employing disulphide bonds; see Bettina Strasser, Veronika Mlitz, Marcela Hermann, Erwin Tschachler, and Leopold Eckhart, "Convergent Evolution of Cysteine-Rich Proteins in Feathers and Hair," *BMC Evolutionary Biology* 15 (2015), 82.
48. See, for example, Gerald Mayr, Michael Pittman, Evan Saitta, Thomas G. Kaye, and Jakob Vinther, "Structure and Homology of *Psittacosaurus* Tail Bristles," *Palaeontology* 59, no. 6 (2016): 793–803.
49. See, for example, Zixiao Yang et al., "Pterosaur Integumentary Structures with Complex Feather-like Branching," *Nature Ecology and Evolution* 3, no. 1 (2019): 24–30; see also comments by David M. Unwin and David M. Martill, "No Protofeathers on Pterosaurs," *Nature Ecology and Evolution* 4, no. 12 (2020): 1590–91, and the authors' reply (1592–93).
50. See Matthew G. Baron, David B. Norman, and Paul M. Barrett, "A New Hypothesis of Dinosaur Relationships and Early Dinosaur Evolution," *Nature* 543, no. 7646 (2017): 501–6; and ensuing critique by Max C. Langer et al., "Untangling the Dinosaur Family Tree," *Nature* 551, no. 7678 (2017): E1–E3, and the authors' reply (E4–E5).
51. See Xing Xu et al., "A Bizarre Maniraptorian Theropod with Preserved Evidence of Membranous Wings," *Nature* 521, no. 7550 (2015): 70–73; and commentary by Kevin Padian, "Dinosaur up in the Air," *Nature* 521, no. 7550 (2015): 40–41.
52. Min Wang, Jingmai K. O'Connor, Xing Xu, and Zhonghe Zhou, "A New Jurassic Scansoriopterygid and the Loss of Membranous Wings in Theropod Dinosaurs," *Nature* 569, no. 7755 (2019): 256–59.
53. Fucheng Zhang, Zhonghe Zhou, Xing Xu, and Xiaolin Wang, "A Juvenile Coelurosaurian Theropod from China Indicates Arboreal Habits," *Naturwissenschaften* 89, no. 9 (2002): 394–98.

54. T. Alexander Dececchi et al., "Aerodynamics Show Membrane-Winged Theropods Were a Poor Gliding Dead-end," *iScience* 23, no. 12 (2020), 101574.
55. Fucheng C. Zhang, Zhonghe Zhou, Xing Xu, Xiaolin Wang, and Corwin Sullivan, "A Bizarre Jurassic Maniraptoran from China with Elongate Ribbon-like Feathers," *Nature* 455, no. 7216 (2008): 1105–8.
56. See, for example, Theresa J. Feo, Daniel J. Field, and Richard O. Prum, "Barb Geometry of Asymmetrical Feathers Reveals a Transitional Morphology in the Evolution of Avian Flight," *Proceedings of the Royal Society of London, Series B: Biological Sciences* 282, no. 1803 (2015), 20142864.
57. Corwin Sullivan, Xing Xu, and Jingmai K. O'Connor, "Complexities and Novelties in the Early Evolution of Avian Flight, as Seen in the Mesozoic Yanliao and Jehol Biotas of Northeast China," *Palaeoworld* 26, no. 2 (2017): 212–29, quotation from p. 214 (my emphasis).
58. Francisco Ortega, Fernando Escaso, and José L. Sanz, "A Bizarre, Humped Carcharodontosauria (Theropoda) from the Lower Cretaceous of Spain," *Nature* 467, no. 7312 (2010): 203–6.
59. Fiann M. Smithwick, Robert Nicholls, Innes C. Cuthill, and Jakob Vinther, "Countershading and Stripes in the Theropod Dinosaur *Sinosauropteryx* Reveal Heterogenous Habitats in Early Cretaceous Jehol Biota," *Current Biology* 27, no. 21 (2017): 3337–43.e2.
60. Oliver W. M. Rauhut, Christian Foth, Helmut Tischlinger, and Mark A. Norell, "Exceptionally Preserved Juvenile Megalosauroid Theropod Dinosaur with Filamentous Integument from the Late Jurassic of Germany," *Proceedings of the National Academy of Sciences of the United States of America* 109, no. 29 (2012): 11746–51.
61. See, for example, Luis M. Chiappe and Ursula B. Göhlich, "Anatomy of *Juraventor starki* (Theropoda: Coelurosauria) from the Late Jurassic of Germany," *Neues Jahrbuch für Geologie und Paläontologie, Abhandlungen* 258, no. 3 (2010): 257–96.
62. Phil R. Bell, Nicolás E. Campione, W. Scott Persons, Philip J. Currie, Peter L. Larson, Darren H. Tanke, and Robert T. Bakker, "Tyrannosaurid Integument Reveals Conflicting Patterns of Gigantism and Feather Evolution," *Biology Letters* 13, no. 6 (2017): 20170092. One should point out, however, that the large *Yutyrannus* did bear prominent integumen-

tary structures; see Xing Xu et al., "A Gigantic Feathered Dinosaur from the Lower Cretaceous of China," *Nature* 484, no. 7392 (2012): 92–95.

63. Xing Xu, Mark A. Norell, Xuewen Kuang, Xiaolin Wang, Qi Zhao, and Chengkai Jia, "Basal Tyrannosauroids from China and Evidence for Protofeathers in Tyrannosauroids," *Nature* 431, no. 7009 (2004): 680–84.
64. Patrick M. O'Connor and Leon P. A. M. Claessens, "Basic Avian Pulmonary Design and Flow-through Ventilation in Non-avian Theropod Dinosaurs," *Nature* 436, no. 7048 (2005): 253–56.
65. Roger B. J. Benson, Richard J. Butler, Matthew T. Carrano, and Patrick M. O'Connor, "Air-filled Postcranial Bones in Theropod Dinosaurs: Physiological Implications and the 'Reptile'-Bird Transition," *Biological Reviews of the Cambridge Philosophical Society* 87, no. 1 (2012): 168–93.
66. Mary Higbee Schweitzer, Wenxia Zheng, Lindsay Zanno, Sarah Werning, and Toshie Sugiyama, "Chemistry Supports the Identification of Gender-specific Reproductive Tissue in *Tyrannosaurus rex*," *Scientific Reports* 6 (2016), 23099; for a more cautionary approach, see Aurore Canonville, Mary H. Schweitzer, and Lindsay Zanno, "Identifying Medullary Bone in Extinct Avemetatarsalians: Challenges, Implications and Perspectives," *Philosophical Transactions of the Royal Society of London, Series B: Biological Sciences* 375, no. 1793 (2020), 20190133.
67. See, for example, Federico Agnolin and Agustin G. Martinelli, "*Guaibasaurus candelariensis* (Dinosauria, Saurischia) and the Early Origin of Avian-like Resting Posture," *Alcheringa* 36, no. 2 (2012): 263–67.
68. Chris L. Organ and Andrew M. Shedlock, "Palaeogenomics of Pterosaurs and the Evolution of Small Genome Size in Flying Vertebrates," *Biology Letters* 5, no. 2 (2009): 47–50.
69. Chris L. Organ, Andrew M. Shedlock, Andrew Meade, Mark Pagel, and Scott V. Edwards, "Origin of Avian Genome Size and Structure in Non-avian Dinosaurs," *Nature* 446, no. 7132 (2007): 180–84.
70. See Enrico L. Rezende, Leonardo D. Bacigalupe, Roberto F. Nespolo, and Francisco Bozinovic, "Shrinking Dinosaurs and the Evolution of Endothermy in Birds," *Science Advances* 6, no. 1 (2020), eaaw4486, which links endothermy to "shrinking dinosaurs."

71. Matt A. White, "The Subarctometatarsus: Intermediate Metatarsus Architecture Demonstrating the Evolution of the Arctometatarsus and Advanced Agility in Theropod Dinosaurs," *Alcheringa* 33, no. 1 (2009): 1–21.
72. P. J. Bishop et al., "Using Step Width to Compare Locomotor Biomechanics between Extinct, Non-avian Theropod Dinosaurs and Modern Obligate Bipeds," *Journal of the Royal Society Interface* 14, no. 132 (2017), 20170276.
73. Bhart-Anjan S. Bhullar, Michael Hanson, Matteo Fabbri, Adam Pritchard, Gabe S. Bever, and Eva Hoffman, "How to Make a Bird Skull: Major Transitions in the Evolution of the Avian Cranium, Paedomorphosis, and the Beak as Surrogate Hand," *Integrative and Comparative Biology* 56, no. 3 (2016): 389–403.
74. Tobin L. Hieronymus and Lawrence M. Witmer, "Homology and Evolution of Avian Compound Rhamphothecae," *Auk* 127, no. 3 (2010): 590–604.
75. Robert V. Hill, Michael D. D'Emic, G. S. Bever, and Mark A. Norell, "A Complex Hyobranchial Apparatus in a Cretaceous Dinosaur and the Antiquity of Avian Paraglossalia," *Zoological Journal of the Linnean Society* 175, no. 4 (2015): 892–909.
76. Phil Senter, "Function in the Stunted Forelimbs of *Mononykus olecranus* (Theropoda), a Dinosaurian Anteater," *Paleobiology* 31, no. 3 (2005): 373–81.
77. Xing Xu, Qingwei Tan, Corwin Sullivan, Fenglu Han, and Dong Xiao, "A Short-Armed Troodontid Dinosaur from the Upper Cretaceous of Inner Mongolia and Its Implications for Troodontid Evolution," *PLoS One* 6, no. 9 (2011), e22916.
78. See Fernando E. Novas, Diego Pol, Juan I. Canale, Juan D. Porfiri, and Jorge O. Calvo, "A Bizarre Cretaceous Theropod Dinosaur from Patagonia and the Evolution of Gondwanan Dromaeosaurids," *Proceedings of the Royal Society of London, Series B: Biological Sciences* 276, no. 1659 (2009): 1101–7; and Xiaoting T. Zheng, Xing Xu, Hailu You, Qi Zhao, and Zhiming Dong, "A Short-Armed Dromaeosaurid from the Jehol Group of China with Implications for Early Dromaeosaurid Evolution,"

Proceedings of the Royal Society of London, Series B: Biological Sciences 277, no. 1679 (2010): 211–17.

79. Andrea Cau et al., "Synchroton Scanning Reveals Amphibious Ecomorphology in a New Clade of Bird-like Dinosaurs," *Nature* 552, no. 7685 (2017): 395–99. Chase D. Brownstein, "*Halszkaraptor escuilliei* and the Evolution of the Paravian Bauplan," *Scientific Reports* 9, no. 1 (2019), 16455 [also correction in 10 (2020), e3333] concurs with the bizarre appearance of *Halszkaraptor* but disputes its inferred semiaquatic mode of life. In turn this ecological niche is reaffirmed by Andrea Cau, "The Body Plan of *Halszkaraptor escuilliei* (Dinosauria, Theropoda) Is Not a Transitional Form along the Evolution of Dromaeosaurid Hypercarnivory," *PeerJ* 8 (2020), e8672, who draws attention to convergences with the semiaquatic spinosaurids.
80. Nicholas R. Longrich and Phillip J. Currie, "*Albertonykus borealis*, a New Alvarezsaur (Dinosauria: Theropoda) from the Early Maastrichtian of Alberta, Canada: Implications for the Systematics and Ecology of the Alvarezsauridae," *Cretaceous Research* 30, no. 1 (2009): 239–52.
81. Robert B. J. Benson, Matthew T. Carrano, and Stephen L. Brusatte, "A New Clade of Archaic Large-Bodied Predatory Dinosaurs (Theropoda: Allosauroidea) that Survived to the Latest Mesozoic," *Naturwissenschaften* 97, no. 1 (2010): 71–78.
82. Evgeny Kurochkin, "Parallel Evolution of Theropod Dinosaurs and Birds," *Entomological Review* 86 (Supplement 1), (2006), S45–S58, quotation from S49.
83. Akinobu Watanabe, Maria Eugenia Leone Gold, Stephen L. Brusatte, Roger B. J. Benson, Jonah Choiniere, Amy Davidson, and Mark A. Norell, "Vertebral Pneumaticity in the Ornithomimosaur *Archaeornithomimus* (Dinosauria: Theropoda) Revealed by Computed Tomography Imaging and Reappraisal of Axial Pneumaticity in Ornithomimosauria," *PLoS One* 10, no. 12 (2015), e0145168.
84. Daria K. Zelenitsky et al., "Feathered Non-Avian Dinosaurs from North America Provide Insight into Wing Origins," *Science* 338, no. 6106 (2012): 510–14; and Aaron J. van der Reest, Alexander P. Wolfe, and Philip J. Currie, "A Densely Feathered Ornithomimid (Dinosauria:

Theropoda) from the Upper Cretaceous Dinosaur Park Formation, Alberta, Canada," *Cretaceous Research* 58 (2016): 108–17.

85. Yaser Saffar Talori, Yun-Fei Liu, Jing-Shan Zhao, Corwin Sullivan, Jingmai K. O'Connor, and Zhi-Heng Li, "Winged Forelimbs of the Small Theropod Dinosaur *Caudipteryx* Could Have Generated Small Aerodynamic Forces during Rapid Terrestrial Locomotion," *Scientific Reports* 8, no. 1 (2018), 17854.
86. Perle Altangerel, Mark A. Norell, Luis M. Chiappe, and James M. Clark, "Flightless Bird from the Cretaceous of Mongolia," *Nature* 362 (1993): 623–26 [correction in *Nature* 363 (1993), 188].
87. See, for example, Jonah N. Choiniere, Xing Xu, James M. Clark, Catherine A. Forster, Yu Guo, and Fenglu Han, "A Basal Alvarezsauroid Theropod from the Early Late Jurassic of Xinjiang, China," *Science* 327, no. 5965 (2010): 571–74.
88. Luis M. Chiappe, Mark A. Norell, and James M. Clark, "The Skull of a Relative of the Stem-Group Bird *Mononykus*," *Nature* 392 (1998): 275–78.
89. M. H. Schweitzer, J. A. Watt, R. Avci, L. Knapp, L. Chiappe, M. Norell, and M. Marshall, "Beta-Keratin Specific Immunological Reactivity in Feather-like Structures of the Cretaceous Alvarezsaurid *Shuvuuia deserti*," *Journal of Experimental Zoology* 285 (1999): 146–57; see also Evan T. Saitta et al., "Preservation of Feather Fibers from the Late Cretaceous Dinosaur *Shuvuuia deserti* Raises Concern about Immunohistochemical Analyses on Fossils," *Organic Geochemistry* 125 (2018): 142–51.
90. See Jonah M. Choiniere et al., "Evolution of Vision and Hearing Modalities in Theropod Dinosaurs," *Science* 372, no. 6542 (2021): 610–13; with commentary by Lawrence M. Witmer, "Making Sense of Dinosaurs and Birds," *Science* 372, no. 6542 (2021): 575–76. See also Michael Hanson, Eva A. Hoffman, Mark A. Norell, and Bhart-Anjan S. Bhullar, "The Early Origin of a Birdlike Inner Ear and the Evolution of Dinosaurian Movement and Vocalization," *Science* 372, no. 6542 (2021): 601–9.
91. Jonah N. Choiniere, James M. Clark, Mark A. Norell, and Xing Xu, "Cranial Osteology of *Haplocheirus sollers* Choiniere et al., 2010 (Therop-

oda: Alvarezsauroidea)," *American Museum Novitates* 3816, no. 3816 (2014): 1–44.

92. Sterling J. Nesbitt, Julia A. Clarke, Alan H. Turner, and Mark A. Norell, "A Small Alvarezsaurid from the Eastern Gobi Desert Offers Insight into Evolutionary Patterns in the Alvarezsauroidea," *Journal of Vertebrate Paleontology* 31, no. 1 (2011): 144–53.
93. X. Xu et al., "A New Feather Type in a Nonavian Theropod and the Early Evolution of Feathers," *Proceedings of the National Academy of Sciences of the United States of America* 106, no. 3 (2009): 832–34.
94. Xing Xu, Yennien Cheng, Xiaolin Wang, and Chunhsiang Chang, "Pygostyle-like Structure from *Beipiaosaurus* (Theropoda, Therizinosauridae) from the Lower Cretaceous Yixian Formation of Liaoning, China," *Acta Geologica Sinica* 77, no. 3 (2003): 294–98.
95. See Stephan Lautenschlager, Lawrence M. Witmer, Perle Altangerel, and Emily J. Rayfield, "Edentulism, Beaks and Biomechanical Innovations in the Evolution of Theropod Dinosaurs," *Proceedings of the National Academy of Sciences of the United States of America* 110, no. 51 (2013): 20657–62.
96. On the vertebral pneumaticity, see David K. Smith, R. Kent Sanders, and Douglas G. Wolfe, "Vertebral Pneumaticity of the North American Therizinosaur *Nothronychus*," *Journal of Anatomy* 238, no. 3 (2021): 598–614.
97. David K. Smith, "The Braincase of the North American Therizinosaurian *Nothronychus mckinleyi* (Dinosauria, Theropoda)," *Journal of Vertebrate Paleontology* 34, no. 3 (2014): 635–46; also the update by David K. Smith et al., "A Re-evaluation of the Basicranial Soft Tissues and Pneumaticity of the Therizinosaurian *Nothronychus mckinleyi* (Theropoda; Maniraptora)," *PLoS One* 13, no. 7 (2018), e0198155.
98. Lindsay E. Zanno, "A Taxonomic and Phylogenetic Re-evaluation of Therizinosauria (Dinosauria: Maniraptora)," *Journal of Systematic Palaeontology* 8, no. 4 (2010): 503–43.
99. See, for example, Amy M. Balanoff and Mark A. Norell, "Osteology of *Khaan mckennai* (Oviraptosauria: Theropoda)," *Bulletin of the American Museum of Natural History* 372 (2012): 1–77.

100. See, for example, Gregory F. Funston, Philip J. Currie, David A. Eberth, Michael J. Ryan, Tsogtbaatar Chinzorig, Demchig Badamgarav, and Nicholas R. Longrich, "The First Oviraptorosaur (Dinosauria: Theropoda) Bonebed: Evidence of Gregarious Behaviour in a Maniraptoran Theropod," *Scientific Reports* 6 (2016), 35782; see also G. F. Funston, P. J. Currie, M. J. Ryan, and Z.-M. Dong, "Birdlike Growth and Mixed-Age Flocks in Avimimids (Theropoda, Oviraptorosauria)," *Scientific Reports* 9 (2019), 18816, giving evidence for mixed-age flocking and birdlike growth.
101. Takanobu Tsuihiji, Lawrence M. Witmer, Mahito Watabe, Rinchen Barsbold, Khishigjav Tsogtbaatar, Shigeru Suzuki, and Purevdorj Khatanbaatar, "New Information on the Cranial Morphology of *Avimimus* (Theropoda: Oviraptorosauria)," *Journal of Vertebrate Paleontology* 37, no. 4 (2017), e1347177.
102. Zhong-He Zhou, Xiao-Lin Wang, Fu-Cheng Zhang, and Xing Xu, "Important Features of *Caudipteryx*—Evidence from Two Nearly Complete New Specimens," *Vertebrata PalAsiatica* 38, no. 4 (2000): 241–54.
103. See Xing Xu, Yen-Nien Cheng, Xiao-Lin Wang, and Chun-Hsiang Chang, "An Unusual Oviraptorosaurian Dinosaur from China," *Nature* 419, no. 6904 (2002): 291–93; and Amy M. Balanoff, Xing Xu, Yoshitsugu Kobayashi, Yusuke Matsufune, and Mark A. Norell, "Cranial Osteology of the Theropod Dinosaur *Incisivosaurus gauthieri* (Theropoda: Oviraptorosauria)," *American Museum Novitates* 3651 (2009): 1–35.
104. Shuo Wang et al., "Heterochronic Truncation of Odontogenesis in Theropod Dinosaurs Provides Insight into the Macroevolution of Avian Beaks," *Proceedings of the National Academy of Sciences of the United States of America* 114, no. 41 (2017): 10930–35, with correction on p. E10506.
105. Nicholas R. Longrich, Philip J. Currie, and Zhi-Ming Dong, "A New Oviraptorid (Dinosauria: Theropoda) from the Upper Cretaceous of Bayan Mandahu, Inner Mongolia," *Palaeontology* 53, no. 5 (2010): 945–60.
106. Rinchen Barsbold, Halszka Osmólska, Mahito Watabe, Philip J. Currie, and Khishigjav Tsogtbaatar, "A New Oviraptorosaur (Dinosauria,

Theropoda) from Mongolia: The First Dinosaur with a Pygostyle," *Acta Palaeontologica Polonica* 45, no. 2 (2006): 97–106.

107. Q. Ji et al., "Two Feathered Dinosaurs from Northeastern China," *Nature* 393 (1998): 753–61; and further critique by Christian Foth, "On the Identification of Feather Structures in Stem-Line Representatives of Birds: Evidence from Fossils and Actuopalaeontology," *Paläontologische Zeitschrift* 86 (2012): 91–102.
108. W. Scott Persons, IV, Philip J. Currie, and Mark A. Norell, "Oviraptorosaur Tail Forms and Functions," *Acta Palaeontologica Polonica* 59, no. 3 (2013): 553–67.
109. Corwin Sullivan, David W. E. Hone, Xing Xu, and Fucheng Zhang, "The Asymmetry of the Carpal Joint and the Evolution of Wing Folding in Maniraptorian Theropod Dinosaurs," *Proceedings of the Royal Society of London, Series B: Biological Sciences* 277, no. 1690 (2010): 2027–33.
110. P. G. Tickle, M. A. Norell, and J. R. Codd, "Ventilatory Mechanics from Maniraptoran Theropods to Extant Birds," *Journal of Evolutionary Biology* 25, no. 4 (2012): 740–47.
111. Junchang Lü et al., "High Diversity of the Ganzhou Oviraptorid Fauna Increased by a New 'Cassowary-like' Crested Species," *Scientific Reports* 7, no. 1 (2017), 6393.
112. Amy M. Balanoff, G. S. Bever, and Mark A. Norell, "Reconsidering the Avian Nature of the Oviraptorosaur Brain (Dinosauria: Theropoda)," *PLoS One* 9, no. 12 (2014), e113559.
113. Derived from study of the endocranial cavity; see Amy M. Balanoff, Mark A. Norell, Aneila V. C. Hogan, and Gabriel S. Bever, "The Endocranial Cavity of Oviraptorosaur Dinosaurs and the Increasingly Complex, Deep History of the Avian Brain," *Brain, Behavior and Evolution* 91, no. 3 (2018): 125–35.
114. Amy M. Balanoff, Jeroen B. Smaers, and Alan H. Turner, "Brain Modularity across the Theropod-Bird Transition: Testing the Influence of Flight on Neuroanatomical Variation," *Journal of Anatomy* 229, no. 2 (2016): 204–14.
115. For a description of not only the egg clutch but enclosed embryos, see Shundong Bi et al., "An Oviraptorid Preserved atop an Embryo-Bearing

Egg Clutch Sheds Light on the Reproductive Biology of Non-avialan Theropod Dinosaurs," *Science Bulletin* 66, no. 9 (2021): 947–54; also James Matthew Clark, Mark Norell, and Luis M. Chiappe, "An Oviraptorid Skeleton from the Late Cretaceous of Ukhaa Tolgod, Mongolia, Preserved in an Avianlike Brooding Position over an Oviraptorid Nest," *American Museum Novitates* 3265 (1999): 1–36.

116. Romain Amiot et al., "δ^{18}O-Derived Incubation Temperatures of Oviraptorosaur Eggs," *Palaeontology* 60, no. 5 (2017): 633–47.
117. See the overview by David J. Varricchio and Frankie D. Jackson, "Reproduction in Mesozoic Birds and Evolution of the Modern Avian Reproductive Mode," *Auk* 133, no. 4 (2016): 654–84.
118. Kohei Tanaka, Darla K. Zelenitsky, and François Therrien, "Eggshell Porosity Provides Insight on Evolution of Nesting in Dinosaurs," *PLoS One* 10, no. 11 (2015), e0142829.
119. Jasmina Wiemann et al., "Dinosaur Origin of Egg Color: Oviraptors Laid Blue-Green Eggs," *PeerJ* 5 (2017), e3706.
120. See Alan Feduccia and Stephen A. Czerkas, "Testing the Neoflightless Hypothesis: Propatagium Reveals Flying Ancestry of Oviraptorosaurs," *Journal of Ornithology* 156, no. 4 (2015): 1067–74. Although a minority view, they identify a propatagium in *Caudipteryx* and argue the oviraptorosaurs are secondarily flightless.
121. T. Maryańska, H. Osmólska, and M. Wolsan, "Avialan Status for Oviraptorosauria," *Acta Palaeontological Polonica* 47 (2002): 97–116.
122. Xing Xu, Qingwei Tan, Jianmin Wang, Xijin Zhao, and Lin Tan, "A Gigantic Bird-like Dinosaur from the Late Cretaceous of China," *Nature* 447, no7146. (2007): 844–47.
123. Waisum Ma, Junyou Wang, Michael Pittman, Qingwei Tan, Lin Tan, Bin Guo, and Xing Xu, "Functional Anatomy of a Giant Toothless Mandible from a Bird-like Dinosaur: *Gigantoraptor* and the Evolution of the Oviraptorosaurian Jaw," *Scientific Reports* 7 (2017), 16247.
124. Junchang Lü, Philip J. Currie, Li Xu, Xingliao Zhang, Hanyong Pu, and Songhai Jia, "Chicken-sized Oviraptorid Dinosaurs from Central China and Their Ontogenetic Implications," *Naturwissenschaften* 100, no. 2 (2013): 165–75.

125. See, for example, Shu'an Ji and Qiang Ji, "*Jinfengopteryx* Compared to *Archaeopteryx*, with Comments on the Mosaic Evolution of Long-tailed Avialan Birds," *Acta Geologica Sinica* 81, no. 3 (2007): 337–43.
126. Zoltán Csiki, Mátyás Vremir, Stephen L. Brusatte, and Mark A. Norell, "An Aberrant Island-Dwelling Theropod Dinosaur from the Late Cretaceous of Romania," *Proceedings of the National Academy of Sciences of the United States of America* 107, no. 35 (2010): 15357–61.
127. Andrea Cau, Tom Brougham, and Darren Naish, "The Phylogenetic Affinities of the Bizarre Late Cretaceous Romanian Theropod *Balaur bondoc* (Dinosauria, Maniraptora): Dromaeosaurid or Flightless Bird?" *PeerJ* 3 (2015), e1032.
128. Xing Xu and Diego Pol, "*Archaeopteryx*, Paravian Phylogenetic Analysis, and the Use of Probability-Based Methods for Palaeontological Datasets," *Journal of Systematic Palaeontology* 12, no. 3 (2014): 323–34.
129. See, for example, Michael S. Y. Lee and Timothy H. Worthy, "Likelihood Reinstates *Archaeopteryx* as a Primitive Bird," *Biology Letters* 8, no. 2 (2012): 299–303; also Martin Kundrát, John Nudds, Benjamin P. Kear, Junchang Lü, and Per Ahlberg, "The First Specimen of *Archaeopteryx* from the Upper Jurassic Mörnsheim Formation of Germany," *Historical Biology* 31, no. 1 (2019): 3–63.
130. See, for example, Mark N. Puttick, Gavin H. Thomas, and Michael J. Benton, "High Rates of Evolution Preceded the Origin of Birds," *Evolution* 68, no. 5 (2014): 1497–510.
131. See, for example, Chunling Gao, Eric M. Morschhauser, David J. Varricchio, Jinyuan Liu, and Bo Zhao, "A Second Soundly Sleeping Dragon: New Anatomical Details of the Chinese Troodontid *Mei long* with Implications for Phylogeny and Taphonomy," *PLoS One* 7, no. 9 (2012), e45203.
132. Phillip J. Currie and Xi-Jin Zhao, "A New Troodontid (Dinosauria, Theropoda) Braincase from the Dinosaur Park Formation (Campanian) of Alberta," *Canadian Journal of Earth Sciences* 30, no. 10 (1993): 2231–47.
133. Takanobu Tsuihiji, Rinchen Barsbold, Mahito Watabe, Khishigjav Tsogtbaatar, Tsogtbaatar Chinzorig, Yoshito Fujiyama, and Shigeru

Suzuki, "An Exquisitely Preserved Troodontid Theropod with New Information on the Palatal Structure from the Upper Cretaceous of Mongolia," *Naturwissenschaften* 101, no. 2 (2014): 131–42. See also Han Hu et al., "Evolution of the Vomer and Its Implications for Cranial Kinesis in Paraves," *Proceedings of the National Academy of Sciences of the United States of America* 116, no. 39 (2019): 19571–78, who identify at best a limited degree of cranial kinesis.

134. Stephen L. Brusatte, Graeme T. Lloyd, Steve C. Wang, and Mark A. Norell, "Gradual Assembly of Avian Body Plan Culminated in Rapid Rates of Evolution across the Dinosaur-Bird Transition," *Current Biology* 24, no. 20 (2014): 2386–92; and informative commentary by Daniel T. Ksepka, "Evolution: A Rapid Flight towards Birds," *Current Biology* 24, no. 21 (2014): R1052–55.
135. Denver W. Fowler, Elizabeth A. Freedman, John B. Scannella, and Robert E. Kambic, "The Predatory Ecology of *Deinonychus* and the Origin of Flapping in Birds," *PLoS One* 6, no. 12 (2011), e28964.
136. Enrique Peñalver et al., "Parasitised Feathered Dinosaurs as Revealed by Cretaceous Amber Assemblages," *Nature Communications* 8, no. 1 (2017), 1924.
137. Alan H. Turner, Peter J. Makovicky, and Mark A. Norell, "Feather Quill Knobs in the Dinosaur *Velociraptor*," *Science* 317, no. 5845 (2017): 1721.
138. Pascal Godefroit, Helena Demuynck, Gareth Dyke, Dongyu Hu, François Escuillié, and Philippe Claeys, "Reduced Plumage and Flight Ability of a New Jurassic Paravian Theropod from China," *Nature Communications* 4 (2013), 1394.
139. Michael S. Y. Lee, Andrea Cau, Darren Naish, and Gareth J. Dyke, "Sustained Miniaturization and Anatomical Innovation in the Dinosaurian Ancestors of Birds," *Science* 345, no. 6196 (2014): 562–66; and commentary by Michael J. Benton, "How Birds Became Birds," 345, no. 6196 (2014): 508–9.
140. See, for example, Rui Pei et al., "Potential for Powered Flight Neared by Most Close Avialan Relatives, but Few Crossed Its Thresholds," *Current Biology* 30, no. 20 (2020): 4033–46; see also the critique by Francisco J. Serrano and Luis M. Chiappe, "Independent Origins of

Powered Flight in Paravian Dinosaurs?" *Current Biology* 31, no. 8 (2021): R370–72; and reply by M. Pittman et al., "Response to Serrano and Chiappe," *Current Biology* 31, no. 8 (2021): R372–73.

141. Phillip L. Manning, David Payne, John Pennicott, Paul M. Barrett, and Roland A. Ennos, "Dinosaur Killer Claws or Climbing Crampons?" *Biology Letters* 2, no. 2 (2006): 110–12.
142. Ashley M. Heers, David B. Baier, Brandon E. Jackson, and Kenneth P. Dial, "Flapping before Flight: High Resolution, Three-Dimensional Skeletal Kinematics of Wings and Legs during Avian Development," *PLoS One* 11, no. 4 (2016), e0153446; also the follow-up: Ashley M. Heers, Stephanie L. Varghese, Leila K. Hatier, and Jeremiah J. Cabrera, "Multiple Functional Solutions during Flightless to Flight-Capable Transitions," *Frontiers in Ecology and Evolution* 8 (2021), 573411.
143. See Xia Wang, Robert L. Nudds, Colin Palmer, and Gareth J. Dyke, "Primary Feather Vane Asymmetry Should Not Be Used to Predict the Flight Capabilities of Feathered Fossils," *Science Bulletin* 62, no. 18 (2017): 1227–28; and Xia Wang, Ho Kwan Tang, and Julia A. Clarke, "Flight, Symmetry and Barb Angle Evolution in the Feathers of Birds and Other Dinosaurs," *Biology Letters* 15, no. 12 (2019), 20190622; also Feo et al., "Barb Geometry of Asymmetrical Feathers," 20142864.
144. John Lees, Terence Garner, Glen Cooper, and Robert Nudds, "Rachis Morphology Cannot Accurately Predict the Mechanical Performance of Primary Feathers (and Therefore Fossil) Feathered Flyers," *Royal Society Open Science* 4, no. 2 (2017), 160927.
145. See, for example, Nicholas R. Longrich, Jakob Vinther, Qingjin Meng, Quangguo Li, and Anthony P. Russell, "Primitive Wing Feather Arrangement in *Archaeopteryx lithographica* and *Anchiornis huxleyi*," *Current Biology* 22, no. 23 (2012): 2262–67.
146. Daniella Schwarz, Martin Kundrát, Helmut Tischlinger, Gareth Dyke, and Ryan M. Carney, "Ultraviolet Light Illuminates the Avian Nature of the Berlin *Archaeopteryx* Skeleton," *Scientific Reports* 9, no. 1 (2019), 6518. They also note a set of fused thoracic vertebrae (the notoarium), again consistent with active flight.
147. See X. Wang, R. L. Nudds, and G. J. Dyke, "The Primary Feather Lengths of Early Birds with Respect to Avian Wing Shape Evolution," *Journal*

of Evolutionary Biology 24, no. 6 (2011): 1226–31; and X. Wang, R. L. Nudds, C. Palmer, and G. J. Dyke, "Size Scaling and Stiffness of Avian Primary Feathers: Implications for the Flight of Mesozoic Birds," *Journal of Evolutionary Biology* 25, no. 3 (2012): 547–55.

148. See Dennis Evangelista, Sharlene Cam, Tony Huynh, Austin Kwong, Homayun Mehrabani, Kyle Tse, and Robert Dudley, "Shifts in Stability and Control Effectiveness during Evolution of Paraves Support Aerial Maneuvering Hypotheses for Flight Origins," *PeerJ* 2 (2014), e632.
149. W. Scott Persons IV and Philip J. Currie, "Dragon Tails: Convergent Caudal Morphology in Winged Archosaurs," *Acta Geologica Sinica* 86, no. 6 (2012): 1402–12.
150. At least eight genera are known, including the recently described *Wulong;* see Ashley W. Proust, Chunling Gao, David J. Varricchio, Jianlin Wu, and Fengjiao Zhang, "A New Microraptorine Theropod from the Jehol Biota and Growth in Early Dromaeosaurids," *Anatomical Record* 303, no. 4 (2020): 963–87.
151. See, for example, Sankar Chatterjee and R. Jack Templin, "Biplane Wing Planform and Flight Performance of the Feathered Dinosaur *Microraptor qui,*" *Proceedings of the National Academy of Sciences of the United States of America* 104, no. 5 (2007): 1576–80.
152. Quanguo Li et al., "Reconstruction of *Microraptor* and the Evolution of Iridescent Plumage," *Science* 335, no. 6073 (2012): 1215–19.
153. These included mammals; see H. C. E. Larsson, D. Hone, T. Dececchi, and P. Currie, "The Winged Non-avian Dinosaur *Microraptor* Fed on Animals: Implications for the Jehol Biota Ecosystem [abstract]," Supplement 2, "Society of Vertebrate Paleontology Annual Meeting 2010, Program and Abstracts," *Journal of Vertebrate Paleontology* 30 (2010), 120A; for birds, see Jingmai M. O'Connor, Zhonghe Zhou, and Xing Xu, "Additional Specimen of *Microraptor* Provides Unique Evidence of Dinosaurs Preying on Birds," *Proceedings of the National Academy of Sciences of the United States of America* 108, no. 49 (2011): 19662–65; for fish, see Lida Xing et al., "Piscivory in the Feathered Dinosaur *Microraptor,*" *Evolution* 67, no. 8 (2013): 2441–45; and for lizards, see Jingmai O'Connor et al., "*Microraptor* with Ingested Lizard Suggests

Non-specialized Digestive Function," *Current Biology* 29, no. 14 (2019), 2423–29.e2.

154. See, for example, Gareth Dyke, Roeland de Kat, Colin Palmer, Jacques van der Kindere, Darren Naish, and Bharathram Ganapathisubramani, "Aerodynamic Performance of the Feathered Dinosaur *Microraptor* and the Evolution of Feathered Flight," *Nature Communications* 4 (2013), e2489; and Dennis Evangelista et al., "Aerodynamic Characteristics of a Feathered Dinosaur Measured Using Physical Models: Effects of Form on Static Stability and Control Effectiveness," *PLoS One* 9, no. 1 (2014), e85203. Similar arguments apply to the related *Sinornithosaurus;* see Sankar Chatterjee and R. J. Templin, "Feathered Coelurosaurs from China: New Light on the Arboreal Origin of Avian Flight," in *Feathered Dragons: Studies on the Transition from Dinosaurs to Birds,* ed. Philip J. Currie, Eva B. Koppelhus, Martin A. Shugar, and Joanna L. Wright (Bloomington: Indiana University Press, 2004), 251–81.
155. Yosef Kiat, Amir Balaban, Nir Sapir, Jingmai Kathleen O'Connor, Min Wang, and Xing Xu, "Sequential Molt in a Feathered Dinosaur and Implications for Early Paravian Ecology and Locomotion," *Current Biology* 30, no. 18 (2020): 3633–38. Notably sequential molting is not seen in birds that are either flightless or where flight is limited (many waterbirds).
156. Xing Xu, Zhonghe Zhou, Xiaolin Wang, Xuewen Kuang, Fucheng Zhang, and Xiangke Du, "Four-Winged Dinosaurs from China," *Nature* 421, no. 6921 (2003): 335–40; and commentary by Richard O. Prum, "Dinosaurs Take to the Air," *Nature* 421, no. 6921 (2003): 323–24.
157. Sang-im Lee, Jooha Kim, Hyungmin Park, Piotr G. Jabłoński, and Haecheon Choi, "The Function of the Alula in Avian Flight," *Scientific Reports* 5 (2015), 9914.
158. Rui Pei, Quanguo Li, Qingjin Meng, Ke-Qin Gao, and Mark A. Norell, "A New Specimen of *Microraptor* (Theropoda: Dromaeosauridae) from the Lower Cretaceous of Western Liaoning, China," *American Museum Novitates* 2014, no. 3821 (2014): 1–28.
159. C. A. Forster, S. D. Sampson, L. M. Chiappe, and D. W. Krause, "The Theropod Ancestry of Birds: New Evidence from the Late Cretaceous

of Madagascar," *Science* 279, no. 5358 (1998): 1915–19; they corrected the name from *Rahona* (a name already assigned to a butterfly) to *Rahonavis* in a letter to the editor in *Science* 280, no. 5361 (1998), 179; also Catherine A. Forster, Patrick M. O'Connor, Luis M. Chiappe, and Alan H. Turner, "The Osteology of the Late Cretaceous Paravian *Rahonavis ostromi* from Madagascar," *Palaeontologia Electronica* 23, no. 2 (2020): 1–75.

160. Dongyu Hu, Lianhai Hou, Lijun Zhang, and Xing Xu, "A Pre-*Archaeopteryx* Troodontid Theropod from China with Long Feathers on the Metatarsus," *Nature* 461, no. 7264 (2009): 640–43, with commentary by Lawrence M. Witmer, "Feathered Dinosaurs in a Tangle," *Nature* 461, no. 7264 (2009): 601–2. Others, however, have suggested *Anchiornis* is closer to *Archaeopterx* and other avialians; see, for example, Rui Pei, Quanguo Li, Qingjin Meng, Mark Norell, and Keqin Gao, "New Specimens of *Anchiornis huxleyi* (Theropoda, Paraves) from the Late Jurassic of Northeastern China," *Bulletin of American Museum of Natural History* 411 (2017): 1–66.

161. Xiaoli Wang, Michael Pittman, Xiaoting Zheng, Thomas G. Kaye, Amanda R. Falk, Scott A. Hartman, and Xing Xu, "Basal Paravian Functional Anatomy Illuminated by High-Detail Body Outline," *Nature Communications* 8 (2017), 14576; arguments against powered flight are given by Yanhong Pan et al., "The Molecular Evolution of Feathers with Direct Evidence from Fossils," *Proceedings of the National Academy of Sciences of the United States of America* 116, no. 8 (2019): 3018–23. So too T. Alexander Dececchi, Hans C. E. Larsson, Michael Pittman, and Michael B. Habib, "Chapter 11: High Flyer or High Fashion? A Comparison of Flight Potential among Small-Bodied Paravians," in "Pennaraptoran Theropod Dinosaurs: Past Progress and New Frontiers," ed. Michael Pittman and Xing Xu, special issue, *Bulletin of the American Museum of Natural History* 440 (2020): 295–320, are even doubtful of an ability to glide.

162. Xiaoting Zheng, Jingmai O'Connor, Xiaoli Wang, Min Wang, Xiaomei Zhang, and Zhonghe Zhou, "On the Absence of Sternal Elements in *Anchiornis* (Paraves) and *Sapeornis* (Aves) and the Complex Early Evolution of the Avian Sternum," *Proceedings of the National*

Academy of Sciences of the United States of America 111, no. 38 (2014): 13900–905; see also critique by Christian Foth (111, no. 50 [2014]: E5334), and the reply by Jingmai O'Connor, Xiaoli Wang, Min Wang, Xiaomei Zhang, and Zhonghe Zhou (111, no. 50 [2014]: E5335).

163. Christian Foth and Oliver W. M. Rauhut, "Re-evaluation of the Haarlem *Archaeopteryx* and the Radiation of Maniraptoran Theropod Dinosaurs," *BMC Evolutionary Biology* 17, no. 1 (2017), 236.
164. Xing Xu et al., "Mosaic Evolution in an Asymmetrically Feathered Dinosaur with Transitional Features," *Nature Communications* 8 (2017), 14972.
165. Dongyu Hu et al., "A Bony-Crested Jurassic Dinosaur with Evidence of Iridescent Plumage Highlights Complexity in Early Paravian Evolution," *Nature Communications* 9, no. 1 (2018), 217.
166. Peter J. Makovicky, Sebastián Apesteguía, and Federico L. Agnolín, "The Earliest Dromaeosaurid Theropod from South America," *Nature* 437, no. 7061 (2005): 1007–11; also Federico A. Gianechini, Peter J. Makovicky, Sebastián Apesteguía, and Ignacio Cerda, "Postcranial Skeletal Anatomy of the Holotype and Referred Specimens of *Buitreraptor gonzalezorum* Makovicky, Apesteguía and Agnolín 2005 (Theropoda, Dromaeosauridae), from the Late Cretaceous of Patagonia," *PeerJ* 6 (2018), e4558.
167. Peter C. Kjærgaard, "'Hurrah for the Missing Link!': A History of Apes, Ancestors and a Crucial Piece of Evidence," *Notes and Records of the Royal Society Journal of the History of Science* 65, no. 1 (2011): 83–98.
168. See, for example, "[Comments on]: The Neanderthal Skull: Its Formation Considered Anatomically," in the *Journal of the Anthropological Society of London* 3 (1865): xv–xix, notably the comments by C. Carter Blake that the newly discovered Neanderthal skull should give no comfort to those pesky "transmutationists" when it came to the "derivation of man from beast" (xvii).
169. Folmer Bokma, Valentijn van den Brink, and Tanja Stadler, "Unexpectedly Many Extinct Hominins," *Evolution* 66, no. 9 (2012): 2969–74.
170. Jeffrey H. Schwartz and Ian Tattersall, "Defining the Genus *Homo*: Early Hominin Species Were as Diverse as Other Mammals," *Science* 349, no. 6251 (2015): 931–32.

171. See, for example, D. Lordkipanidze et al., "A Complete Skull from Dmanisi, Georgia, and the Evolutionary Biology of Early *Homo,*" *Science* 342, no. 6156 (2013): 326–31.
172. See, for example, Robert A. Foley, "Mosaic Evolution and the Pattern of Transitions in the Hominin Lineage," *Philosophical Transactions of the Royal Society of London, Series B: Biological Sciences* 371, no. 1698 (2016), 20150244.
173. See the characteristically insightful article by Matt Cartmill, "Primate Origins, Human Origins, and the End of Higher Taxa," *Evolutionary Anthropology* 21, no. 6 (2012): 208–20; also comment by Ian Tattersall, "Higher Taxa: an Alternate Perspective" (221–23) and the author's reply (172–73).
174. See the overview by Lee R. Berger, "The Mosaic Nature of *Australopithecus sediba*. Introduction," *Science* 340, no. 6129 (2013): 163. For a more skeptical stance on the phylogenetic position of this hominin, see Andrew Du and Zeresenay Alemseged, "Temporal Evidence Shows *Australopithecus sediba* Is Unlikely to Be the Ancestor of *Homo,*" *Science Advances* 5 (2019), eaav9038.
175. Steven E. Churchill et al., "The Upper Limb of *Australopithecus sediba,*" *Science* 340, no. 6129 (2013), 1233477; also Thomas R. Rein, Terry Harrison, Kristian J. Carlson, and Katerina Harvati, "Adaptation to Suspensory Locomotion in *Australopithecus sediba,*" *Journal of Human Evolution* 104 (2017): 1–12.
176. Job M. Kibii, Steven E. Churchill, Peter Schmid, Kristian J. Carlson, Nichelle D. Reed, Darryl J. de Ruiter, and Lee R. Berger, "A Partial Pelvis of *Australopithecus sediba,*" *Science* 333, no. 6048 (2011): 1407–11. These authors also note that the birth canal would be sufficiently wide to accommodate a large head even though the brain size is still that of an australopithecine. See also Natalie M. Laudicina, Frankee Rodriguez, and Jeremy M. DeSilva, "Reconstructing Birth in *Australopithecus sediba,*" *PLoS One* 14, no. 9 (2019), e0221871.
177. Tracy L. Kivell, Job M. Kibii, Steven E. Churchill, Peter Schmid, and Lee R. Berger, "*Australopithecus sediba* Hand Demonstrates Mosaic Evolution of Locomotor and Manipulative Abilities," *Science* 333, no. 6048 (2011): 1411–17; also Christopher J. Dunmore et al., "The Posi-

tion of *Australopithecus sediba* within Fossil Hominin Hand Use Diversity," *Nature Ecology and Evolution* 14, no. 7 (2020): 911–18.

178. Peter Schmid, Steven E. Churchill, Shahed Nalla, Eveline Weissen, Kristian J. Carlson, Darryl J. de Ruiter, and Lee R. Berger, "Mosaic Morphology in the Thorax of *Australopithecus sediba*," *Science* 340, no. 6129 (2013), e1234598.
179. William H. Kimbel and Brian Villmoare, "From *Australopithicus* to *Homo:* The Transition That Wasn't," *Philosophical Transactions of the Royal Society of London, Series B: Biological Sciences* 371, no. 1698 (2016), 20150248.
180. See Lee R. Berger et al., "*Homo naledi,* a New Species of the Genus *Homo* from the Dinaledi Chamber, South Africa," *eLife* 4 (2015), e09560; and John Hawks et al., "New Fossil Remains of *Homo naledi* from the Lesedi Chamber, South Africa," *eLife* 6 (2017), e24232.
181. Jessie L. Robbins et al., "Providing Context to the *Homo naledi* Fossils: Constraints from Flowstones on the Age of Sediment Deposits in Rising Star Cave, South Africa," *Chemical Geology* 567 (2021), 120108.
182. Mana Dembo, "The Evolutionary Relationships and Age of *Homo naledi:* An Assessment Using Dated Bayesian Phylogenetic Methods," *Journal of Human Evolution* 97 (2016): 17–26.
183. Elen M. Feuerriegel, David J. Green, Christopher S. Walker, Peter Schmid, John Hawks, Lee R. Berger, and Steven E. Churchill, "The Upper Limb of *Homo naledi,*" *Journal of Human Evolution* 104 (2017): 155–73.
184. Tracey L. Kivell et al., "The Hand of *Homo naledi,*" *Nature Communications* 6 (2015), 8431.
185. W. E. H. Harcourt-Smith et al., "The Foot of *Homo naledi,*" *Nature Communications* 6 (2015), 84321; Damiano Marchi et al., "The Thigh and Leg of *Homo naledi,*" *Journal of Human Evolution* 104 (2017): 174–204.
186. Heather M. Garvin, Marina C. Elliott, Lucas K. Delezene, John Hawks, Steven E. Churchill, Lee R. Berger, and Trenton W. Holliday, "Body Size, Brain Size, and Sexual Dimorphism in *Homo naledi* from the Dinaledi Chamber," *Journal of Human Evolution* 111 (2017): 119–38. With respect to endocasts, see Ralph L. Holloway, Shawn D. Hurst,

Heather M. Garvin, P. Thomas Schoenemann, William B. Vanti, Lee R. Berger, and John Hawks, "Endocast Morphology of *Homo naledi* from the Dinaledi Chamber, South Africa," *Proceedings of the National Academy of Sciences of the United States of America* 115, no. 22 (2018): 5738–43; and commentary by Stephen Montgomery, "Hominin Brain Evolution: The Only Way Is Up?" *Current Biology* 28, no. 14 (2018): R788–90.

187. For an overview, Philip L. Reno, "Genetic and Developmental Basis for Parallel Evolution and Its Significance for Hominoid Evolution," *Evolutionary Anthropology* 23, no. 5 (2014): 188–200.
188. See the review by Diego Sustaita, Emmanuelle Pouydebat, Adriana Manzano, Virginia Abdala, Fritz Hertel, and Anthony Herrel, "Getting a Grip on Tetrapod Grasping: Form, Function, and Evolution," *Biological Reviews* 88, no. 2 (2013): 380–405.
189. Adriana S. Manzano, Virginia Abdala, and Anthony Herrel, "Morphology and Function of the Forelimb in Arboreal Frogs: Specializations for Grasping Ability?" *Journal of Anatomy* 213, no. 3 (2008): 296–307.
190. See, for example, Allison L. Machnicki, Linda B. Spurlock, Karen B. Strier, Philip L. Reno, and C. Owen Lovejoy, "First Steps in Bipedality in Hominids: Evidence from the Atelid and Proconsulid Pelvis," *PeerJ* 4 (2016), e1521.
191. See, for example, Yohannes Haile-Selassie, Beverly Z. Saylor, Alan Deino, Naomi E. Levin, Mulugeta Alene, and Bruce M. Latimer, "A New Hominin Foot from Ethiopia Shows Multiple Pliocene Bipedal Adaptations," *Nature* 483, no. 7391 (2012): 565–69; and Jeremy M. DeSilva, Kenneth G. Holt, Steven E. Churchill, Kristian J. Carlson, Christopher S. Walker, Bernhard Zipfel, and Lee R. Berger, "The Lower Limb and Mechanics of Walking in *Australopithecus sediba*," *Science* 340, no. 6129 (2013), 1232999.
192. Bernhard Zipfel, Jeremy M. DeSilva, Robert S. Kidd, Kristian J. Carlson, Steven E. Churchill, and Lee R. Berger, "The Foot and Ankle of *Australopithecus sediba*," *Science* 333, no. 6048 (2011): 1417–20.
193. See, for example, Susan C. Antón, Richard Potts, and Leslie C. Aiello, "Evolution of Early *Homo:* An Integrated Biological Perspective," *Science* 345, no. 6192 (2014), 1236828.

194. See, for example, Robert Foley and Marta Mirasón Lahr, "On Stony Ground: Lithic Technology, Human Evolution, and the Emergence of Culture," *Evolutionary Anthropology* 12, no. 3 (2003): 109–22.
195. See the appropriately titled Ron Shimelmitz, Steven L. Kuhn, Arthur J. Jelinek, Avraham Ronen, Amy E. Clark, and Mina Weinstein-Evron, "'Fire at Will': The Emergence of Habitual Fire Use 350,000 Years Ago," *Journal of Human Evolution* 77 (2014): 196–203. It is, however, possible that fire was being used at a much earlier stage; see, for example, Sarah Hlubik, Russell Cutts, David R. Braun, Francesco Berna, Craig S. Feibel, and John W. K. Harris, "Hominin Fire Use in the Okote Member at Koobi Fora, Kenya: New Evidence for the Old Debate," *Journal of Human Evolution* 133 (2019): 214–29.
196. Particularly early, perhaps, is a ca. 500 kyr engraved shell from Java; see Josephine C. A. Joordens et al., "*Homo erectus* at Trinil on Java Used Shells for Tool Production and Engraving," *Nature* 518, no. 7538 (2015): 228–31. For somewhat younger (ca. 300–350 kyr) engravings that include one with regular spacing, see Dietrich Mania and Ursula Mania, "Deliberate Engravings on Bone Artefacts of *Homo erectus*," *Rock Art Research* 5, no. 2 (1988): 91–107, which includes discussion (95–104) and authors' reply (104–7). Other telling evidence comes in the form of spears from the Schöningen site in Germany; see, for example, Nicholas J. Conard, Jordi Serangeli, Utz Böhner, Britt M. Starkovich, Christopher E. Miller, Brigitte Urban, and Thijs Van Kolfschoten, "Excavations at Schöningen and Paradigm Shifts in Human Evolution," *Journal of Human Evolution* 89 (2015): 1–17; as well as evidence for throwing sticks: Nicholas J. Conard, Jordi Serangeli, Gerlinde Bigga, and Veerle Rots, "A 300,000-Year-Old Throwing Stick from Schöningen, Northern Germany, Documents the Evolution of Human Hunting," *Nature Ecology and Evolution* 4, no. 5 (2020): 690–93.
197. See, for example, Nick Blegen, "The Earliest Long-Distance Obsidian Transport: Evidence from the ~200 ka Middle Stone Age Sibilo School Road Site, Baringo, Kenya," *Journal of Human Evolution* 103 (2017): 1–19.
198. See, for example, the evidence from the Mediterranean documented by Duncan Howitt-Marshall and Curtis Runnels, "Middle Pleistocene

Sea-Crossings in the Eastern Mediterranean?" *Journal of Anthropological Archaeology* 42 (2016): 140–53.

199. See, for example, the evidence for pigment use at least 350 kyr ago by Lawrence S. Barham, "Systematic Pigment Use in the Middle Pleistocene of South-Central Africa," *Current Anthropology* 43, no. 1 (2002): 181–90.
200. S. Bruno, "The Multi-use of Ochre in Prehistory," *Human Evolution* 23, no. 3 (2008): 233–39.
201. This may include the use of specularite documented by Ian Watts, Michael Chazan, and Jayne Wilkins, "Early Evidence for Brilliant Ritualized Display: Specularite Use in the Northern Cape (South Africa) between ~500 and ~300 Ka," *Current Anthropology* 57, no. 3 (2016): 287–310.
202. See, for example, Tammy Hodgskiss, "Cognitive Requirements for Ochre Use in the Middle Stone Age at Sibudu, South Africa," *Cambridge Archaeological Journal* 24, no. 3 (2014): 405–28.
203. See, for example, Mary C. Stiner, Ran Barkai, and Avi Gopher, "Cooperative Hunting and Meat Sharing 400-200 kya at Qesem Cave, Israel," *Proceedings of the National Academy of Sciences of the United States of America* 106, no. 32 (2009): 13207–212; Mary C. Stiner, Avi Gopher, and Ran Barkai, "Hearth-side Socioeconomics, Hunting and Paleoecology during the Late Lower Paleolithic at Qesem Cave, Israel," *Journal of Human Evolution* 60, no. 2 (2011): 213–33.
204. See, for example, the article and discussion on the Berekhat Ram: F. D'Errico and A. Nowell, "A New Look at the Berekhat Ram Figurine: Implications for the Origins of Symbolism," *Cambridge Archaeological Journal* 10, no. 1 (200): 123–46 (discussion: 146–57; 157–67). See also Tan-Tan figurines in Robert G. Bednarik, "A Figurine from the African Acheulian," *Current Anthropology* 44, no. 3 (2003): 405–13.
205. Although certainly not construed in this way. On the use of fire in the Qesem Cave, see Ran Barkai, Jordi Rosell, Ruth Blasco, and Avi Gopher, "Fire for a Reason: Barbecue at Middle Pleistocene Qesem Cave, Israel," *Current Anthropology* 58, no. S16 (2017): S314–28. They observe sophisticated behaviors that are enigmatically ahead of the

"expected" hominin time line, which is congruent with the exploration here of "hidden" realities.

206. Eudald Carbonell and Maria Mosquera, "The Emergence of a Symbolic Behaviour: The Sepulchral Pit of Sima de los Huesos, Sierra de Atapuerca, Burgos, Spain," *Comptes Rendus Palevol* 5, no. 1–2 (2006): 155–60. Discovery of a quartzite hand ax might add force to this interpretation. See also Arantza Aranburu, Juan Luis Arsuaga, and Nohemi Sala, "The Stratigraphy of the Sima de los Huesos (Atapuerca, Spain) and Implications for the Origin of the Fossil Hominin Accumulation," *Quaternary International* 433, pt. A (2017): 5–21; they point to but do not confirm deliberate accumulation.
207. Paul H. G. M. Dirks et al., "Geological and Taphonomic Context for the New Hominin Species *Homo naledi* from the Dinaledi Chamber, South Africa," *eLife* 4 (2015), e09561. But for a more critical view, see Aurore Val, "Deliberate Body Disposal by Hominins in the Dinaledi Chamber, Cradle of Humankind, South Africa?" *Journal of Human Evolution* 96 (2016): 145–48, and the authors' reply (149–53).
208. Francois Durand, "Naledi: An Example of How Natural Phenomena Can Inspire Metaphysical Assumptions," *HTS Teologiese Studies/Theological Studies* 73, no. 3 (2017), a4507; also Detleve L. Tönsing, "*Homo faber* or *Homo credente*? What Defines Humans, and What Could *Homo naledi* Contribute to this Debate?" *HTS Teologiese Studies/Theological Studies* 73, no. 3 (2017), e4495.
209. As a mere taster, see, for example, evidence for seafaring: George Ferentinos, Maria Gkioni, Maria Geraga, and George Papatheodorou, "Early Seafaring Activity in the Southern Ionian Islands, Mediterranean Sea," *Journal of Archaeological Science* 39, no. 7 (2012): 2167–76; sophisticated and altruistic health care: Penny Spikins, Andy Needham, Barry Wright, Calvin Dytham, Maurizio Gatta, and Gail Hitchens, "Living to Fight Another Day: The Ecological and Evolutionary Significance of Neanderthal Healthcare," *Quaternary Science Reviews* 217, no. 1 (2019): 98–118; and Penny Spikins, Andy Needham, Lorna Tilley, and Gail Hitchens, *World Archaeology* 50, no. 3 (2018): 384–403; hafted tools that rely on a sophisticated technology: Ilaria Degano

et al., "Hafting of Middle Paleolithic Tools in Latium (Central Italy): New Data from Fossellone and Sant'Agostino Caves," *PLoS One* 14, no. 6 (2019), e0213473; and Marcel J. L. T. Niekus et al., "Middle Paleolithic Complex Technology and a Neandertal Tar-Backed Tool from the Dutch North Sea," *Proceedings of the National Academy of Sciences of the United States of America* 116, no. 44 (2019): 22081–87; with commentary by João Zilhão, "Tar Adhesives, Neandertals, and the Tyranny of the Discontinuous Mind," *Proceedings of the National Academy of Sciences of the United States of America* 116, no. 44 (2019): 21966–68. However, Patrick Schmidt, Maxime Rageot, Matthias Blessing, and Claudio Tennie, "The Zandmotor Data Do Not Resolve the Question Whether Middle Paleolithic Birch Tar Making Was Complex or Not," *Proceedings of the National Academy of Sciences of the United States of America* 117, no. 9 (2019): 4456–57, queried the inferred degree of technological sophistication, and the reply by Paul R. B. Kozowyk, Geeske H. J. Langejans, Gerrit L. Dusseldorp, and Marcel J. L. Th. Niekus, "Interpretation of Paleolithic Adhesive Production: Combining Experimental and Paleoenvironmental Information," *Proceedings of the National Academy of Sciences of the United States of America* 117, no. 9 (2020): 4458–59. For wooden "diggers" that employ the exceptionally hard boxwood, see Biancamaria Aranguren et al., "Wooden Tools and Fire Technology in the Early Neanderthal Site of Poggetti Vecchi (Italy)," *Proceedings of the National Academy of Sciences of the United States of America* 115, no. 9 (2018): 2054–59; and for shell tools, see Katerina Douka and Enza Elena Spinapolice, "Neanderthal Shell Tool Production: Evidence from Middle Palaeolithic Italy and Greece," *Journal of World Prehistory* 25 (2012): 45–79. Highly competent hunting ranged from aurochs to rhino, along with a penchant for the choice bits; see Mark White, Paul Pettitt, and Danielle Schreve, "Shoot First, Ask Questions Later: Interpretative Narratives of Neanderthal Hunting," *Quaternary Science Reviews* 140 (2016): 1–20. The diet was diverse; the evidence includes the ingenious identification of food fragments in dental calculi; see, for example, Robert C. Power et al., "Dental Calculus Indicates Widespread Plant Use within the Stable Neanderthal Dietary Niche," *Jour-*

nal of Human Evolution 119 (2018): 27–41. This included marine resources, which evidently extended to skin diving; see Paola Villa, Sylvain Soriano, Luca Pollarolo, Carlo Smriglio, Mario Gaeta, Massimo D'Orazio, Jacopo Conforti, and Carlo Tozzi, "Neandertals on the Beach: Use of Marine Resources at Grotta dei Moscerini (Latium, Italy)," *PLoS One* 15, no. 1 (2020), e0226690. See also Erik Trinkhaus, Mathilde Samsel, and Sébastien Villotte, "External Auditory Exostoses among Western Eurasian Late Middle and Late Pleistocene Humans," *PLoS One* 14, no. 8 (2019), e0220464, on bony protrusions in the ear (exostoses) that might point to swimming (aka surfer's ear). Agile prey were included and even mushrooms as well as perhaps nets to put them in; see Bruce L. Hardy et al., "Impossible Neanderthals? Making String, Throwing Projectiles and Catching Small Game during Marine Isotope Stage 4 (Abri du Maras, France)," *Quaternary Science Reviews* 82 (2013): 23–40; the latter was supported by the discovery of three-ply fibers made from conifer bark; see B. L. Hardy, M.-H. Moncel, C. Kerfant, M. Lebon, L. Bellot-Gurlet, and N. Mélard, "Direct Evidence of Neanderthal Fibre Technology and Its Cognitive and Behavioral Implications," *Scientific Reports* 10, no. 1 (2020), 4889. They also used ocher for decoration of objects; see, for example, Marco Peresani, Marian Vanhaeren, Ermanno Quaggiotto, Alain Queffelec, and Francesco d'Errico, "An Ochered Fossil Marine Shell from the Mousterian of Fumane Cave, Italy," *PLoS One* 8, (2013), e68572; Marin Cârciumaru, Elena-Cristina Nițu, and Ovidiu Cîrstina, "A Geode Painted with Ochre by the Neanderthal Man," *Comptes Rendus Palevol.* 14, no. 1 (2015): 31–41; and Dirk L. Hoffmann, Diego E. Angelucci, Valentín Villaverde, Josefina Zapata, and João Zilhão, "Symbolic Use of Marine Shells and Mineral Pigments by Iberian Neandertals 115,000 Years Ago," *Science Advances* 4, no. 2 (2018), eaar5255. The latter report dated the find at approximately 115 kyr. Perhaps it also was a dietary supplement; see Carlos M. Duarte, "Red Ochre and Shells: Clues to Human Evolution," *Trends in Ecology and Evolution* 29, no. 10 (2014): 560–65, not to mention manganese dioxide employed to decrease the ignition temperature when making fires; see Peter J. Heyes, Konstantinos Anastasakis, Wiebren de Jong, Annelies van Hoesel, Wil Roebroeks, and Marie Soressi, "Selection and Use of Manganese Dioxide by Neanderthals," *Scientific Reports* 6 (2016), 22159.

210. See, for example, the overview by Paola Villa and Wil Roebroeks, "Neandertal Demise: An Archaeological Analysis of the Modern Human Superiority Complex," *PLoS One* 9, no. 4 (2014), e96424.
211. Jacques Jaubert et al., "Early Neanderthal Constructions Deep in Bruniquel Cave in Southwestern France," *Nature* 534, no. 7605 (2016): 111–14; and commentary by Marie Soressi, "Neanderthals Built Underground," *Nature* 534, no. 7605 (2016): 43–44.
212. Robert H. Gargett et al., "Grave Shortcomings: The Evidence for Neandertal Burial," *Current Anthropology* 30, no. 2 (1989): 157–90. For the opposing side see, for example, William Rendu et al., "Evidence Supporting an Intentional Neandertal Burial at La Chapelle-aux-Saints," *Proceedings of the National Academy of Sciences of the United States of America* 111, no. 1 (2014): 81–86; and a general overview by Paul Pettitt, *The Palaeolithic Origins of Human Burial* (London: Routledge, 2011), who in addition to discussing La Ferrassie provides strong evidence for intentional burial and possible grave goods in La Grotte du Régourdou. For an update of the celebrated burials at Shanidar, see Emma Pomeroy et al., "Issues of Theory and Method in the Analysis of Paleolithic Mortuary Behavior: A View from Shanidar Cave," *Evolutionary Anthropology* 29, no. 5 (2020): 263–79.
213. See the monograph by D. Peyrony, "La Ferrassie. Mousterien-Perigordien-Aurignacien," *Préhistoire* 3 (1934): 1–92, with evidence also for a slab with engraved "cups" on the underside; also Asier Gómez-Olivenci, Rolf Quam, Nohemi Salag, Morgane Bardey, James C. Ohman, and Antoine Balzeau, "La Ferrassie 1: New Perspectives on a 'Classic' Neandertal," *Journal of Human Evolution* 117 (2018): 13–32. A detailed analysis of the child burial, again with an east-west orientation, is given by Antoine Balzeau et al., "Pluridisciplinary Evidence for Burial for the La Ferrassie 8 Neandertal Child," *Scientific Reports* 10, no. 1 (2020), 21230.
214. On the first count, see, for example, Hélène Rougier et al., "Neandertal Cannibalism and Neandertal Bones Used as Tools in Northern

Europe," *Scientific Reports* 6 (2016), 29005; for contrast, see David W. Frayer, Jakov Radovčić, and Davorka Radovčić, "Krapina and the Case for Neandertal Symbolic Behavior," *Current Anthropology* 61, no. 6 (2020): 713–31, with commentaries (721–26) and authors' reply (726–27), which places cannibalism in the context of a wider cultural backdrop.

215. See, for example, C. Tuniz, F. Bernardini, I. Turk, L. Dimkaroski, L. Mancini, and D. Dreossi, "Did Neanderthals Play Music? X-ray Computed Micro-tomography of the Divje Babe 'Flute,'" *Archaeometry* 54, no. 3 (2012): 581–90; see also the subsequent overview by Matija Turk, Ivan Turk, and Marcel Otte, "The Neanderthal Musical Instrument from Divje Babe I Cave (Slovenia): A Critical Review of the Discussion," *Applied Sciences (Switzerland)* 10, no. 4 (2020), e1226, who support the musical interpretation.

216. See, for example, evidence from Gibraltar by Joaquín Rodríguez-Vidal et al., "A Rock Engraving Made by Neanderthals in Gibraltar," *Proceedings of the National Academy of Sciences of the United States of America* 111, no. 37 (2014): 13301–306. Also with respect to a cave bear bone: Ana Majkic, Francesco d'Errico, Stefan Milošević, and Vesna Stepanchuk, "Sequential Incisions on a Cave Bear Bone from the Middle Paleolithic of Pešturina Cave, Serbia," *Journal of Archaeological Method and Theory* 25, no. 1 (2018): 69–116; raven wing bone: Ana Majkic, Sarah Evans, Vadim Stepanchuk, Alexander Tsvelykh, and Francesco d'Errico, "A Decorated Raven Bone from the Zaskalnaya VI (Kolosovskaya) Neanderthal Site, Crimea," *PLoS One* 12, no. 3 (2017), e0173435; and stones: Ana Majkić, Francesco d'Errico, and Vadim Stepanchuk, "Assessing the Significance of Palaeolithic Engraved Cortexes. A Case Study from the Mousterian Site of Kiik-Koba, Crimea," *PLoS One* 13, no. 5 (2018), e0195049. Although in the former case a symbolic interpretation may not be the best default assumption, in the case of the bird bone the regularity of the notches and their mode of implementation are striking.

217. Radiometric dating of carbonate crusts above stencils and the like by D. L. Hoffmann et al., "U-Th Dating of Carbonate Crusts Reveals Neandertal Origin of Iberian Cave Art," *Science* 359, no. 6378 (2018):

912–15; and commentary by Tim Appenzeller, "Europe's First Artists Were Neandertals," *Science* 359, no. 6378 (2018): 852–53. See, however, Randall White et al., "Still No Archaeological Evidence That Neanderthals Created Iberian Cave Art," *Journal of Human Evolution* 144 (2020), 102640, for a much more skeptical view; and response by Dirk L. Hoffmann et al., *Journal of Human Evolution* 144 (2020), 102810.

218. On this potential "proto-figurine," see Jean-Claude Marquet and Michel Lorblanchet, "A Neanderthal Face? The Proto-Figurine from La Roche-Cotard, Langeais (Indreet-Loire, France)," *Antiquity* 77 (2003): 661–70.
219. See, for example, Matteo Romandini, Marco Peresani, Véronique Laroulandie, Laure Metz, Andreas Pastoors, Manuel Vaquero, and Ludovic Slimak, "Convergent Evidence of Eagle Talons Used by Late Neanderthals in Europe: A Further Assessment on Symbolism," *PLoS One* 9, no. 7 (2014), e101278; and Davorka Radovčić, Ankica Oros Sršen, Jakov Radovčić, and David W. Frayer, "Evidence for Neandertal Jewelry: Modified White-Tailed Eagle Claws at Krapina," *PLoS One* 10, no. 3 (2015), e0119802, with the latter specifically suggesting use as jewelry.
220. See, for example, Clive Finlayson et al., "Birds of a Feather: Neanderthal Exploitation of Raptors and Corvids," *PLoS One* 7, no. 9 (2012), e45927; and also Stewart Finlayson, Geraldine Finlayson, Francisco Giles Guzman, and Clive Finlayson, "Neanderthals and the Cult of the Sun Bird," *Quaternary Science Reviews* 217 (2019): 217–24; in addition, see Matteo Romandini, Ivana Fiore, Monica Gala, Martina Cestari, Giuseppe Guida, Antonio Tagliacozzo, and Marco Peresania, "Neanderthal Scraping and Manual Handling of Raptors Wing Bones: Evidence from Fumane Cave. Experimental Activities and Comparison," *Quaternary International* 421 (2016): 154–72.
221. For example, consistent with language are both J. F. Hoffecker, "The Complexity of Neanderthal Technology," *Proceedings of the National Academy of Sciences of the United States of America* 115, no. 9 (2018): 1959–61 (a commentary on Aranguren et al., "Stratigraphy of the Sima de los Huesos," 5–21) and the structure of the hyoid bone. See Ruggero D'Anastasio et al., "Micro-biomechanics of the Kebara 2

Hyoid and Its Implications for Speech in Neanderthals," *PLoS One* 8, (2014), e82261.

222. See, for example, Karen Ruebens, Shannon J. P. McPherron, and Jean-Jacques Hublin, "On the Local Mousterian Origin of the Châtelperronian: Integrating Typo-technological, Chronostratigraphic and Contextual Data," *Journal of Human Evolution* 86 (2015): 55–91.
223. Melissa A. Toups, Andrew Kitchen, Jessica E. Light, and David L. Reed, "Origin of Clothing Lice Indicates Early Clothing Use by Anatomically Modern Humans in Africa," *Molecular Biology and Evolution* 28, no. 1 (2011): 29–32.
224. Marie Soressi et al., "Neandertals Made the First Specialized Bone Tools in Europe," *Proceedings of the National Academy of Sciences of the United States of America* 110, no. 35 (2013): 14186–90.
225. Mark Collard, Lia Tarle, Dennis Sandgathe, and Alexander Allan, "Faunal Evidence for a Difference in Clothing Use between Neanderthals and Early Modern Humans in Europe," *Journal of Anthropological Archaeology* 44 (2016): 235–46. On the other hand, discovery of juvenile leopard remains in a French cave (La Grotte de l'Hortus) in association with a Neanderthal encampment might hint at a fashion statement far ahead of its time? See the monograph by Henry de Lumley, *La grotte moustérienne de l'Hortus (Valflaunès, Hérault)*, Études Quaternaires, Memoire 1 (Aix-en-Provence, France: Université de Provence, 1972).
226. Duilio Garofoli, "Cognitive Archaeology without Behavioral Modernity: An Eliminativist Attempt," *Quaternary International* 405, pt. A (2016): 125–35.
227. See Duilio Garofoli, "Ornamental Feathers without Mentalism: A Radical Enactive View on Neanderthal Body Adornment," in *Embodiment, Enaction, and Culture: Investigating the Constitution of the Shared World*, ed. Christoph Durt, Thomas Fuchs, and Christian Tewes (Cambridge, MA: MIT Press, 2017), 279–305.
228. Mark Nielsen, Michelle C. Langley, Ceri Shipton, and Rohan Kapitány, "*Homo neanderthalensis* and the Evolutionary Origins of Ritual in *Homo sapiens*," *Philosophical Transactions of the Royal Society of London, Series B: Biological Sciences* 375, no. 1805 (2020), 20190424.

229. Simona Petru, "I Remember. Differences between the Neanderthals and Modern Human Mind," *Documenta Praehistorica* 44 (2017): 402–14.
230. See, for example, Frederick L. Coolidge and Thomas Wynn, "A Cognitive and Neuropsychological Perspective on the Châtelperronian," *Journal of Anthropological Research* 60, no. 1 (2004): 55–73.
231. See, for example, C. Philip Beaman, "Modern Cognition in the Absence of Working Memory: Does the Working Memory Account of Neandertal Cognition Work?" *Journal of Human Evolution* 52, no. 6 (2007): 702–6; and reply from Frederick L. Coolidge and Thomas Wynn, "The Working Memory Account of Neandertal Cognition—How Phonological Storage Capacity May Be Related to Recursion and the Pragmatics of Modern Speech," *Journal of Human Evolution* 52, no. 6 (2007): 707–10.
232. See Dan Dediu and Stephen C. Levinson, "Neanderthal Language Revisited: Not Only Us," in "The Evolution of Language," ed. Chris Petkov and William-Marslen-Wilson, special issue, *Current Opinion in Behavioral Sciences* 21 (2018): 49–55.
233. See Mercedes Conde-Valverde et al., "Neanderthals and *Homo sapiens* Had Similar Auditory and Speech Capacities," *Nature Ecology and Evolution* 5, no. 5 (2021): 609–15.
234. Brian Hayden, "Neanderthal Social Structure?" *Oxford Journal of Archaeology* 31, no. 1 (2012): 1–26.
235. See the overview by Thomas Wynn, Karenleigh A. Overmann, and Frederick L. Coolidge, "The False Dichotomy: A Refutation of the Neandertal Indistinguishability Claim," *Journal of Anthropological Sciences* 94 (2016): 201–21.
236. In support of differences see, for example, Eiluned Pearce, Chris Stringer, and R. I. M. Dunbar, "New Insights into Differences in Brain Organization between Neanderthals and Anatomically Modern Humans," *Proceedings of the Royal Society of London, Series B: Biological Sciences* 280, no. 1758 (2013), 20130168; versus similarities as argued by Marcia S. Ponce de León, Thibaut Bienvenu, Takeru Akazawa, and Christoph P. E. Zollikofer, "Brain Development Is Similar in Neanderthals and Modern Humans," *Current Biology* 26, no.14 (2016): R665–66. Yet other studies, including one looking at the parietal lobe, see differences but more of degree; see Ana Sofia Pereira-Pedro, Emiliano Bruner,

Philipp Gunz, and Simon Neubauer, "A Morphometric Comparison of the Parietal Lobe in Modern Humans and Neanderthals," *Journal of Human Evolution* 142 (2020), 102770.

237. See Sarah Traynor, Alia N. Gurtov, Jess Hutton Senjem, and John Hawks, "Assessing Eye Orbits as Predictors of Neandertal Group Size," *American Journal of Physical Anthropology* 157, no. 4 (2015): 680–83; and the subsequent discussion in Robin Dunbar, Eiluned Pearce, and Chris Stringer in *American Journal of Physical Anthropology* 159, no. 2 (201): 358–60, and the authors' reply (361). A reemphasis on differences in cortical structure is found in Antonio García-Tabernero, Angel Peña-Melián, and Antonio Rosas, "Primary Visual Cortex in Neandertals as Revealed from the Occipital Remains from the El Sidrón Site, with Emphasis on the New SD-2300 Specimen," *Journal of Anatomy* 233, no. 1 (2018): 33–45.

238. See, for example, documentation from Krapina in Carles Lalueza Fox and David W. Frayer, "Non-dietary Marks in the Anterior Dentition of the Krapina Neanderthals," *International Journal of Osteoarchaeology* 7, no. 2 (1997): 133–49.

239. Almudena Estalrrich and Antonio Rosas, "Handedness in Neandertals from the El Sidrón (Asturias, Spain): Evidence from Instrumental Striations with Ontogenetic Inferences," *PLoS One* 8, no. 5 (2013), e62797.

240. Kristin L. Krueger, Peter S. Ungar, Debbie Guatelli-Steinberg, Jean-Jacques Hublin, Alejandro Pérez-Pérez, Erik Trinkaus, and John C. Willman, "Anterior Dental Microwear Textures Show Habitat-Driven Variability in Neandertal Behavior," *Journal of Human Evolution* 105 (2017): 13–23.

241. Emiliano Bruner and Atsushi Iriki, "Extending Mind, Visuospatial Integration, and the Evolution of the Parietal Lobes in the Human Genus," *Quaternary International* 405, pt. A (2016): 98–110.

242. Marco Langbroek, "Ice Age Mentalists: Debating Neurological and Behavioural Perspectives on the Neandertal and Modern Mind," *Journal of Anthropological Sciences* 92 (2014): 288–89. See also a subsequent commentary by Duilio Garofoli, "Neandertal Three Hands: Epistemological Foundations and a Theory of Visual Impedance," *Journal of Anthropological Sciences* 93 (2015): 185–90.

243. Marco Langbroek, "Trees and Ladders: A Critique of the Theory of Human Cognitive and Behavioural Evolution in Palaeolithic Archaeology," *Quaternary International* 270 (2012): 4–14, with commentary in Duilio Garofoli in *Quaternary International* 299 (2013): 116–18, and the author's reply (119–20).

CHAPTER 5

1. Robert Wokler, "Perfectible Apes in Decadent Cultures: Rousseau's Anthropology Revisited," *Daedalus* 107, Summer (1978): 107–34.
2. See, for example, Michael Tomasello, *Becoming Human: A Theory of Ontogeny* (Cambridge, MA: Harvard University Press, 2019). Very many of the points I discuss in this chapter are covered by Tomasello, but as will be apparent his effectively recapitulatory arguments that human cognition is an ontogenetic extension of the ape counterparts are ultimately built on the promissory notes of materialism.
3. Jordi Vallverdú et al., "Slime Mould: The Fundamental Mechanisms of Biological Cognition," *BioSystems* 165 (2018): 57–70, and corrigendum, 166 (2019): 66; also Jules Smith-Ferguson and Madeleine Beekman, "Who Needs a Brain? Slime Moulds, Behavioural Ecology and Minimal Cognition," *Adaptive Behavior* 28, no. 6 (2020): 465–78.
4. Nicholas P. Money, "Hyphal and Mycelial Consciousness: The Concept of the Fungal Mind," *Fungal Biology* 125, no. 4 (2021): 257–59.
5. Derek C. Penn, Keith J. Holyoak, and Daniel J. Povinelli, "Darwin's Mistake: Explaining the Discontinuity between Human and Nonhuman Minds," *Behavioral and Brain Sciences* 31, no. 2 (2008): 109–30, with a collection of critiques (130–53), and the authors' response, "Darwin's Triumph: Explaining the Uniqueness of the Human Mind without a *Deus ex Machina*," 31, no. 2 (2008): 153–69, followed by collected references (169–78). Many authors explore similar territory in one way or another; another useful overview on the question of human uniqueness includes Sara J. Shettleworth, "Modularity, Comparative Cognition and Human Uniqueness," *Philosophical Transactions of the Royal Society of London, Series B: Biological Sciences* 367, no. 1603 (2012): 2794–802.
6. See Jerome Bruner's response, "*Homo sapiens*, a Localized Species," *Behavioral and Brain Sciences* 28, no. 5 (2005): 694–95, to Michael Toma-

sello, Malinda Carpenter, Josep Call, Tanya Behne, and Henrike Moll, "Understanding and Sharing Intentions: The Origins of Cultural Cognition," *Behavioral and Brain Sciences* 28, no. 5 (2005): 675–91.

7. Francesco Ferretti, "The Social Brain Is Not Enough: On the Importance of the Ecological Brain for the Origin of Language," *Frontiers in Psychology* 7 (2016), 1138, https://doi.org/10.3389/fpsyg.2016.01138.
8. I have not been able to locate the actual source, although Dobzhansky repeatedly makes reference to human uniqueness. Yet, not surprisingly, this is widely quoted. For instance, see David Premack and Ann James Premack, "Why Animals Have Neither Culture nor History," in *Companion Encyclopedia of Anthropology: Humanity, Culture and Social Life*, ed. Tim E. Ingold (Routledge, 1994), 350–65, p. 350n11.
9. Thomas Suddendorf, *The Gap: The Science of What Separates Us from Other Animals* (New York: Basic Books, 2013), 278.
10. And if the 4,300-year-old stone tools are anything to go by, no change in their lithics either; see Julio Mercader, Huw Barton, Jason Gillespie, Jack Harris, Steven Kuhn, Robert Tyler, and Christophe Boesch, "4,300-Year-Old Chimpanzee Sites and the Origins of Percussive Stone Technology," *Proceedings of the National Academy of Sciences of the United States of America* 104, no. 9 (2007): 3043–48. For an update, see T. Proffitt, M. Haslam, J. F. Mercader, C. Boesch, and L. V. Luncz, "Revisiting Panda 100, the First Archaeological Chimpanzee Nut-Cracking Site," *Journal of Human Evolution* 124 (2018): 117–39. Importantly they conclude that these chimpanzee lithics have little bearing on the earliest tools made by hominins.
11. Premack and Premack, "Why Animals," 350–65.
12. Jonathan Marks, "The Biological Myth of Human Evolution," *Contemporary Social Science* 7, no. 2 (2012): 139–57.
13. See, for example (with reference to the underpinnings of logical reasoning), Nicolò Cesana-Arlotti, Ana Martín, Ernő Téglás, Liza Vorobyova, Ryszard Cetnarski, and Luca L. Bonatti, "Precursors of Logical Reasoning in Preverbal Human Infants," *Science* 359, no. 6381 (2018): 1263–66; with commentary by Justin Halberda, "Logic in Babies," *Science* 359, no. 6381 (2018): 1214–15.

14. Rosemary A. Varley and Michael Siegal, "Evidence for Cognition without Grammar from Causal Reasoning and 'Theory of Mind' in an Agrammatic Aphasic Patient," *Current Biology* 10, no. 12 (2000): 723–25; and Rosemary A. Varley, Michael Siegal, and S. C. Want, "Severe Impairment in Grammar Does Not Preclude Theory of Mind," *Neurocase* 7, no. 6 (2001): 489–93.
15. As just one example of chirpy self-confidence, pertaining to continuity in cultures, see Andrew Whiten, Robert A. Hinde, Kevin N. Laland, and Christopher B. Stringer, "Culture Evolves," *Philosophical Transactions of the Royal Society of London, Series B: Biological Sciences* 366, no. 1567 (2011): 938–48; see also Paul Carron, "Ape Imagination? A Sentimentalist Critique of Frans de Waal's Gradualist Theory of Human Morality," *Biology and Philosophy* 33 (2018), 22, https://doi.org/10.1007/s10539-018-9632-4.
16. Daniel Povinelli and Jennifer Vonk, "Chimpanzee Minds: Suspiciously Human?" *Trends in Cognitive Sciences* 7, no. 4 (2003): 157–60; see also Daniel John Povinelli, "Behind the Ape's Appearance: Escaping Anthropocentrism in the Study of Other Minds," *Daedalus* 133, no. 1 (2004): 29–41.
17. An overview by Derek C. Penn and Daniel J. Povinelli, "The Human Enigma," in *The Oxford Handbook of Thinking and Reasoning*, ed. Keith J. Holyoak and Robert G. Morrison (New York: Oxford University Press, 2012), 529–42.
18. Stefano Ghirlanda, Johan Lind, and Magnus Enquist, "Memory for Stimulus Sequences: A Divide between Humans and Other Animals?" *Royal Society Open Science* 4, no. 6 (2017), 161011, https://doi.org/10.1098/rsos.161011.
19. Joël Fagot and Anaïs Maugard, "Analogical Reasoning in Baboons (*Papio papio*): Flexible Encoding of the Source Relation Depending on the Target Relation," *Learning and Behavior* 41, no. 3 (2013): 229–37.
20. Joël Fagot and Roger K. R. Thompson, "Generalized Relational Matching by Guinea Baboons (*Papio papio*) in Two-by-Two-Item Analogy Problems," *Psychological Science* 22, no. 10 (2011): 1304–309.
21. Yet in trials involving prospective tool selection with three male and three female Goffin's cockatoos, only the latter succeeded, even as one

male (Doolittle) plugged on through fifty-three training sessions. Even so, the females were hardly fast off the mark; see I. Beinhauer, T. Bugnyar, and A. M. I. Auersperg, "Prospective but Not Retrospective Tool Selection in the Goffin's Cockatoo (*Cacatua goffiniana*)," *Behaviour* 156, no. 5–8 (2019): 633–59.

22. So in a study to show mental representations in New Caledonian crows there were three sets of ingenious experiments of increasing difficulty. Eleven crows reported for duty; in experiment 1, four succeeded quickly, but four more required many more trials. These eight mental athletes then tackled experiment 2, but took between twenty-five and sixty-nine trials. One then dropped out, so the final hurdle was only offered to three crows, with success in fewer than twenty trials. It is not quite clear, however, what the total number of trials was; more importantly, these procedures refer to stick tools, which are part and parcel of the everyday life of these crows. Substituting stones led to a much weaker performance. Romana Gruber et al., "New Caledonian Crows Use Mental Representations to Solve Metatool Problems," *Current Biology* 29, no. 4 (2019): 686–92; with commentary by Thomas Bugnyar, "Tool Use: New Caledonian Crows Engage in Mental Planning," *Current Biology* 29, no. 6 (2019): R200–R202. See also the critical overview by Jennifer Vonk, "Sticks and Stones: Associative Learning Alone?" *Learning and Behavior* 48, no. 3 (2020): 277–78.
23. See the trenchant overview by Alex Thornton and Dieter Lukas, "Individual Variation in Cognitive Performance: Developmental and Evolutionary Perspectives," *Philosophical Transactions of the Royal Society of London, Series B: Biological Sciences* 367, no. 1603 (2012): 2773–83.
24. See, for example, the trials and tribulations of Sherman, Lana, and their fellow chimpanzees in Michael J. Beran, Mattea S. Rossettie, and Audrey E. Parrish, "Trading Up: Chimpanzees (*Pan troglodytes*) Show Self-Control through their Exchange Behavior," *Animal Cognition* 19, no. 1 (2016): 109–21.
25. Krist Vaesen, "Chimpocentrism and Reconstructions of Human Evolution," *Studies in the History and Philosophy of Science, Part C: Studies in History and Philosophy of Biological and Biomedical Sciences* 45, no. 1 (2014): 12–21.

26. See, for example, Louise Barrett, "A Better Kind of Continuity," *Southern Journal of Philosophy* 53, Supplement 1 (2015): 28–49.
27. That the horse deserves most of the credit is amusingly relayed by Matt Cartmill, "Human Uniqueness and Theoretical Content in Paleoanthropology," *International Journal of Primatology* 11 (1990): 173–91.
28. See the overview by Jesse M. Bering, "A Critical Review of the 'Enculturation Hypothesis': The Effects of Human Rearing on Great Ape Social Cognition," *Animal Cognition* 7, no. 4 (2004): 201–12.
29. See Josep Call and Michael Tomasello, "The Effect of Humans on the Cognitive Development of Apes," in *Reaching into Thought: The Minds of the Great Apes,* ed. Anne E. Russon, Kim A. Bard, and Sue Taylor Parker (Cambridge: Cambridge University Press, 1996), 371–403.
30. Louise Barrett, "Why Brains Are Not Computers, Why Behaviorism Is Not Satanism, and Why Dolphins Are Not Aquatic Apes," *Behavior Analyst* 39, no. 1 (2016): 9–23, p. 10.
31. Christopher David Bird and Nathan John Emery, "Rooks Use Stones to Raise the Water Level to Reach a Floating Worm," *Current Biology* 19, no. 16 (2009): 1410–14; and commentary by Alex H. Taylor and Russell D. Gray, "Animal Cognition: Aesop's Fable Flies from Fiction to Fact," *Current Biology* 19, no. 16 (2009): R731–32.
32. See, for example, studies using the cognitive poster-child of New Caledonian crows by Sarah A. Jelbert, Alex H. Taylor, Lucy G. Cheke, Nicola S. Clayton, and Russell D. Gray, "Using the Aesop's Fable Paradigm to Investigate Causal Understanding of Water Displacement by New Caledonian Crows," *PLoS One* 9, no. 3 (2014), e92895, https://doi.org/10.1371/journal.pone.0092895; and Corina J. Logan, Sarah A. Jelbert, Alexis J. Breen, Russell D. Gray, and Alex H. Taylor, "Modifications to the Aesop's Fable Paradigm Change New Caledonian Crow Performances," *PLoS One* 9, no. 7 (2014), e103049, https://doi.org/10.1371/journal.pone.0103049.
33. Stefano Ghirlanda and Johan Lind, "'Aesop's Fable' Experiments Demonstrate Trial-and-Error Learning in Birds, but No Causal Understanding," *Animal Behaviour* 123 (2017): 239–47. Along the same theme, see the results with Western scrub-jays: Corina J. Logan, "Behavioral Flexibility and Problem Solving in an Invasive Bird," *PeerJ* 4 (2016), e1707,

https://doi.org/10.7717/peerj.1975; with kea: R. Schwing, F. Weiss, A. Tichy, and G. Gajdon, "Testing the Causal Understanding of Water Displacement by Kea (*Nestor notabilis*)," *Behaviour* 156, no. 5–8 (2019): 447–78; and with azure-winged magpies: Yigui Zhang, Cong Yu, Lixin Chen, and Zhongqiu Li, "Performance of Azure-Winged Magpies in Aesop's Fable Paradigm," *Scientific Reports* 11 (2021), 804, https://doi.org/10.1038/s41598-020-80452-5. All point to a failure by the birds to grasp causalities.

34. P. D. Neilands, S. A. Jelbert, A. J. Breen, M. Schiestl, and A. H. Taylor, "How Insightful Is 'Insight'? New Caledonian Crows Do Not Attend to Object Weight during Spontaneous Stone Dropping," *PLoS One* 11, no. 12 (2016), e0167419, https://doi.org/10.1371/journal.pone.0167419.
35. See the meta-analysis by Laura Hennefield, Hyesung G. Hwang, Sara J. Weston, and Daniel J. Povinelli, "Meta-analytic Techniques Reveal That Corvid Causal Reasoning in the Aesop's Fable Paradigm Is Driven by Trial-and-Error Learning," *Animal Cognition* 21, no. 6 (2018): 735–48.
36. Alex H. Taylor et al., "Of Babies and Birds: Complex Tool Behaviours Are Not Sufficient for the Evolution of the Ability to Create a Novel Causal Intervention," *Proceedings of the Royal Society of London, Series B: Biological Sciences* 281 (2014), 20140837, https://doi.org/10.1098/rspb.2014.0837.
37. See, for example, Bernd Heinrich and Thomas Bugnyar, "Testing Problem Solving in Ravens: String-Pulling to Reach Food," *Ethology* 111, no. 10 (2005): 962–76.
38. See, for example, Alex H. Taylor, Felipe S. Medina, Jennifer C. Holzhaider, Lindsay J. Hearne, Gavin R. Hunt, and Russell D. Gray, *PLoS One* 5, no. 2 (2010), e9345, https://doi.org/10.1371/journal.pone.0009345; also with crossed-strings where the Goffin's cockatoos never got the hang of things, see Birgit Wakonig, Alice M. Auersperg, and Mark O'Hara, "String-Pulling in the Goffin's Cockatoo (*Cacatua goffiniana*)," *Learning and Behavior* 49, no. 1 (2021): 124–36.
39. Alex H. Taylor, Brenna Knaebe, and Russell D. Gray, "An End to Insight? New Caledonian Crows Can Spontaneously Solve Problems without Planning Their Actions," *Proceedings of the Royal Society of London, Series B: Biological Sciences* 279, no. 1749 (2012): 4977–81; also commentary by

Amanda M. Seed and Neeltje J. Boogert, "Animal Cognition: An End to Insight?" *Current Biology* 23, no. 2 (2013): R67–R69.

40. For a balanced appraisal, see Amanda Seed and Richard Byrne, "Animal Tool-Use," *Current Biology* 20, no. 23 (2010): R1032–39.
41. See, for example, the trials with macaws, where the birds can be trained to undertake particular tasks, but as the experimenters upped the ante, the cognitive blinds came down. Laurie O'Neill, Anthony Picaud, Jana Maehner, Manfred Gahr, and Auguste M. P. von Bayern, "Two Macaw Species Can Learn to Solve an Optimised Two-Trap Problem, but without Functional Causal Understanding," *Behavior* 156, no. 5–8 (2019): 691–720.
42. See, for example, the overview by Derek C. Penn and Daniel J. Povinelli, "Causal Cognition in Human and Nonhuman Animals: A Comparative, Critical Review," *Annual Review of Psychology* 58 (2007): 97–118; also Jennifer Vonk and Francys Subiaul, "Do Chimpanzees Know What Others Can and Cannot Do? Reasoning about 'Capability,'" *Animal Cognition* 12 (2009): 267–86.
43. See, for example, the overview by Juiane Bräuer, Daniel Hanus, Simone Pika, Russell Gray, and Natalie Uomini, "Old and New Approaches to Animal Cognition: There Is Not 'One Cognition,'" *Journal of Intelligence* 8, no. 3 (2020), 28, https://dx.doi.org/10.3390/jintelligence8030028; also the previously cited essay by Barrett, "Why Brains Are Not Computers."
44. See, for example, Ken Cheng, "Cognition Beyond Representation: Varieties of Situated Cognition in Animals," *Comparative Cognition and Behavior Reviews* 13 (2018): 1–20; but also critiques such as by Stephen D. Goldinger, Megan H. Papesh, Anthony S. Barnhart, and Michael C. Hout, "The Poverty of Embodied Cognition," *Psychonomic Bulletin and Review* 23 (2016): 959–78.
45. David Premack, "Why Humans Are Unique: Three Theories," *Perspectives on Psychological Science* 5, no. 1 (2010): 22–32. This specific example of how an animal sees things is only one section of a much wider survey.
46. See Hjalmar S. Kühl et al., "Chimpanzee Accumulative Stone Throwing," *Scientific Reports* 6 (2016), 22219, https://doi.org/10.1038/srep22219.
47. Michael J. Beran and Audrey E. Parrish, "Non-human Primate Token Use Shows Possibilities but Also Limitations for Establishing a Form of

Currency," *Philosophical Transactions of the Royal Society of London, Series B: Biological Sciences* 376, no. 1819 (2021): 20190675, https://doi.org/10.1098/rstb.2019.0675.

48. See, for example, Alicia P. Melis, Brian Hare, and Michael Tomasello, "Engineering Cooperation in Chimpanzees: Tolerance Constraints on Cooperation," *Animal Behaviour* 72, no. 2 (2006): 275–86.
49. As Michael Tomasello, Malinda Carpenter, Josep Call, Tanya Behne, and Henrike Moll remark in "Understanding and Sharing Intentions: The Origins of Cultural Cognition," *Behavioral and Brain Sciences* 28 (2005): 675–91, p. 675; with commentaries (691–721), the authors' reply (721–27), and collected references (727–35).
50. See the exploration of the Sartrean theme by Jesse M. Bering, "Why Hell Is Other People: Distinctively Human Psychological Suffering," *Review of General Psychology* 12, no. 1 (2008): 1–8.
51. Marks, "Biological Myth of Human Evolution."
52. For an early overview, see C. M. Heyes, "Theory of Mind in Nonhuman Primates," *Behavioral and Brain Sciences* 21, no. 1 (1998): 101–14, with extended commentary (115–134), the author's reply (134–44), and collected references (144–48).
53. See the overview by Krist Vaesen, "The Cognitive Bases of Human Tool Use," *Behavioral and Brain Sciences* 35, no. 4 (2012): 203–18, with critiques (218–44), the author's response (244–52), and collected references (252–62).
54. See the wide-ranging review of the discontinuities that divide human and animal by Penn and Povinelli, "Human Enigma."
55. This curiosity, *Flatland: A Romance of Many Dimensions* (London: Seeley & Co., 1884), was originally published under the ironic pseudonym of A. Square. This book is more than simply an exploration of how two-dimensional beings would "encounter" a third dimension, but also an acute social critique.
56. See V. Petkov, ed., *Space and Time: Minkowski's Papers on Relativity*, trans. Fritz Lewertoff and Vesselin Petkov (Montreal: Minkowski Institute Press, 2012); also Peter Louis Galison's overview in "Minkowski's Space-Time: From Visual Thinking to the Absolute World," *Historical Studies in the Physical Sciences* 10 (1979): 85–121.

57. Catherine Crockford, Roman M. Wittig, Roger Mundry, and Klaus Zuberbühler, "Wild Chimpanzees Inform Ignorant Group Members of Danger," *Current Biology* 22, no. 2 (2012): 142–46, and commentary by Robert M. Seyfarth and Dorothy L. Cheney, "Animal Cognition: Chimpanzee Alarm Calls Depend on What Others Know," *Current Biology* 22, no. 2 (2012): R51–R52.
58. Alicia P. Melis, Brian Hare, and Michael Tomasello, "Chimpanzees Recruit the Best Collaborators," *Science* 311, no. 5765 (2006): 1297–300.
59. Annie Bissonnette, Mathias Franz, Oliver Schülke, and Julia Ostner, "Socioecology, but Not Cognition, Predicts Male Coalitions across Primates," *Behavioral Ecology* 25, no. 4 (2014): 794–801.
60. See Craig B. Stanford's *Chimpanzee and Red Colobus: The Ecology of Predator and Prey* (Cambridge, MA: Harvard University Press, 1998); also John C. Mitani and David P. Watts, "Why Do Chimpanzees Hunt and Share Meat?" *Animal Behaviour* 61, no. 5 (2001): 915–24.
61. Anke F. Bullinger, Emily Wyman, Alicia P. Melis, and Michael Tomasello, "Coordination of Chimpanzees (*Pan troglodytes*) in a Stag Hunt Game," *International Journal of Primatology* 32 (2011): 1296–310.
62. See Michael Tomasello, "Why Don't Apes Point?" in *Roots of Human Sociality: Culture, Cognition and Interaction*, ed. Nicholas J. Enfield and Stephen C. Levinson (Oxford: Berg, 2006), 506–30.
63. Henrike Moll and Michael Tomasello, "Cooperation and Human Cognition: The Vygotskian Intelligence Hypothesis," *Philosophical Transactions of the Royal Society of London, Series B: Biological Sciences* 362, no. 1480 (2007): 639–48.
64. See, for example, Kevin Laland, Clive Wilkins, and Nicky Clayton, "The Evolution of Dance," *Current Biology* 26, no. 1 (2016): R5–R9.
65. Ralf Stoecker, "Why Animals Can't Act," *Inquiry (Oslo)* 52, no. 3 (2009): 255–71.
66. That bees can "dance" is not in dispute, but this is only one way by which information is conveyed, and the dance is often ignored in favor of other indications of suitable localities; see Christoph Grüter, M. Sol Balbuena, and Walter M. Farina, "Informational Conflicts Created by the Waggle Dance," *Proceedings of the Royal Society of London, Series B: Biological Sciences* 275, no. 1640 (2008): 1321–27.

67. This point and its implications for the gulf between ourselves and animals is frequently made. See, for example, Daniel J. Povinelli's *World without Weight: Perspectives on an Alien Mind* (Oxford: Oxford University Press, 2012).
68. Jonathan Redshaw and Thomas Suddendorf, "Children's and Apes' Preparatory Responses to Two Mutually Exclusive Possibilities," *Current Biology* 26, no. 13 (2016): 1758–62, and commentary by Amanda M. Seed and Katherine L. Dickerson, "Future Thinking: Children but Not Apes Consider Multiple Possibilities," *Current Biology* 26, no. 13 (2016): R525–27.
69. See, for example, a case involving baboons by Timothy M. Flemming, Roger K. R. Thompson, and Joël Fagot, "Baboons, Like Humans, Solve Analogy by Categorical Abstraction of Relations," *Animal Cognition* 16, no. 3 (2013): 519–24.
70. This is a much rehearsed area, but see, for example, Derek C. Penn and Daniel J. Povinelli, "On the Lack of Evidence That Non-human Animals Possess Anything Remotely Resembling a 'Theory of Mind,'" *Philosophical Transactions of the Royal Society of London, Series B: Biological Sciences* 362, no. 1480 (2007): 731–44; and Derek C. Penn and Daniel J. Povinelli, "Causal Cognition in Human and Nonhuman Animals: A Comparative, Critical Review," *Annual Review of Psychology* 58 (2007): 97–118.
71. See the overview by Philippe Rochat and Dan Zahavi, "The Uncanny Mirror: A Re-framing of Mirror Self-experience," *Consciousness and Cognition* 20 (2011): 204–13.
72. M. van Buuren, A. Auersperg, G. Gajdon, S. Tebbich, and A. von Bayern, "No Evidence of Mirror Self-recognition in Keas and Goffin's Cockatoos," *Behaviour* 156, no. 5–8 (2019): 763–81, who also address Goffin's cockatoos.
73. See the crisp overview by Eric Saidel, "Through the Looking Glass, and What We (Don't) Find There," *Biology and Philosophy* 31 (2016): 335–52.
74. Tom Suddendorf and Michael Corballis, "The Evolution of Foresight: What Is Mental Time Travel and Is It Unique to Humans?" *Behavioral and Brain Sciences* 30 (2007): 299–313.
75. See, for example, Catherine Crockford, Roman M. Wittig, Roger Mundry, and Klaus Zuberbühler, "Wild Chimpanzees Inform Ignorant

Group Members of Danger," *Current Biology* 22, no. 2 (2012): 142–46, and commentary by Robert M. Seyfarth and Dorothy L. Cheney, "Animal Cognition: Chimpanzee Alarm Calls Depend on What Others Know," *Current Biology* 22, no. 2 (2012): R51–R52.

76. See Anke F. Bullinger, Alicia P. Melis, and Michael Tomasello, "Chimpanzees (*Pan troglodytes*) Instrumentally Help but Do Not Communicate in a Mutualistic Cooperative Task," *Journal of Comparative Psychology* 128, no. 3 (2014): 251–60.
77. Joan P. Silk et al., "Chimpanzees Are Indifferent to the Welfare of Unrelated Group Members," *Nature* 437, no. 7063 (2005): 1357–59.
78. Giulia Sirianni, Roger Mundry, and Christophe Boesch, "When to Choose Which Tool: Multidimensional and Conditional Selection of Nut-Cracking Hammers in Wild Chimpanzees," *Animal Behaviour* 100 (2015): 152–65.
79. See, for example, Daniel J. Povinelli, Jesse M. Bering, and Steve Giambrone, "Toward a Science of Other Minds: Escaping the Argument by Analogy," *Cognitive Science* 24, no. 3 (2000): 509–41.
80. See, for example, Daniel J. Povinelli and Jennifer Vonk, "Chimpanzee Minds: Suspiciously Human?" *Trends in Cognitive Sciences* 7, no. 4 (2003): 157–60.
81. See, for example, Luca Surian, Stefania Caldi, and Dan Sperber, "Attribution of Beliefs by 13-Month-Old Infants," *Psychological Science* 18, no. 7 (2007): 580–86; and Renée Baillargeau, Rose M. Scott, and Zijing He, "False-Belief Understanding in Infants," *Trends in Cognitive Sciences* 14, no. 3 (2010): 110–18.
82. The arguments go back and forth; see, for example, B. Hare, "From Hominoid to Hominid Mind: What Changed and Why?" *Annual Review of Anthropology* 40 (2011): 293–309; Josep Call and Michael Tomasello, "Does the Chimpanzee Have a Theory of Mind? 30 Years Later," *Child Development* 70, no. 5 (1999): 381–95; Daniel J. Povinelli and Jennifer Vonk, "We Don't Need a Microscope to Explore the Chimpanzee's Mind," *Mind and Language* 19, no. 1 (2004): 1–28; and Cecilia Heyes, "Apes Submentalise," *Trends in Cognitive Sciences* 21, no. 1 (2017): 1–2, as well as her earlier and telling overview: Cecilia Heyes, "Theory of Mind in Nonhuman Primates," *Behavioral and Brain Sciences* 21, no. 1

(1998): 101–14. More recently there has been the trenchant overview by Daniel J. Horschler, Evan L. MacLean, and Laurie R. Santos, "Do Non-Human Primates Really Represent Others' Beliefs?" *Trends in Cognitive Sciences* 24, no. 8 (2020): 594–605, and follow-up remarks, "Advancing Gaze-Based Research on Primate Theory of Mind," *Trends in Cognitive Sciences* 24, no. 8 (2020): 778–79, on the employment of so-called anticipatory looking that, although hailed as a key development in the recognition of nonhuman false belief, appears to show serious weaknesses.

83. Kristin Andrews, "Apes Track False Beliefs but Might Not Understand Them," *Learning and Behavior* 46, no. 1 (2018): 3–4; see also, for example, Drew C. W. Marticorena, April M. Ruiz, Cora Mukerji, Anna Goddu, and Laurie R. Santos, "Monkeys Represent Others' Knowledge but Not Their Beliefs," *Developmental Science* 14, no. 6 (2011): 1406–16.
84. Juliane Kaminski, Joseph Call, and Michael Tomasello, "Chimpanzees Know What Others Know, but Not What They Believe," *Cognition* 109, no. 2 (2008): 224–34.
85. Katja Karg, Martin Schmelz, Josep Call, and Michael Tomasello, "Differing Views: Can Chimpanzees Do Level 2 Perspective-Taking?" *Animal Cognition* 19, no. 3 (2016): 555–64; also the trenchant remarks by Daniel J. Povinelli and Jesse M. Bering, "The Mentality of Apes Revisited," *Current Directions in Psychological Science* 11, no. 4 (2002): 115–19.
86. See David Buttelmann, Malinda Carpenter, Josep Call, and Michael Tomasello, "Chimpanzees, *Pan troglodytes,* Recognize Successful Actions, but Fail to Imitate Them," *Animal Behaviour* 86 (2013): 755–61.
87. Eric M. Anderson, Yin-Juei Chang, Susan Hespos, and Dedre Gentner, "Comparison within Pairs Promotes Analogical Abstraction in Three-Month-Olds," *Cognition* 176 (2018): 74–86.
88. Douglas Hofstadter and Emmanuel Sander, *Surfaces and Essences: Analogy as the Fuel and Fire of Thinking* (New York: Basic Books, 2013). As others have, they also point out that human analogical thinking is very far from most formulations of artificial cognition. If you have a few minutes to spare, turn to the index and check out the entry "angel stung by bumblebee."
89. Yes, the chimpanzee Sarah is famous for her capacity for a variety of analogical thinking, but her earlier experience of dealing with tokens was

evidently an essential prerequisite for the former capacity and has effectively no parallel among other chimpanzees. See the chapter by David L. Oden, Roger K. R. Thompson, and David Premack, "Can an Ape Reason Analogically? Comprehension and Production of Analogical Problems by Sarah, a Chimpanzee (*Pan troglodytes*)," in *The Analogical Mind: Perspectives from Cognitive Science,* ed. Dedre Gentner, Keith J. Holyoak, and Boicho N. Kokinov (Cambridge, MA: MIT Press, 2001), 471–98.

90. Daphna Buchsbaum, Sophie Bridgers, Deena Skolnick Weisberg, and Alison Gopnik, "The Power of Possibility: Causal Learning, Counterfactual Reasoning, and Pretend Play," *Philosophical Transactions of the Royal Society of London, Series B: Biological Sciences* 367, no. 1599 (2012): 2202–12. Also, in the more general context of human cognition, see the overview by Michael Tomasello and Hannes Rakoczy, "What Makes Human Cognition Unique? From Individual to Shared to Collective Intentionality," *Mind and Language* 18, no. 2 (2003): 121–47.
91. See the chapter by Juan-Carlos Gómez and Beatriz Martín-Andrade, "Fantasy Play in Apes," in *The Nature of Play: Great Apes and Humans,* ed. Anthony D. Pellegrini and Peter K. Smith (New York: Guilford Press, 2005), 139–72.
92. From Robert Browning's poem "A Grammarian's Funeral: Shortly after the Revival of Learning in Europe."
93. See, for example, Gema Martin-Ordas, Dorthe Berntsen, and Josep Call, "Memory for Distant Past Events in Chimpanzees and Orangutans," *Current Biology* 23, no. 15 (2013): 1438–41.
94. For a balanced overview, see Seed and Byrne, "Animal Tool-Use."
95. Jordi Pladevall, Natacha Mendes, David Riba, Miquel Llorente, and Federica Amici, "No Evidence of What-Where-When Memory in Great Apes (*Pan troglodytes, Pan paniscus, Pongo abelii,* and *Gorilla gorilla*)," *Journal of Comparative Psychology* 134, no. 2 (2020): 252–61.
96. As William A. Roberts asks, "Are Animals Stuck in Time?" *Psychological Bulletin* 128, no. 3 (2002): 473–89.
97. Not that this is cognitively trivial; see Cecilia Heyes, "Simple Minds: A Qualified Defence of Associative Learning," *Philosophical Transactions of the Royal Society of London, Series B: Biological Sciences* 367, no. 1603 (2012): 2695–703; and Anthony Dickinson, "Associative Learning and

Animal Cognition," *Philosophical Transactions of the Royal Society of London, Series B: Biological Sciences* 367, no. 1603 (2012): 2733–42. Experiments that can disentangle associative learning from more cognitively sophisticated sorts of learning are demanding, but one hallmark of the former is that allowing more time to help solve a problem is of no benefit; see Johan Lind, Sofie Lönnberg, Tomas Persson, and Magnus Enquist, "Time Does Not Help Orangutans *Pongo abelii* Solve Physical Problems," *Frontiers in Psychology* 8 (2017), 161, https://doi.org/10.3389/fpsyg.2017.00161.

98. See Johan Lind, "What Can Associative Learning Do for Planning?" *Royal Society Open Science* 5 (2018), 180778, https://doi.org/10.1098/rsos.180778.
99. See, for example, Thomas Suddendorf and Michael C. Corballis, "The Evolution of Foresight: What Is Mental Time Travel, and Is It Unique to Humans?" *Behavioral and Brain Sciences* 30, no. 3 (2007): 299–313, with commentaries (313–35), the authors' reply (335–45), and collected references (345–51); and Thomas Suddendorf and Michael C. Corballis, "Behavioural Evidence for Mental Time Travel in Nonhuman Animals," *Behavioural Brain Research* 215, no. 2 (2010): 292–98.
100. A comment made by Mathias Osvath and Peter Gärdenfors, "Oldowan Culture and the Evolution of Anticipatory Cognition," *Lund University Cognitive Science* 122 (2005): 1–16, in their much wider overview of the emergence of anticipatory cognition.
101. Metaphors respectively of the savage winters that were the beginning of the end for two of the many monsters of European history, Napoleon and Hitler, and the words from the Christmas carol "Good King Wenceslas."
102. Sen Cheng, Markus Werning, and Thomas Suddendorf, "Dissociating Memory Traces and Scenario Construction in Mental Time Travel," *Neuroscience and Biobehavioral Reviews* 60 (2016): 82–89.
103. See the chapter by Jonathan Redshaw and Adam Bulley, "Future-Thinking in Animals: Capacities and Limits," in *The Psychology of Thinking about the Future,* ed. Gabriele Oettingen, A. Timur Sevincer, Peter M. Gollwitzer, Francis T. Anderson, and Cristina M. Atance (New York: Guilford Press, 2018), 31–51.

104. See the chapter by Ulric Neisser, "Memory with a Grain of Salt," in *Memory: An Anthology*, ed. Harriet Harvey Wood and A. S. Byatt (London: Chatto and Windus, 2008), 80–88.
105. Michael C. Corballis, "Mental Time Travel and the Shaping of Language," *Experimental Brain Research* 192, no. 3 (2009): 553–60.
106. See the overview by Jonathan Redshaw, "Does Metarepresentation Make Human Mental Time Travel Unique?" *Wiley Interdisciplinary Reviews. Cognitive Science* 5, no. 5 (2014): 519–31.
107. See, for example, the overview by William A. Roberts, "Evidence for Future Cognition in Animals," *Learning and Motivation* 43 (2012): 169–80. Overall he is skeptical that animals show genuine "mental time travel" and points out that in the majority of purported examples it refers to species with a strong tendency to cache food and so with a premium on memory and thus would be trait specific rather than domain general.
108. See Donald Davidson, "The Emergence of Thought," *Erkenntnis* 51 no. 1 (1999): 7–17.
109. For a crisp overview of so-called Morgan's canon, see Thomas R. Zentall, "Morgan's Canon: Is It Still a Useful Rule of Thumb?" *Ethology* 124, no. 7 (2018): 449–57.
110. But note that, as ever, it is context specific. Although rescue is seen among ants, where the risk of burial is high (especially in sand), in environments where the risk is much lower the ants show no such propensity to leap into action. See Krzysztof Miler, Bakhtiar Effendi Yahya, and Marcin Czarnoleski, "Pro-social Behaviour of Ants Depends on Their Ecological Niche—Rescue Actions in Species from Tropical and Temperate Regions," *Behavioural Processes* 144 (2017): 1–4.
111. See, for example, Lindsay P. Schwartz, Alan Silberberg, Anna H. Casey, David N. Kearns, and Burton Slotnick, "Does a Rat Release a Soaked Conspecific Due to Empathy?" *Animal Cognition* 20, no. 2 (2017): 299–308; and Kelsey A. Heslin and Michael F. Brown, "No Preference for Prosocial Helping Behavior in Rats with Concurrent Social Interaction Opportunities," *Learning and Behavior* (2021), https://doi.org/10.3758/s13420-021-00471-8.

112. Marco Vasconcelos, Karen Hollis, Elise Nowbahari, and Alex Kacelnik, "Pro-sociality without Empathy," *Biology Letters* 8, no. 6 (2012): 910–12.
113. See the overview by Ana Pérez-Manrique and Antoni Gomila, "The Comparative Study of Empathy: Sympathetic Concern and Empathic Perspective-Taking in Non-human Animals," *Biological Reviews* 93, no. 1 (2018): 248–69.
114. See Kate Arnold and A. Whiten, "Post-conflict Behaviour of Wild Chimpanzees (*Pan troglodytes schweinfurthii*) in the Budongo Forest, Uganda," *Behaviour* 138, no. 5 (2001): 649–90.
115. See Christophe Boesch, "New Elements of a Theory of Mind in Wild Chimpanzees," *Behavioral and Brain Sciences* 15, no. 1 (1992): 149–50.
116. For the case of children versus chimpanzees, see Julia Ulber, Katharina Hamann, and Michael Tomasello, "Young Children, but Not Chimpanzees, Are Averse to Disadvantageous and Advantageous Inequities," *Journal of Experimental Child Psychology* 155 (2017): 48–66; also Robert Hepach, Leïla Benziad, and Michael Tomasello, "Chimpanzees Help Others with What They Want; Children Help Them with What They Need," *Developmental Science* 23, no. 3 (2020), e12922, https://doi.org/10.1111/desc.12922.
117. Sarah A. Jelbert, Puja J. Singh, Russell D. Gray, and Alex H. Taylor, "New Caledonian Crows Rapidly Solve a Collaborative Problem without Cooperative Cognition," *PLoS One* 10, no. 8 (2015), e0133253.
118. Although they are generous after solicitation; see Toshisada Nishida and Linda A. Turner, "Food Transfer between Mother and Infant Chimpanzees of the Mahale Mountains National Park, Tanzania," *International Journal of Primatology* 17, no. 6 (1996): 947–68.
119. But sometimes they will allow the infants access to nutritious food; see Ari Ueno and Tetsuro Matsuzawa, "Food Transfer between Chimpanzee Mothers and Their Infants," *Primates* 45, no. 4 (2004): 231–39.
120. Yvonne Rekers, Daniel B. M. Haun, and Michael Tomasello, "Children, but Not Chimpanzees, Prefer to Collaborate," *Current Biology* 21, no. 20 (2011): 1756–58.
121. See, for example, Claudio Tennie, Keith Jensen, and Josep Call, "The Nature of Prosociality in Chimpanzees," *Nature Communications* 7 (2016), 13915, https://doi.org/10.1038/ncomms13915; also critique by

Alicia P. Melis, Jan M. Engelmann, and Felix Warneken, "Correspondence: Chimpanzee Helping Is Real, Not a Byproduct," *Nature Communications* 9, no. 1 (2018): 615, https://doi.org/10.1038/s41467-017-02321-6, and the authors' reply (616); also Anke F. Bullinger, Judith M. Burkart, Alicia P. Melis, and Michael Tomasello, "Bonobos, *Pan paniscus*, Chimpanzees, *Pan troglodytes*, and Marmosets, *Callithrix jacchus*, Prefer to Feed Alone," *Animal Behaviour* 85, no. 1 (2013): 51–60.

122. See Anke F. Bullinger, Alicia P. Melis, and Michael Tomasello, "Chimpanzees, *Pan troglodytes,* Prefer Individual over Collaborative Strategies towards Goals," *Animal Behaviour* 82 (2011): 1135–41.
123. See experiments conducted by Alicia P. Melis, Anna-Claire Schneider, and Michael Tomasello, "Chimpanzees, *Pan troglodytes,* Share Food in the Same Way after Collaborative and Individual Food Acquisition," *Animal Behaviour* 82 (2011): 485–93. Yes, chimpanzees can help to obtain food, but this only occurs if the partner makes a racket, so food exchange is never spontaneous. See Alicia P. Melis, Felix Warneken, Keith Jensen, Anna-Claire Schneider, Josep Call, and Michael Tomasello, "Chimpanzees Help Conspecifics Obtain Food and Non-food Items," *Proceedings of the Royal Society of London, Series B: Biological Sciences* 278, no. 1710 (2011): 1405–13.
124. See, for example, Grace E. Fletcher, Felix Warneken, and Michael Tomasello, "Differences in Cognitive Processes Underlying the Collaborative Activities of Children and Chimpanzees," *Cognitive Development* 27 (2012): 136–53.
125. With reference to chimpanzees and orangutans, see Katja Liebal and Federico Rossano, "The Give and Take of Food Sharing in Sumatran Orang-utans, *Pongo abelii,* and Chimpanzees, *Pan troglodytes,*" *Animal Behaviour* 133 (2017): 91–100; and more widely among the primates, see Federica Amici, Elisabetta Visalberghi, and Josep Call, "Lack of Prosociality in Great Apes, Capuchin Monkeys and Spider Monkeys: Convergent Evidence from Two Different Food Distribution Tasks," *Proceedings of the Royal Society of London, Series B: Biological Sciences* 281, no. 1793 (2014): 20141699, https://doi.org/10.1098/rspb.2014.1699.
126. Michael Newton, *Savage Girls and Wild Boys: A History of Feral Children* (London: Faber and Faber, 2002).

127. See Katherine A. Cronin, Evelien De Groot, Jeroen Stevens, "Bonobos Show Limited Social Tolerance in a Group Setting: A Comparison with Chimpanzees and a Test of the Relational Model," *Folia Primatologica* 86, no. 3 (2015): 167–77.
128. Jingzhi Tan, Suzy Kwetuenda, and Brian Hare, "Preference or Paradigm? Bonobos Show No Evidence of Other-Regard in the Standard Prosocial Choice Task," *Behaviour* 152, no. 3–4 (2015): 521–44.
129. Christopher Krupenye and Brian Hare, "Bonobos Prefer Individuals That Hinder Others over Those That Help," *Current Biology* 28, no. 2 (2018): 280–85.e5; and commentary by J. Kiley Hamlin, "Social Behavior: Bonobos Are Nice but Prefer Mean Guys," *Current Biology* 28, no. 4 (2018): R164–66.
130. Megan L. Lambert, Jorg J. M. Massen, Amanda M. Seed, Thomas Bugnyar, and Katie E. Slocombe, "An 'Unkindness' of Ravens? Measuring Prosocial Preferences in *Corvus corax*," *Animal Behaviour* 123 (2017): 383–93.
131. Catherine Hobaiter, Anne Marijke Schel, Kevin Langergraber, and Klaus Zuberbühler, "'Adoption' by Maternal Siblings in Wild Chimpanzees," *PLoS One* 9, no. 8 (2014), e103777, https://doi.org/10.1371/journal.pone.0103777.
132. See, for example, Dora Biro, Tatyana Humle, Kathelijne Koops, Claudia Sousa, Misato Hayashi, and Tetsuro Matsuzawa, "Chimpanzee Mothers at Bossou, Guinea Carry the Mummified Remains of Their Dead Infants," *Current Biology* 20, no. 8 (2010): R351–52; also James R. Anderson, Alasdair Gillies, and Louise C. Lock, "*Pan* Thanatology," *Current Biology* 20, no. 8 (2010): R349–51.
133. Derek C. Penn, "How Folk Psychology Ruined Comparative Psychology," in *Animal Thinking: Contemporary Issues in Comparative Cognition*, ed. Randolf Menzel and Julia Fischer (Cambridge, MA: MIT Press, 2011), 235–65. This chapter complements Penn et al., "Darwin's Mistake," with a brisk survey of the various anthropomorphic pitfalls that repeatedly confound the cognitive studies that presuppose animals actually *understand* the rules of engagement.
134. Keith Jensen, Brian Hare, Josep Call, and Michael Tomasello, "What's in It for Me? Self-Regard Precludes Altruism and Spite in Chimpanzees,"

Proceedings of the Royal Society of London, Series B: Biological Sciences 273, no.1589 (2006): 1013–21; and Keith Jensen, Brian Hare, Josep Call, and Michael Tomasello, "Chimpanzees Are Vengeful but Not Spiteful," *Proceedings of the National Academy of Sciences of the United States of America* 104, no. 32 (2007): 13046–50.

135. See Katrin Riedl, Keith Jensen, Josep Call, and Michael Tomasello, "No Third-Party Punishment in Chimpanzees," *Proceedings of the National Academy of Sciences of the United States of America* 109, no. 37 (2012): 14824–29.

136. Patricia Kanngiesser, Federico Rossano, Ramona Frickel, Anne Tomm, and Michael Tomasello, "Children, but Not Great Apes, Respect Ownership," *Developmental Science* 23, no. 1 (2020), e12842, https://doi.org/10.1111/desc.12842.

137. See, for example, Ingrid Kaiser, Keith Jensen, Josep Call, and Michael Tomasello, "Theft Is an Ultimatum Game: Chimpanzees and Bonobos Are Insensitive to Unfairness," *Biology Letters* 8, no. 6 (2012): 942–45; and Jan M. Engelmann, Jeremy B. Clift, Esther Herrmann, and Michael Tomasello, "Social Disappointment Explains Chimpanzees' Behaviour in the Inequity Aversion Task," *Proceedings of the Royal Society of London, Series B: Biological Sciences* 284, no. 1861 (2017): 20171502, https://doi.org/10.1098/rspb.2017.1502.

138. In her compelling book, Margaret Visser, *The Geometry of Love: Space, Time, Mystery and Meaning in an Ordinary Church* (London: Viking, 2001), where she continues, "but that every human being seeks and hopes to find" (93).

139. Daniel B. M. Haun, Yvonne Rekers, and Michael Tomasello, "Children Conform to the Behavior of Peers; Other Great Apes Stick with What They Know," *Psychological Science* 25, no. 12 (2014): 2160–67.

140. See the overview by Juliane Bräuer and Daniel Hanus, "Fairness in Non-human Primates?" *Social Justice Research* 25, no. 3 (2012): 256–76.

141. Alicia P. Melis, Patricia Grocke, Josefine Kalbitz, and Michael Tomasello, "One for You, One for Me: Humans' Unique Turn-Taking Skills," *Psychological Science* 27, no. 7 (2016): 987–96; also Alejandro Sánchez-Amaro, Shona Duguid, Josep Call, and Michael Tomasello,

"Chimpanzees and Children Avoid Mutual Defection in a Social Dilemma," *Evolution and Human Behaviour* 40, no. 1 (2019): 46–54, a comparison of chimpanzees and children in terms of when to cooperate and when to defect (prisoners' dilemma), where only the latter engage in active cooperation.

142. Specifically moving stones to assist in the righting of the shell; see P. J. Weldon and D. L. Hoffman, "Unique Form of Tool-Using in Two Gastropod Molluscs (Trochidae)," *Nature* 256, no. 5520 (1975): 720–21.
143. Gavin R. Hunt, Russell D. Gray, and Alex H. Taylor, "Why Is Tool Use Rare in Animals?" in *Tool Use in Animals: Cognition and Ecology*, ed. Crickette M. Sanz, Josep Call, and Christophe Boesch (Cambridge: Cambridge University Press, 2013), 89–118.
144. See the overview by I. Teschke et al., "Did Tool-Use Evolve with Enhanced Physical Cognitive Abilities?" *Philosophical Transactions of the Royal Society of London, Series B: Biological Sciences* 368, no. 1630 (2013): 20120418, https://doi.org/10.1098/rstb.2012.0418. Less surprisingly there is a strong correlation between brain size (at least in Australian birds) and play, especially social play; see Gisela Kaplan, "Play Behaviour, not Tool Using, Relates to Brain Mass in a Sample of Birds," *Scientific Reports* 10, no. 1 (2020), 20437, https://doi.org/10.1038/s41598-020-76572-7.
145. With respect to the general question of animal intelligences, see, for example, the ruminations of E. M. Macphail and J. J. Bolhuis, "The Evolution of Intelligence: Adaptive Specializations versus General Process," *Biological Reviews of the Cambridge Philosophical Society* 76 (2001): 341–64.
146. See Joanna H. Wimpenny, Alex A. S. Weir, Lisa Clayton, Christian Rutz, and Alex Kacelnik, "Cognitive Processes Associated with Sequential Tool Use in New Caledonian Crows," *PLoS One* 4, no. 8 (2009), e6471, https://doi.org/10.1371/journal.pone.0006471; also A. M. P. Bayern et al., "Compound Tool Construction by New Caledonian Crows," *Scientific Reports* 8, no. 1 (2018), 15676, https://doi.org/10.1038/s41598-018-33458-z.
147. Michael Haslam, "'Captivity Bias' in Animal Tool Use and Its Implications for the Evolution of Hominin Technology," *Philosophical*

Transactions of the Royal Society of London, Series B: Biological Sciences 368, no. 1630 (2013), 20120421; also Ken Cheng and Richard W. Byrne, "Why Human Environments Enhance Animal Capacities to Use Objects: Evidence from Keas (*Nestor notabilis*) and Apes (*Gorilla gorilla, Pan paniscus, Pongo abelii, Pongo pygmaeus*)," *Journal of Comparative Psychology* 132, no. 4 (2018): 419–26.

148. Alex A. S. Weir, Jackie Chappell, and Alex Kacelnik, "Shaping of Hooks in New Caledonian Crows," *Science* 297, no. 5583 (2002): 981; also Alex A. S. Weir and Alex Kacelnik, "A New Caledonian Crow (*Corvus moneduloides*) Creatively Re-designs Tools by Bending or Unbending Aluminium Strips," *Animal Cognition* 9, no. 4 (2006): 317–34.
149. A phrase ultimately derived from Daniel Dennett. Sara Shettleworth, "Clever Animals and Killjoy Explanations in Comparative Psychology," *Trends in Cognitive Sciences* 14 (2010): 477–81.
150. Christian Rutz, Shoko Sugasawa, Jessica E. M. van der Wal, Barbara C. Klump, and James J. H. St Clair, "Tool Bending in New Caledonian Crows," *Royal Society Open Science* 3, (2016), 160439, https://doi.org/10.1098/rsos.160439.
151. Jennifer C. Holzhaider, Gavin R. Hunt, and Russell D. Gray, "The Development of Pandanus Tool Manufacture in Wild New Caledonian Crows," *Behaviour* 147 (2010): 553–86.
152. See, for example, Ben Kenward, Christian Rutz, Alex A. Weir, and Alex Kacelnik, "Development of Tool Use in New Caledonian Crows: Inherited Action Patterns and Social Influences," *Animal Behaviour* 72 (2006): 1329–43.
153. Jennifer C. Holzhaider, M. D. Sibley, A. H. Taylor, P. J. Singh, R. D. Gray, and G. R. Hunt, "The Social Structure of New Caledonian Crows," *Animal Behaviour* 81, no. 1 (2011): 83–92; and Gavin R. Hunt, Jennifer C. Holzhaider, and Russell D. Gray, "Prolonged Parental Feeding in Tool-Using New Caledonian Crows," *Ethology* 118, no. 5 (2012): 423–30.
154. Hiroshi Matsui et al., "Adaptive Bill Morphology for Enhanced Tool Manipulation in New Caledonian Crows," *Scientific Reports* 6 (2016), 22776, https://doi.org/10.1038/srep22776.

155. Albeit with "monocular tool control." Antone Martinho III, Zackory T. Burns, Auguste M. P. von Bayern, and Alex Kacelnik, "Monocular Tool Control, Eye Dominance, and Laterality in New Caledonian Crows," *Current Biology* 24, no. 24 (2014): 2930–34.
156. W. C. McGrew, "Is Primate Tool Use Special? Chimpanzee and New Caledonian Crow Compared," *Philosophical Transactions of the Royal Society of London, Series B: Biological Sciences* 368, no. 1630 (2013): 20120422.
157. Susana Carvalho, Eugénia Cunha, Cláudia Sousa, and Tetsuro Matsuzawa, "Chaînes Opératoires and Resource-Exploitation Strategies in Chimpanzee (*Pan troglodytes*) Nut Cracking," *Journal of Human Evolution* 55, no. 1 (2008): 148–63; and Susana Carvalho, Dora Biro, William C. McGrew, and Tetsuro Matsuzawa, "Tool-Composite Reuse in Wild Chimpanzees (*Pan troglodytes*): Archaeologically Invisible Steps in the Technological Evolution of Early Hominins?" *Animal Cognition* 12, Supplement (2009), S103–S104.
158. In principle maybe, but although the common ancestor of bonobo-chimpanzee-human almost certainly used tools, most likely stone tools evolved convergently among the chimpanzees. See Michael Haslam, "On the Tool Use Behavior of the Bonobo-Chimpanzee Last Common Ancestor, and the Origins of Hominine Stone Tool Use," *American Journal of Primatology* 76, no. 10 (2014): 910–18.
159. See Sirianni et al., "When to Choose Which Tool," 152–65.
160. Tatyana Humle and Tetsuro Matsuzawa, "Ant-Dipping among the Chimpanzees of Bossou, Guinea, and Some Comparisons with Other Sites," *American Journal of Primatology* 58, no. 3 (2002): 133–48.
161. Crickette Sanz, Josep Call, and David Morgan, "Design Complexity in Termite-Fishing Tools of Chimpanzees (*Pan troglodytes*)," *Biology Letters* 5, no. 3 (2009): 293–96.
162. J. D. Pruetz, P. Bertolani, K. Boyer Ontl, S. Lindshield, M. Shelley, and E. G. Wessling, "New Evidence on the Tool-Assisted Hunting Exhibited by Chimpanzees (*Pan troglodytes*) in a Savannah Habitat at Fongoli, Sénégal," *Royal Society Open Science* 2, no. 4 (2015), 140507, https://doi.org/10.1098/rsos.140507.

163. Crickette M. Sanz, Caspar Schöning, and David B. Morgan, "Chimpanzees Prey on Army Ants with Specialized Tool Set," *American Journal of Primatology* 72, no. 1 (2010): 17–24.
164. R. Adriana Hernandez-Aguilar, Jim Moore, and Travis Rayne Pickering, "Savanna Chimpanzees Use Tools to Harvest the Underground Storage Organs of Plants," *Proceedings of the National Academy of Sciences of the United States of America* 104, no. 49 (2007): 19210–13.
165. Gaku Ohashi, "Pestle-Pounding and Nut-Cracking by Wild Chimpanzees at Kpala, Liberia," *Primates* 56, no. 2 (2015): 113–17.
166. See Thibaud Gruber, Klaus Zuberbühler, Fabrice Clément, and Carel van Schaik, "Apes Have Culture but May Not Know That They Do," *Frontiers in Psychology* 6 (2015), 91, https://doi.org/10.3389/fpsyg.2015.00091.
167. For an overview that picks apart conservatism, functional fixedness, and conformity, see Thibaud Gruber, "Great Apes Do Not Learn Novel Tool Use Easily: Conservatism, Functional Fixedness, or Cultural Influence?" *International Journal of Primatology* 37 (2016): 296–316.
168. Lydia V. Luncz and Christophe Boesch, "Tradition over Trend: Neighboring Chimpanzee Communities Maintain Differences in Cultural Behavior Despite Frequent Immigration of Adult Females," *American Journal of Primatology* 76, no. 7 (2014): 649–57.
169. Nor need social conformity be the only determinant—at least in the case of marmosets: Mario B. Pesendorfer, Tina Gunhold, Nicola Schiel, Antonio Souto, Ludwig Huber, and Friederike Range, "The Maintenance of Traditions in Marmosets: Individual Habit, Not Social Conformity? A Field Experiment," *PLoS One* 4, no. 2 (2009), e4472. The maintenance of traditions reflects habits acquired associated with the less cognitively demanding pathway of associative learning.
170. See Thibaud Gruber, Martin N. Muller, Pontus Strimling, Richard Wrangham, and Klaus Zuberbühler, "Wild Chimpanzees Rely on Cultural Knowledge to Solve an Experimental Honey Acquisition Task," *Current Biology* 19, no. 21 (2009): 1806–10.
171. On the Tai chimpanzees, see Lydia V. Luncz, Giulia Sirianni, Roger Mundry, and Christophe Boesch, "Costly Culture: Differences in Nut-

Cracking Efficiency between Wild Chimpanzee Groups," *Animal Behaviour* 137 (2018): 63–73.

172. See Daniel J. Povinelli and Derek C. Penn, "Through a Floppy Tool Darkly: Toward a Conceptual Overthrow of Animal Alchemy," in *Tool Use and Causal Cognition*, ed. Teresa McCormack, Christoph Hoerl, and Stephen Butterfill (Oxford: Oxford University Press, 2011), 69–88; also see the follow-up report by Daniel J. Povinelli and Scott H. Frey, "Constraints on the Exploitation of the Functional Properties of Objects in Expert Tool-Using Chimpanzees (*Pan troglodytes*)," *Cortex* 82 (2016): 11–23.
173. E. E. Furlong, K. J. Boose, and S. T. Boysen, "Raking It In: The Impact of Enculturation on Chimpanzee Tool Use," *Animal Cognition* 11, no. 1 (2008): 83–97.
174. What Lambros Malafouris calls a "tectonoetic awareness" in "Between Brains, Bodies and Things: Tectonoetic Awareness and the Extended Self," *Philosophical Transactions of the Royal Society of London, Series B: Biological Sciences* 363, no. 1499 (2008): 1993–2002; see also his contribution to a discussion in Lambros Malafouris, "Select Prosthetic Gestures: How the Tool Shapes the Mind," *Behavioral and Brain Sciences* 35, no. 4 (2012): 230–31.
175. S. H. Johnson-Frey, "What's So Special about Human Tool Use?" *Neuron* 39 (2003): 201–4.
176. Lambros Malafouris, "Knapping Intentions and the Marks of the Mental," in *The Cognitive Life of Things: Recasting the Boundaries of the Mind*, ed. Lambros Malafouris and Colin Renfrew (Cambridge: Macdonald Institute for Archaeological Research, 2010), 13–22.
177. See evidence from 2.34 Ma Lokalalei C by Anne Delagnes and Hélène Roche, "Late Pliocene Hominid Knapping Skills: The Case of Lokalalei 2C, West Turkana, Kenya," *Journal of Human Evolution* 48, no. 5 (2005): 455–72.
178. Lambros Malafouris, "Creative Thinking: The Feeling of and for Clay," *Pragmatics and Cognition* 22 (2014): 140–58; also his earlier work in "Beads for a Plastic Mind: The 'Blind Man's Stick' (BMS) Hypothesis and the Active Nature of Material Culture," *Cambridge Archaeological Journal* 18, no. 3 (2008): 401–14.

179. Michael Muthukrishna and Joseph Heinrich, "Innovation in the Collective Brain," *Philosophical Transactions of the Royal Society of London, Series B: Biological Sciences* 371, no. 1690 (2016): 20150192.
180. Bennett G. Galef, "Culture in Animals?" in *The Question of Animal Culture,* ed. Kevin N. Laland and Bennett G. Galef (Cambridge, MA: Harvard University Press, 2009), 232–46. This writer also has trenchant things to say about so-called teaching among animals.
181. Kim Hill, "Animal 'Culture'?" in *Question of Animal Culture,* 269–87.
182. See G. L. Vale, E. G. Flynn, and R. L. Kendal, "Cumulative Culture and Future Thinking: Is Mental Time Travel a Prerequisite to Cumulative Cultural Evolution?" *Learning and Motivation* 43 (2012): 220–30.
183. Claudio Tennie and Carel P. van Schaik, "Spontaneous (Minimal) Ritual in Non-human Great Apes?" *Philosophical Transactions of the Royal Society of London, Series B: Biological Sciences* 375, no. 1805 (2020): 20190423. They are hard pressed to find any convincing examples of ritual in chimpanzees.
184. Susan Perry, "Are Nonhuman Primates Likely to Exhibit Cultural Capacities Like Those of Humans, and If So, How Can This Be Demonstrated?" in *Question of Animal Culture,* 247–68.
185. See, for example, Gillian L. Vale et al., "Why Do Chimpanzees Have Diverse Behavioral Repertoires yet Lack More Complex Cultures? Invention and Social Information Use in a Cumulative Task," *Evolution and Human Behavior* 42, no. 3 (2021): 247–58. They show that individual chimpanzees can innovate but have no clue how to pass on the necessary information. The contrasts between primates and children are also telling; see L. G. Dean, R. L. Kendal, S. J. Schapiro, B. Thierry, and K. N. Laland, "Identification of the Social and Cognitive Processes Underlying Human Cumulative Culture," *Science* 335, no. 6072 (2012): 1114–18; also commentary by Robert Kurzban and H. Clark Barrett, "Origins of Cumulative Culture," *Science* 335, no. 6072 (2012): 1056–57.
186. See Claudio Tennie, Josep Call, and Michael Tomasello, "Push or Pull: Imitation vs. Emulation in Great Apes and Human Children," *Ethology* 112, no. 12 (2006): 1159–69.

187. Nor does it appear to be innate with a genetic basis; see the commentary by Cecilia Heyes, "Imitation: Not in Our Genes," *Current Biology* 26, no. 10 (2016): R412–14; see also Cecilia Heyes, "*Homo imitans?* Seven Reasons Why Imitation Couldn't Possibly Be Associative," *Philosophical Transactions of the Royal Society of London, Series B: Biological Sciences* 371, no. 1686 (2016): 20150069, https://doi.org/10.1098/rstb.2015.0069.
188. See Ina C. Uzgiris, "Two Functions of Imitation during Infancy," *International Journal of Behavioral Development* 4, no. 1 (1981): 1–12.
189. B. M. Waller, A. Misch, J. Whitehouse, and E. Herrmann, "Children, but Not Chimpanzees, Have Facial Correlates of Determination," *Biology Letters* 10, no. 3 (2014), 20130974.
190. Helen Wasielewski, "Imitation Is Necessary for Cumulative Cultural Evolution in an Unfamiliar, Opaque Task," *Human Nature* 25, no. 1 (2014): 161–79.
191. See, for example, Claudio Tennie, Josep Call, and Michael Tomasello, "Ratcheting Up the Ratchet: On the Evolution of Cumulative Culture," *Philosophical Transactions of the Royal Society of London, Series B: Biological Sciences* 364, no. 1528 (2009): 2405–15.
192. For an overview see Richard W. Byrne and Lisa G. Rapaport, "What Are We Learning from Teaching?" *Animal Behaviour* 82, no. 5 (2011): 1207–11; and critique by Alex Thornton and Katherine McAuliffe, "Teaching Can Teach Us a Lot," 83, no. 4 (2011), e6–e9, and authors' reply (e1–e3).
193. Joah R. Madden, Hans-Joerg P. Kunc, Sinead English, and Tim H. Clutton-Brock, "Why Do Meerkat Pups Stop Begging?" *Animal Behaviour* 78, no. 1 (2009): 85–89.
194. As with chimpanzees; see, for example, Dora Biro et al., "Cultural Innovation and Transmission of Tool Use in Wild Chimpanzees: Evidence from Field Experiments," *Animal Cognition* 6, no. 4 (2003): 213–23.
195. See, for example, Sarah J. Davis, Gillian L. Vale, Steven J. Schapiro, Susan P. Lambeth, and Andrew Whiten, "Foundations of Cumulative Culture in Apes: Improved Foraging Efficiency through Relinquishing and Combining Witnessed Behaviours in Chimpanzees (*Pan troglodytes*)," *Scientific Reports* 6 (2016), 35953. Although their emphasis is

on the supposed rudiments of cumulative culture, the glacial progress is self-evident. See also Claudio Tennie, Josep Call, and Michael Tomasello, "Untrained Chimpanzees (*Pan troglodytes schweinfurthii*) Fail to Imitate Novel Actions," *PLoS One* 7, no. 8 (2012), e41548.

196. See, for example, Elouise Leadbeater, Nigel E. Raine, and Lars Chittka, "Social Learning: Ants and the Meaning of Teaching," *Current Biology* 16, no. 9 (2006): R323–25; Thomas O. Richardson, Philippa A. Sleeman, John M. McNamara, Alasdair I. Houston, and Nigel R. Franks, "Teaching with Evaluation in Ants," *Current Biology* 17, no. 17 (2007): 1520–26; and Elizabeth L. Franklin and Nigel R. Franks, "Individual and Social Learning in Tandem-Running Recruitment by Ants," *Animal Behaviour* 84, no. 2 (2012): 361–68.
197. Premack and Premack, "Why Animals," 350–65.
198. For example, in the context of how to collaborate to coordinate decisions the contrast between children and chimpanzees is profound; see Shona Duguid, Emily Wyman, Sebastian Gruenseisen, and Michael Tomasello, "The Strategies Used by Chimpanzees (*Pan troglodytes*) and Children (*Homo sapiens*) to Solve a Simple Coordination Problem," *Journal of Comparative Psychology* 134, no. 4 (2020): 401–11.
199. See, for example, Michael Tomasello and Malinda Carpenter, "Shared Intentionality," *Developmental Science* 10, no. 1 (2007): 121–25; and Sidney Strauss and Margalit Ziv, "Teaching Is a Natural Cognitive Ability for Humans," *Mind, Brain, and Education* 6, no. 4 (2012): 186–96.
200. Cecilia M. Heyes and Chris D. Frith, "The Cultural Evolution of Mind Reading," *Science* 344, no. 6190 (2014), 1243091.
201. See the sparkling essay by Stephen R. L. Clark, "The Evolution of Language: Truth and Lies," *Philosophy* 75, no. 3 (2000): 401–21.
202. Robert Boyd, Peter J. Richerson, and Joseph Henrich, "The Cultural Niche: Why Social Learning Is Essential for Human Adaptation," *Proceedings of the National Academy of Sciences of the United States of America* 108, Supplement 2 (2011): 10918–25.
203. Cecilia Heyes, "Simple Minds: A Qualified Defence of Associative Learning," *Philosophical Transactions of the Royal Society of London, Series B: Biological Sciences* 367, no. 1603 (2012), 2181–91.

204. Cecilia Heyes, "Born Pupils? Natural Pedagogy and Cultural Pedagogy," *Perspectives on Psychological Science* 11, no. 2 (2016): 280–95, quotation is on p. 292.
205. Referring specifically to the monkey larynx: Tecumseh Fitch, Bart de Boer, Neil Mathur, and Asif A. Ghazanfar, "Monkey Vocal Tracts Are Speech-Ready," *Science Advances* 2, no. 12 (2016): e1600723; with commentary by Philip Lieberman, "Comment on 'Monkey Vocal Tracts Are Speech-Ready,'" 3, no. 7 (2017), and the authors' reply (e1701859).
206. As relayed by Denis Diderot; see, for example, Mary Efrosini Gregory, *Evolutionism in Eighteenth-Century French Thought* (New York: Peter Lang, 2008), 196.
207. See, for example, John W. Pilley, "Border Collie Comprehends Sentences Containing a Prepositional Object, Verb, and Direct Object," *Learning and Motivation* 44, no. 4 (2013): 229–40; and J. Kaminsky et al., "Word Learning in a Domestic Dog: Evidence for 'Fast Mapping,'" *Science* 304, no. 5677 (2004): 1682–83.
208. See, for example, Ulrike Griebel and D. Kimbrough Oller, "Vocabulary Learning in a Yorkshire Terrier: Slow Mapping of Spoken Words," *PLoS One* 7, no. 2 (2012), e30182.
209. Sebastian Tempelmann, Juliane Kaminski, and Michael Tomasello, "Do Domestic Dogs Learn Words Based on Humans' Referential Behaviour?" *PLoS One* 9, no. 3 (2014), e91014.
210. See, for example, Juliane Kaminski, Martina Neumann, Juliane Bräuer, and Josep Call, "Dogs, *Canis familiaris*, Communicate with Humans to Request but Not to Inform," *Animal Behaviour* 82, no. 4 (2011): 651–58.
211. Katharina C. Kirchhofer, Felizitas Zimmermann, Juliane Kaminski, and Michael Tomasello, "Dogs (*Canis familiaris*), but Not Chimpanzees (*Pan troglodytes*), Understand Imperative Pointing," *PLoS One* 7, no. 2 (2012), e30913.
212. Krista Macpherson and William A. Roberts, "Do Dogs (*Canis familiaris*) Seek Help in an Emergency?" *Journal of Comparative Psychology* 120, no. 2 (2006): 113–19; see also Linda Jaasma, Isabelle Kamm, Annemie Ploeger, and Mariska E. Kret, "The Exceptions That Prove the Rule? Spontaneous Helping Behaviour towards Humans in Some

Domestic Dogs," *Applied Animal Behaviour Science* 224 (2020), 104941, which is scarcely more encouraging.

213. See the overview by Dorothy L. Cheney and Robert M. Seyfarth, "Précis of *How Monkeys See the World*," *Behavioral and Brain Sciences* 15, no. 1 (1992): 135–47, with extended commentary (147–71), reply (172–78), and collected references (178–82).
214. See Nicholas Ducheminsky, S. Peter Henzi, and Louise Barrett, "Responses of Vervet Monkeys in Large Troops to Terrestrial and Aerial Predator Alarm Calls," *Behavioral Ecology* 25, no. 6 (2014): 1474–84; and James L. Fuller and Marina Cords, "Multiple Functions and Signal Concordance of the *pyow* Loud Call of Blue Monkeys," *Behavioral Ecology and Sociobiology* 71, no. 1 (2017), 1–15.
215. See, for example, Camille Coye, Karim Ouattara, Malgorzata E. Arlet, Alban Lemasson, and Klaus Zuberbühler, "Flexible Use of Simple and Combined Calls in Female Campbell's Monkeys," *Animal Behaviour* 141 (2018): 171–81.
216. Karim Ouattara, Alban Lemasson, and Klaus Zuberbühler, "Campbell's Monkeys Concatenate Vocalizations into Context-Specific Call Sequences," *Proceedings of the National Academy of Sciences of the United States of America* 106, no. 51 (2009): 22026–31.
217. As Julia Fischer and Tabitha Price remark in their overview of primate vocalizations, "signallers communicate, but they do not communicate that they communicate" (29). "Meaning, Intention, and Inference in Primate Vocal Communication," *Neuroscience and Biobehavioral Reviews* 82 (2017): 22–31.
218. See, for example, the thoughtful essay revolving around the hermit thrush by the musicologist Emily L. Doolittle, "'Hearken to the Hermit-Thrush': A Case Study in Interdisciplinary Listening," *Frontiers in Psychology* 11 (2020), 613510; see also Hollis Taylor, "Connecting Interdisciplinary Dots: Songbirds, 'White Rats' and Human Exceptionalism," *Social Science Information* 52, no. 2 (2013): 287–306. Both make interesting reading but are as burdened with as much ideological baggage as the rest of us.
219. See, for example, Patricia K. Kuhl, "Human Speech and Birdsong: Communication and the Social Brain," *Proceedings of the National*

Academy of Sciences of the United States of America 100, no. 17 (2003): 9645–46.

220. See Nigel I. Mann, Kimberly A. Dingess, and P. J. B. Slater, "Antiphonal Four-Part Synchronized Chorusing in a Neotropical Wren," *Biology Letters* 2, no. 1 (2006): 1–4.
221. See, for example, Michael D. Beecher and Eliot A. Brenowitz, "Functional Aspects of Song Learning in Songbirds," *Trends in Ecology and Evolution* 20, no. 3 (2005): 143–49.
222. See, for example, Johan D. Bolhuis, "Evolution Cannot Explain How Minds Work," *Behavioural Processes* 117 (2015): 82–91.
223. Rindy C. Anderson, William A. Searcy, Susan Peters, Melissa Hughes, Adrienne L. DuBois, and Stephen Nowicki, "Song Learning and Cognitive Ability Are Not Consistently Related in a Songbird," *Animal Cognition* 20, no. 2 (2017): 309–20; also Adrienne L. Du Bois, Stephen Nowicki, Susan Peters, Karla D. Rivera-Caceres, and William A. Searcy, "Song Is Not a Reliable Signal of General Cognitive Ability in a Songbird," *Animal Behaviour* 137 (2018): 205–13.
224. See, for example, Arik Kershenbaum, Ann E. Bowles, Todd M. Freeberg, Dezhe Z. Jin, Adriano R. Lameira, and Kirsten Bohn, "Animal Vocal Sequences: Not the Markov Chains We Thought They Were," *Proceedings of the Royal Society of London, Series B: Biological Sciences* 281, no. 1792 (2014): 20141370; also Gabriël J. L. Beckers, Johan J. Bolhuis, Kazuo Okanoya, and Robert C. Berwick, "Birdsong Neurolinguistics: Songbird Context-Free Grammar Claim Is Premature," *NeuroReport* 23, no. 3 (2012): 139–45.
225. See, for example, Caroline A. A. van Heijningen, Jos de Visser, Willem Zuidema, and Carel ten Cate, "Simple Rules Can Explain Discrimination of Putative Recursive Syntactic Structures by a Songbird Species," *Proceedings of the National Academy of Sciences of the United States of America* 106, no. 48 (2009): 20538–43.
226. Yoshimasa Seki, Kenta Suzuki, Ayumi M. Osawa, and Kazuo Okanoya, "Songbirds and Humans Apply Different Strategies in a Sound Sequence Discrimination Task," *Frontiers in Psychology* 4 (2013), 447.
227. See Johan J. Bolhuis et al., "Meaningful Syntactic Structure in Songbird Vocalizations?" *PLoS Biology* 16, no. 6 (2018), e2005157; and

subsequent critiques by Simon W. Townsend, Sabrina Engesser, Sabine Stoll, Klaus Zuberbühler, and Balthasar Bickel, "Compositionality in Animals and Humans," *PLoS Biology* 16, no. 6 (2018), e2006425; and Toshitaka N. Suzuki, David Wheatcroft, and Michael Griesser, "Call Combinations in Birds and the Evolution of Compositional Syntax," *PLoS Biology* 16, no. 6 (2018), e2006532; also Robert C. Berwick, Kazuo Okanoya, Gabriel J. L. Beckers, and Johan J. Bolhuis, "Songs to Syntax: The Linguistics of Birdsong," *Trends in Cognitive Sciences* 15, no. 3 (2011): 113–21.

228. See Johan J. Bolhuis, Gabriel J. L. Beckers, Marinus A. C. Huybregts, Robert C. Berwick, and Martin B. H. Everaert, "The Slings and Arrows of Comparative Linguistics," *PLoS Biology* 16, no. 9 (2018), e3000019.
229. See the trenchant remarks by S. Baron-Cohen, "How Monkeys Do Things with 'Words,'" *Behavioral and Brain Sciences* 15 (1992): 148–49.
230. See Thomas C. Scott-Phillips, "Meaning in Animal and Human Communication," *Animal Cognition* 18, no. 3 (2015): 801–15, and "Nonhuman Primate Communication, Pragmatics, and the Origins of Language," *Current Anthropology* 56, no. 1 (2015): 56–80, including the commentaries.
231. A point made in a wide-ranging discussion by Michael D. Beecher, "Why Are No Animal Communication Systems Simple Language?" *Frontiers in Psychology* 12 (2021), e602635.
232. Marc D. Hauser and Jeffrey Watumull, "The Universal Generative Faculty: The Source of Our Expressive Power in Language, Mathematics, Morality, and Music," *Journal of Neurolinguistics* 43, no. B (2017): 78–94. They put this in the context of a "Universal Generative Faculty," but what traction this concept can bear remains to be seen. On the other hand, their strictures on the limitations of artificial intelligence may provide a suitable corrective in some quarters.
233. See Thomas C. Scott-Phillips, James Gurney, Alasdair Ivens, Stephen P. Diggle, and Roman Popat, "Combinatorial Communication in Bacteria: Implications for the Origins of Linguistic Generativity," *PLoS One* 9, no. 4 (2014), e95929.
234. See Thomas C. Scott-Phillips and Richard A. Blythe, "Why Is Combinatorial Communication Rare in the Natural World, and Why Is Lan-

guage an Exception to This Trend?" *Journal of the Royal Society Interface* 10, no. 88 (2013), https://doi.org/10.1098/rsif.2013.0520.

235. See M. Winkler, L. L. Mueller, A. D. Friederici, and C. Männel, "Infant Cognition Includes the Potentially Human-Unique Ability to Encode Embedding," *Science Advances* 4 (2018), eaar8334.
236. David Premack, "Human and Animal Cognition: Continuity and Discontinuity," *Proceedings of the National Academy of Sciences of the United States of America* 104, no. 35 (2007): 13861–67, p. 13865. His discussion, of course, deals with far more than this racy example of recursion, and provides a valuable survey of how similar, and how different, are animal and human minds.
237. Michael C. Corballis, "Recursion, Language, and Starlings," *Cognitive Science* 31, no. 4 (2007): 697–704.
238. Timothy Q. Gentner, Kimberly M. Fenn, Daniel Margoliash, and Howard C. Nusbaum, "Recursive Syntactic Pattern Learning by Songbirds," *Nature* 440, no. 7088 (2006): 1204–207; with commentary by G. F. Marcus, "Language-Startling Starlings," *Nature* 440, no. 7088 (2006): 1117–18.
239. See, for example, the overview by Michael C. Corballis, "Language and Episodic Sharing," in *Animal Thinking: Contemporary Issues in Comparative Cognition*, ed. Randolf Menzel and Julia Fischer (Cambridge, MA: MIT Press, 2011), 175–85, with a particular emphasis on the centrality of mental time travel.
240. Thomas Suddendorf, Donna Rose Addis, and Michael C. Corballis, "Mental Time Travel and the Shaping of the Human Mind," *Philosophical Transactions of the Royal Society of London, Series B: Biological Sciences* 364, no. 1521 (2009): 1317–24; also Tok Thompson, "The Ape That Captured Time: Folklore, Narrative, and the Human-Animal Divide," *Western Folklore* 69 (2010): 395–420.
241. Charles Yang, "Ontogeny and Phylogeny of Language," *Proceedings of the National Academy of Sciences of the United States of America* 110, no. 16 (2013): 6324–27.
242. See, for example, Johan J. Bolhuis, Ian Tattersall, Noam Chomsky, and Robert C. Berwick, "How Could Language Have Evolved?" *PLoS Biology* 12, no. 8 (2014), e1001934.

243. M. Siegal, R. Varley, and S. C. Want, "Mind over Grammar: Reasoning in Aphasia and Development," *Trends in Cognitive Sciences* 5, no. 7 (2001): 296–301.
244. See, for example, the resounding "Gavagai" of David Premack, "'Gavagai!' or the Future History of the Animal Language Controversy," *Cognition* 19, no. 3 (1985): 207–96.
245. See, for example, Esther Herrmann and Michael Tomasello, "Apes' and Children's Understanding of Cooperative and Competitive Motives in a Communicative Situation," *Developmental Science* 9, no. 5 (2006): 518–29.
246. See his compelling book: Paul Oppenheimer, *Evil and the Demonic: A New Theory of Monstrous Behaviour* (New York: New York University Press, 1999).
247. Massimo Piattelli-Palmarini, "Normal Language in Abnormal Brains," *Neuroscience and Biobehavioral Reviews* 81, no. 8 (2017): 188–93. More generally, and extending beyond cases of hydrocephaly, to other extraordinary cases such as the absence of a hemisphere, etc., see Michael Nahm, David Rousseau, and Bruce Greyson, "Discrepancy between Cerebral Structure and Cognitive Functioning: A Review," *Journal of Nervous and Mental Disease* 205 (2017): 967–72.
248. For a brief account of this bizarre story, see Judith Schalansky's *Pocket Atlas of Remote Islands: Fifty Islands I Have Not Visited and Never Will* (London: Penguin, 2012), 120–21; also Damien Personnaz, "Ile Rapa: L'invraisemblable histoire de Marc Liblin," *Agora Vox*, August 9, 2010, https://www.agoravox.fr/culture-loisirs/voyages/article/ile-rapa-l-invraisemblable-79470.
249. See, for example, Kurt Hammerschmidt and Julia Fischer, "Baboon Vocal Repertoires and the Evolution of Primate Vocal Diversity," *Journal of Human Evolution* 126 (2019): 1–13.
250. See, for example, Michael C. Corballis, "From Hand to Mouth: The Gestural Origins of Language," in *Language Evolution*, ed. Morten H. Christiansen and Simon Kirby (Oxford: Oxford University Press, 2003), 201–18.
251. Véronique Izard, Coralie Sann, Elizabeth S. Spelke, and Arlette Streri, "Newborn Infants Perceive Abstract Numbers," *Proceedings of the*

National Academy of Sciences of the United States of America 106, no. 25 (2009): 10382–85.

252. Rosemary Varley, Nicolai J. C. Klessinger, Charles A. J. Romanowski, and Michael Siegal, "Agrammatic but Numerate," *Proceedings of the National Academy of Sciences of the United States of America* 102, no. 9 (2005): 3519–24, with commentary by Elizabeth M. Brannon, "The Independence of Language and Mathematical Reasoning," 102, no. 9 (2005): 3177–78; and Nicolai Klessinger, Marcin Szczerbinski, and Rosemary Varley, "Algebra in a Man with Severe Aphasia," *Neuropsychologia* 45, no. 8 (2007): 1642–48.
253. See, for example, evidence for numerosity in fish: Christian Agrillo, Marco Dadda, Giovanna Serena, and Angelo Bisazza, "Do Fish Count? Spontaneous Discrimination of Quantity in Female Mosquitofish," *Animal Cognition* 11, no. 3 (2008): 495–503; and in bees: Scarlett R. Howard, Aurore Avarguès-Weber, Jair E. Garcia, Andrew D. Greentree, and Adrian G. Dyer, "Numerical Ordering of Zero in Honey Bees," *Science* 360, no. 6393 (2018): 1124–26; with commentary by Andreas Nieder, "Honey Bees Zero in on the Empty Set," *Science* 360, no. 6393 (2018): 1069–70.
254. See, for example, Frederick L. Coolidge and Karenleigh A. Overmann, "Numerosity, Abstraction, and the Emergence of Symbolic Thinking," *Current Anthropology* 53, no. 2 (2012): 204–12, with comments (212–19), reply (219–23), and references (223–25); also further discussion by Caleb Everett, "Without Language, No Distinctly Human Numerosity: A Reply to Coolidge and Overmann," *Current Anthropology* 54, no. 1 (2013): 81–82; and a response from the authors: "On the Nature of Numerosity and the Role of Language in Developing Number Concepts: A Reply to Everett," *Current Anthropology* 54, no. 1 (2013): 83–84.
255. Christian Agrillo, Laura Piffer, Angelo Bisazza, and Brian Butterworth, "Evidence for Two Numerical Systems That Are Similar in Humans and Guppies," *PLoS One* 7, no. 2 (2012), e31923.
256. Eugene Wigner, "The Unreasonable Effectiveness of Mathematics in the Natural Sciences," *Communications on Pure and Applied Mathematics* 13, no. 1 (1960): 1–14.

257. See my overview in Simon Conway Morris, "It All Adds Up . . . Or Does It? Numbers, Mathematics and Purpose," *Studies in the History and Philosophy of Science, Part C: Studies in History and Philosophy of Biological and Biomedical Sciences* 58 (2016): 117–22.
258. David C. Burr, Marco Turi, and Giovanni Anobile, "Subitizing but Not Estimation of Numerosity Requires Attentional Resources," *Journal of Vision* 10, no. 6 (2010), 20, https://psycnet.apa.org/doi/10.1167/10.6.20.
259. See, for example, Andreas Nieder and Earl K. Miller, "Coding of Cognitive Magnitude: Compressed Scaling of Numerical Information in the Primate Prefrontal Cortex," *Neuron* 37, no. 1 (2003): 149–57; also commentary by Stanislas Dehaene, "The Neural Basis of the Weber-Fechner Law: A Logarithmic Mental Number Line," *Trends in Cognitive Sciences* 7, no. 4 (2003): 145–47.
260. As per Israel Kleiner, "Thinking the Unthinkable: The Story of Complex Numbers (with a Moral)," *The Mathematical Teacher* 81, no. 7 (1988): 583–92.
261. See, for example, Andreas Nieder, "Supramodal Numerosity Selectivity of Neurons in Primate Prefrontal and Posterior Parietal Cortices," *Proceedings of the National Academy of Sciences of the United States of America* 109, no. 29 (2012): 11860–65.
262. See, for example, Stanislas Dehaene and Laurent Cohen, "Cultural Recycling of Cortical Maps," *Neuron* 56, no. 2 (2007): 384–98.
263. For a critical and intelligent overview see Rafael E. Núñez, "Is There Really an Evolved Capacity for Number?" *Trends in Cognitive Sciences* 21, no. 6 (2017): 409–24; also his response (404–5) to a critique by A. Nieder, "Number Faculty Is Rooted in Our Biological Heritage," *Trends in Cognitive Sciences* 21, no. 6 (2017): 403–4.
264. See, for example, Helen de Cruz and Johan de Smedt, "Mathematical Symbols as Epistemic Actions," *Synthese* 190 (2012): 3–19.
265. Richard Hamming, "The Unreasonable Effectiveness of Mathematics," *American Mathematical Monthly* 87, no. 2 (1980): 81–90, p. 84.
266. Denise Schmandt-Besserat, "The Token System of the Ancient Near East: Its Role in Counting, Writing, the Economy and Cognition," in *The Archaeology of Measurement: Comprehending Heaven, Earth and*

Time in Ancient Societies, ed. Iain Morley and Colin Renfrew (Cambridge: Cambridge University Press, 2010), 27–34.

267. See the overview by Tetsuro Matsuzawa, "Symbolic Representation of Number in Chimpanzees," *Current Opinion in Neurobiology* 19 (2009): 92–98.
268. For an overview, see M. Hauser, "Our Chimpanzee Mind," *Nature* 437, no. 7055 (2005): 60–63.
269. Jan Lonnemann et al., "Differences in Arithmetic Performance between Chinese and German Adults Are Accompanied by Differences in Processing of Non-symbolic Numerical Magnitude," *PLoS One* 12, no. 4 (2017), e0174991.
270. See, for example, Pierre Pica, Cathy Lemer, Véronique Izard, and Stanislas Dehaene, "Exact and Approximate Arithmetic in an Amazonian Indigene Group," *Science* 306, no. 5695 (2004): 499–503; and Stanislas Dehaene, Véronique Izard, Elizabeth Spelke, and Pierre Pica, "Log or Linear? Distinct Intuitions of the Number Scale in Western and Amazonian Indigene Cultures," *Science* 320, no. 5880 (2008): 1217–20.
271. See, for example, Ilka Diester and Andreas Nieder, "Semantic Associations between Signs and Numerical Categories in the Prefrontal Cortex," *PLoS Biology* 5, no. 11 (2007), e294.
272. Rafael Núñez, D. Doan, and Anastasia Nikoulina, "Squeezing, Striking, and Vocalizing: Is Number Representation Fundamentally Spatial?" *Cognition* 120, no. 2 (2011): 225–35.
273. See the critical overview by Titia Gebuis, Roi Cohen Kadosh, and Wim Gevers, "Sensory-Integration System Rather Than Approximate Number System Underlies Numerosity Processing: A Critical Review," *Acta Psychologica* 171 (2016): 17–35.
274. Michael Tarsitano, "Route Selection by a Jumping Spider (*Portia labiata*) during the Locomotory Phase of a Detour," *Animal Behaviour* 72, no. 6 (2006): 1437–42. Similar arguments apply to the sophistication of archerfish, whose shooting would otherwise do credit to a Spitfire pilot; see, for example, the overview by Stefan Schuster, "Hunting in Archerfish—an Ecological Perspective on a Remarkable Combination of Skills," *Journal of Experimental Biology* 221, no. 24 (2018), jeb.159723.

275. Patricia Gray, Bernie Krause, Jelle Atema, Roger Payne, Carol Krumhansl, and Luis Baptista, "The Music of Nature and the Nature of Music," *Science* 291, no. 5501 (2001): 52–54.
276. From the Keats poem "Ode to a Nightingale" (1819).
277. Evelyn Cheesman, *Who Stand Alone* (London: Geoffrey Bles, 1965), 76.
278. For an overview, see Aniruddh D. Patel, "The Evolutionary Biology of Musical Rhythm: Was Darwin Wrong?" *PLoS Biology* 12, no. 3 (2014), e1001821 (with correction on e1001873).
279. Quoted in Anthony Boden, ed., *Stars in a Dark Night: The Letters of Ivor Gurney to the Chapman Family* (Stroud, UK: Sutton, 2004), 220; and taken from Ivor Gurney, "The Springs of Music," *The Musical Quarterly* 8 (1922): 319–22.
280. See, for example, Charles P. Nicklin, James D. Darling, and Meagan E. Jones, "Humpback Whale Songs: Do They Organize Males during the Breeding Season?" *Behaviour* 143, no. 9 (2006): 1051–101; also Eduardo Mercado and Stephen Handel, "Understanding the Structure of Humpback Whale Songs," *Journal of the Acoustical Society of America* 132, no. 5 (2012): 2947–50.
281. W. Tecumseh Fitch, "Four Principles of Bio-musicology," *Philosophical Transactions of the Royal Society of London, Series B: Biological Sciences* 370, no. 1664 (2015): 2014091, https://doi.org/10.1098/rstb.2014.0091.
282. See Emily Doolittle and Bruno Gingras, "Zoomusicology," *Current Biology* 25, no. 19 (2015): R819–20.
283. See Valérie Dufour, Cristian Pasquaretta, Pierre Gayet, and Elisabeth H. M. Sterck, "The Extraordinary Nature of Barney's Drumming: A Complementary Study of Ordinary Noise Making in Chimpanzees," *Frontiers in Neuroscience* 11 (2017), 2, https://doi.org/10.3389/fnins.2017.00002.
284. Margaret Wilson and Peter Cook, "Rhythmic Entrainment: Why Humans Want to, Fireflies Can't Help It, Pet Birds Try, and Sea Lions Have to Be Bribed," *Psychonomic Bulletin and Review* 23, no. 6 (2016): 1647–59.
285. See Wilbert Zarco, Hugo Merchant, Luis Prado, and Juan Carlos Mendez, "Subsecond Timing in Primates: Comparison of Interval Production between Human Subjects and Rhesus Monkeys," *Journal of*

Neurophysiology 102, no. 6 (2009): 3191–202; also H. Honing et al., "Rhesus Monkeys (*Macaca mulatta*) Detect Rhythmic Groups in Music, but Not the Beat," *PLoS One* 7, no. 12 (2012), e51369.

286. See Yuko Hattori, Masaki Tomonaga, and Tetsuro Matsuzawa, "Spontaneous Synchronized Tapping to an Auditory Rhythm in a Chimpanzee," *Scientific Reports* 3 (2013), 1566, https://doi.org/10.1038/srep01566.

287. See Aniruddh D. Patel, John R. Iversen, Micah R. Bregman, and Irena Schulz, "Experimental Evidence for Synchronization to a Musical Beat in a Nonhuman Animal," *Current Biology* 19, no. 10 (2009): 827–30. For a description of drumming by male palm cockatoos, see Robert Heinsohn, Christina N. Zdenek, Ross B. Cunningham, John A. Endler, and Naomi E. Langmore, "Tool-Assisted Rhythmic Drumming in Palm Cockatoos Shares Key Elements of Human Instrumental Music," *Science Advances* 3, no. 6 (2017), e1602399. All the birds drummed in different ways, and they never combined forces.

288. As Matt Cartmill points out in "Convergent? Minds? Some Questions about Mental Evolution," *Interface Focus* 7 (2017), 2016.0125, it may not be a coincidence that such parrots are also adept at vocal mimicry, although why this link exists is not obvious. More generally his discussion is an exhilarating overview of many issues raised in this chapter.

289. See, for example, John Bispham, "'Music' Means Nothing if We Don't Know What It Means," *Journal of Human Evolution* 50, no. 5 (2006): 587–93.

290. For an overview, see Javier DeFelipe, "The Evolution of the Brain, the Human Nature of Cortical Circuits, and Intellectual Creativity," *Frontiers in Neuroanatomy* 5 (2011), 29, https://doi.org/10.3389/fnana.2011.00029. I am skeptical that the cortical differences he draws attention to will go very far in explaining our cognitive uniqueness.

291. For an overview, see R. I. M. Dunbar and Susanne Schultz, "Why Are There So Many Explanations for Primate Brain Evolution?" *Philosophical Transactions of the Royal Society of London, Series B: Biological Sciences* 372, no. 1727 (2017): 2016.0244, https://doi.org/10.1098/rstb.2016.0244.

292. For example, Dunbar's number that posits the cognitive limits associated with human group sizes; see Patrik Lindenfors, Andreas Wartel, and Johan Lind, "'Dunbar's Number' Deconstructed," *Biology Letters* 7, no. 5 (2021): 2021.0158, https://doi.org/10.1098/rsbl.2021.0158.
293. See, for example, Pamela Lyon, "The Cognitive Cell: Bacterial Behavior Reconsidered," *Frontiers in Microbiology* 6 (2015), 264, https://doi.org/10.3389/fmicb.2015.00264. For a more circumspect approach, see Fermin C. Fulda, "Natural Agency: The Case of Bacterial Cognition," *Journal of the American Philosophical Association* 3, no. 1 (2017): 69–90.
294. See, for example, with respect to *Paramecium,* Harvard L. Armus, Amber R. Montgomery, and Jenny L. Jellison, "Discrimination Learning in Paramecia (*P. caudatum*)," *Psychological Record* 56 (2006): 489–98.
295. See, for example, František Baluška, Stefano Mancuso, and Dieter Volkmann, eds., *Communication in Plants: Neuronal Aspects of Plant Life* (New York: Springer, 2006).
296. Critical overview by Charles I. Abramson and Harrington Wells, "An Inconvenient Truth: Some Neglected Issues in Invertebrate Learning," *Perspectives on Behavior Science* 41, no. 2 (2018): 395–416.
297. See, for example, the overview by Nicholas Toth and Kathy Schick, "An Overview of the Cognitive Implications of the Oldowan Industrial Complex," *Azania* 53, no. 1 (2018): 3–39.
298. See, for example, Brian Hayden, "What Were They Doing in the Oldowan?" *Lithic Technology* 33 (2008): 105–39; and Cristina Lemorini et al., "Old Stones' Song: Use-Wear Experiments and Analysis of the Oldowan Quartz and Quartzite Assemblage from Kanjera South (Kenya)," *Journal of Human Evolution* 72 (2014): 10–25. The latter authors argue that Oldowan lithics were employed in woodworking, including spear production, and perhaps twine production as well, collectively pointing to the beginning of a hunter-gatherer existence.
299. See Adrián Arroyo, Satoshi Hirata, Tetsuro Matsuzawa, and Ignacio de la Torre, "Nut Cracking Tools Used by Captive Chimpanzees (*Pan troglodytes*) and Their Comparison with Early Stone Age Percussive Artefacts from Olduvai Gorge," *PLoS One* 11, no. 11 (2016), e0166788.

300. See, for example, Erella Hovers, "Invention, Reinvention and Innovation: Makings of Oldowan Lithic Technology," in *Origins of Human Innovation and Creativity*, ed. Scott A. Elias (Amsterdam: Elsevier, 2012), 51–68.
301. Raymond Corbey, Adam Jagich, Krist Vaesen, and Mark Collard, "The Acheulean Handaxe: More Like a Bird's Song Than a Beatles' Tune?" *Evolutionary Anthropology* 25, no. 1 (2016): 6–19.
302. See, for example, Aldo Faisal, Dietrich Stout, Jan Apel, and Bruce Bradley, "The Manipulative Complexity of Lower Paleolithic Stone Toolmaking," *PLoS One* 5, no. 11 (2010), e13718; E. E. Hecht et al., "Acquisition of Paleolithic Toolmaking Abilities Involves Structural Remodeling to Inferior Frontoparietal Regions," *Brain Structure and Function* 220, no. 4 (2015): 2315–31; and Dietrich Stout, Erin Hecht, Nada Khreisheh, Bruce Bradley, and Thierry Chaminade, "Cognitive Demands of Lower Paleolithic Toolmaking," *PLoS One* 10, no. 4 (2015), e0121804.
303. Sophie A. de Beaune, "The Invention of Technology: Prehistory and Cognition," *Current Anthropology* 45, no. 2 (2004): 139–62, with commentaries (151–56) and author reply (156–58).
304. See, for example, Terrence Twomey, "The Cognitive Implications of Controlled Fire Use by Early Humans," *Cambridge Archaeological Journal* 23, no. 1 (2013): 113–28.
305. See the chapter by Robert A. Dielenberg, "The Comparative Psychology of Human Uniqueness: A Cognitive Behavioral Review," in *The Psychology of Human Behavior*, ed. R. G. Bednarik (Hauppauge, NY: Nova Science, 2013), 111–82. The author provides an outstanding overview of animal cognition.
306. See the chapter by Jennifer Vonk and Daniel J. Povinelli, "Similarity and Difference in the Conceptual Systems of Primates: The Unobservability Hypothesis," in *Comparative Cognition: Experimental Explorations of Animal Intelligence*, ed. Edward A. Wasserman and Thomas R. Zentall (Oxford: Oxford University Press, 2006), 363–87.
307. See Daniel L. Everett, *Don't Sleep, There Are Snakes: Life and Language in the Amazonian Jungle* (New York: Pantheon, 2008).
308. Jean Molino, "Toward an Evolutionary Theory of Language," in *The Origins of Music*, ed. Nils L. Wallin, Björn Merker, and Steven Brown (Cambridge, MA: MIT Press, 2000), 165–76, p. 175.

309. See, for example, Owen Barfield, *History in English Words,* 2nd ed. (London: Faber and Faber, 1956); also *Poetic Diction: A Study in Meaning,* 2nd ed. (London: Faber and Faber, 1952); and *The Rediscovery of Meaning, and Other Essays* (see especially "Part Two: Meaning, Language and Imagination") (Middletown, CT: Wesleyan University Press, 1977). Michael Vincent Di Fuccia, *Owen Barfield: Philosophy, Poetry and Theology* (Eugene, OR: Cascade Books, 2016), provides an excellent overview.
310. Cited in his obituary in the *Daily Telegraph,* June, 20, 2017, p. 27.
311. See, for example, Edward F. Kelly, Emily Williams Kelley, Adam Crabtree, Alan Gauld, Michael Grosso, and Bruce Greyson, *Irreducible Mind: Toward a Psychology for the 21st Century* (Lanham, MD: Rowman & Littlefield; 2007). One of several clues that this approach will be far more fruitful is the case of terminal lucidity. This is a very peculiar aspect of cognitive recall found in some patients suffering from dementias (and other cerebral disorders that can include lifelong cognitive impairments), which typically emerges shortly before their death. For an overview, see Michael Nahm, Bruce Greyson, Emily Williams Kelly, and Erlendur Haraldsson, "Terminal Lucidity: A Review and a Case Collection," *Archives of Gerontology and Geriatrics* 55, no. 1 (2012): 138–42; as well as the extraordinary accounts given by Michael Nahm and Bruce Greyson in "The Death of Anna Katharina Ehmer: A Case Study in Terminal Lucidity," *Omega (Westport)* 68, no. 1 (2013–2014): 77–87.

CHAPTER 6

1. See Simon Conway Morris, "*Typhloesus wellsi* (Melton and Scott, 1973), a Bizarre Metazoan from the Carboniferous of Montana, U. S. A," *Philosophical Transactions of the Royal Society of London, Series B: Biological Sciences* 327, no. 1242 (1990): 595–624.
2. See my *jeu d'esprit:* Simon Conway Morris, "Aliens Like Us?" *Astronomy and Geophysics* 46, no. 4 (2005), 4.24–4.26.
3. See Christopher Rose and Gregory Wright, "Inscribed Matter as an Energy-Efficient Means of Communication with an Extraterrestrial Civilization," *Nature* 431, no. 7004 (2004): 47–49; and commentary by

Woodruff T. Sullivan 3rd, "Astrobiology: Message in a Bottle," *Nature* 431, no. 7004 (2004): 27–28.

4. Jacob Haqq-Misra and Ravi Kumar Kopparapu, "On the Likelihood of Non-terrestrial Artifacts in the Solar System," *Acta Astronautica* 72 (2012): 15–20.
5. Or pessimism—no, not nuclear Armageddon or death by global warming but spiraling health costs. See Alistair V. W. Geoffrey, W. Guy, and Jimmy D. Bell, "The Intelligence Paradox; Will ET Get the Metabolic Syndrome? Lessons from and for Earth," *Nutrition and Metabolism* 11 (2014): 34, https://doi.org/10.1186/1743-7075-11-34.
6. See the informative piece by Eric M. Jones, "Where Is Everybody? An Account of Fermi's Question," *Physics Today* 38, no. 8 (1985): 11–13. Also see Robert H. Gray, "The Fermi Paradox Is Neither Fermi's Nor a Paradox," *Astrobiology* 15, no. 3 (2015): 195–99; he points out that that was not quite what Fermi meant—he may have been thinking more about the possibilities of interstellar travel. The laurels for the initiation of this paradox belong as much to people such as Michael H. Hart, "An Explanation for the Absence of Extraterrestrials on Earth," *Quarterly Journal of the Royal Astronomical Society* 16 (1975): 128–35.
7. Stephen Webb, *If the Universe Is Teeming with Aliens . . . Where Is Everybody? Seventy-five Solutions to the Fermi Paradox and the Problem of Extraterrestrial Life* (New York: Springer, 2015); also Milan M. Ćirković's *The Great Silence: The Science and Philosophy of Fermi's Paradox* (New York: Oxford University Press, 2018).
8. David Brin, "The 'Great Silence': The Controversy Concerning Extraterrestrial Intelligent Life," *Quarterly Journal of the Royal Astronomical Society* 24, no. 3 (1983): 283–309. Although it shows its age, Brin's article is a comprehensive overview that still sets the scene.
9. See, for example, the overview by Edd Wheeler, "The 'WOW' Signal, Drake Equation, and Exoplanet: Considerations," *Journal of the British Interplanetary Society* 67 (2014): 412–17. It was certainly very anomalous, but scans in the direction of Sagittarius drew a blank; see Robert H. Gray and Simon Ellingsen, "A Search for Periodic Emissions at the WOW Locale," *Astrophysical Journal* 578 (2002): 967–71. A more recent survey was no more successful but did detect a potential signal outside the

region of interest. The latter, however, may have been from a satellite, and the authors concluded "the original Wow signal was due to a man-made source, such as an aircraft or spacecraft" (7); see G. R. Harp, R. H. Gray, J. Richards, G. S. Shostak, and J. C. Tarter, "An ATA Search for a Repetition of the Wow Signal," *Astronomical Journal* 160, no. 4 (2020), 162, https://doi.org/10.3847/1538-3881/aba58f.

10. I discount unidentified flying objects (UFOs); they certainly exist, but they have nothing to do with extraterrestrials in any traditional sense. In chapter 5, I more than hinted that the world around us is interpenetrated by orthogonal realities, and UFOs are part of that story. What is clear is that the physics these objects employ is utterly "impossible"; see, for example, Kevin H. Knuth, Robert M. Powell, and Peter A. Reali, "Estimating Flight Characteristics of Anomalous Unidentified Aerial Vehicles," *Entropy* 21, no. 10 (2019), 939, https://doi.org/10.3390/e21100939.
11. James Benford, "Looking for Lurkers: Co-orbiters as SETI Observables," *Astronomical Journal* 158, no. 4 (2019), 150, https://doi.org/10.3847/1538-3881/ab3e35.
12. Paul Davies gives us a wish list of things we might look for in "Footprints of Alien Technology," *Acta Astronautica* 73 (2012): 250–57. See also Seth Shostak, "SETI: The Argument for Artefact Searches," *International Journal of Astrobiology* 19, no. 6 (2020): 456–61.
13. For an overview, see, for example, K. D. Olum, "Conflict between Anthropic Reasoning and Observation," *Analysis* 64, no. 1 (2004): 1–8. For critiques, see Dien Ho and Bradley Monton, "Anthropic Reasoning Does Not Conflict with Observation," *Analysis* 65, no. 1 (2005): 42–45; and Milan M. Ćirković, "Too Early? On the Apparent Conflict of Astrobiology and Cosmology," *Biology and Philosophy* 21, no. 3 (2006): 369–79. Also see Milan M. Ćirković, "Fermi's Paradox—the Last Challenge for Copernicanism?" *Serbian Astronomical Journal* 178 (2009): 1–20; and Milan M. Ćirković, "Fermi Paradox Is a Daunting Problem—under Whatever Label," *Astrobiology* 16, no. 10 (2016): 737–40; with a response by Robert H. Gray, "The So-Called Fermi Paradox Is Misleading, Flawed, and Harmful," *Astrobiology* 16, no. 10 (2016): 741–43.

14. For a characteristically optimistic overview, see "The Astrobiology Primer v2.0," ed. Shawn D. Domagal-Goldman and Katherine E. Wright, *Astrobiology* 16, no. 8 (2016): 561–653. But not everybody agrees; see, for example, Howard Smith, "Alone in the Universe," *Zygon* 51, no. 2 (2016): 497–519. For a further dose of pessimism, this should do nicely: Leonard Ornstein, "Extraterrestrial Intelligence: The Debate Continues. A Biologist Looks at the Numbers," *Physics Today* 35, no. 3 (1982): 27–31.
15. Irwin and colleagues suggest perhaps 100 million (10^8) planets have the potential to house complex life. Louis N. Irwin, Abel Méndez, Alberto G. Fairén, and Dirk Schulze-Makuch, "Assessing the Possibility of Biological Complexity on Other Worlds, with an Estimate of the Occurrence of Complex Life in the Milky Way Galaxy," *Challenges* 5, no. 159 (2014): 159–74.
16. See Rodrigo Ramirez, Marco A. Gómez-Muñoz, Roberto Vázquez, and Patricia G. Núñez, "New Numerical Determination of Habitability in the Galaxy: The SETI Connection," *International Journal of Astrobiology* 17, no. 1 (2018): 34–43.
17. See Pratika Dayal, Charles Cockell, Ken Rice, and Anupam Mazumdar, "The Quest for Cradles of Life: Using the Fundamental Metallicity Relation to Hunt for the Most Habitable Type of Galaxy," *Astrophysical Journal Letters* 810, no. 1 (2015), L2, https://doi.org/10.1088/2041-8205/810/1/L2. For a more pessimistic view, see Daniel P. Whitmire, "The Habitability of Large Elliptical Galaxies," *Monthly Notices of the Royal Astronomical Society* 494, no. 2 (2020): 3048–52.
18. See Erik Zackrisson, Per Calissendorff, Juan González, Andrew Benson, Anders Johansen, and Markus Janson, "Terrestrial Planets across Space and Time," *Astrophysical Journal* 833, no. (2016), 214, https://doi.org/10.3847/1538-4357/833/2/214.
19. For a balanced overview, see, for example, Nathalie A. Cabrol, "Alien Mindscapes—A Perspective on the Search for Extraterrestrial Intelligence," *Astrobiology* 16, no. 9 (2016): 661–76.
20. See, for example, Keith B. Wiley, "The Fermi Paradox, Self-Replicating Probes, and the Interstellar Transportation Band Width," *arXiv*, November 26, 2011, https://arxiv.org/abs/1111.6131. For a more cautionary

outlook that argues that these putative probes will be inevitably subject to error propagation, leading to ultimate error catastrophe, see Axel Kowald, "Why Is There No von Neumann Probe on Ceres? Error Catastrophe Can Explain the Fermi-Hart Paradox," *Journal of the British Interplanetary Society* 68, no. 12 (2015): 383–88, along with the critique by Stephen Ashworth, "Von Neumann Probes, Errors and the Absence of Evidence for Extraterrestrial Civilisations," *Journal of the British Interplanetary Society* 69, no. 4 (2016): 148–50, with a response from Axel Kowald (150–52).

21. See, for example, Anthony R. Martin and Alan Bond, "Is Mankind Unique?—The Lack of Evidence for Extraterrestrial Intelligence," *Journal of the British Interplanetary Society* 36, no. 5 (1983): 223–25.
22. An analogy employed by Jason T. Wright, Shubham Kanodia, and Emily G. Lubar, "How Much SETI Has Been Done? Finding Needles in the n-Dimensional Cosmic Haystack," *Astronomical Journal* 156, no. 6 (2018), 260, https://dx.doi.org/10.3847/1538-3881/aae099, suggesting the "space" searched to date is about equivalent to the volume of a small swimming pool as against that of Earth's oceans.
23. Concerning the observations of nearby 692 stars, see J. Emilio Enriquez et al., "The Breakthrough Listen Search for Intelligent Life: 1.1–1.9 GHz Observations of 692 Nearby Stars," *Astrophysical Journal* 849, no. 2 (2017), 104, https://doi.org/10.3847/1538-4357/aa8d1b. In a similar vein, see the investigation of 14 exoplanetary systems in Jean-Luc Margot et al., "A Search for Technosignatures from 14 Planetary Systems in the Kepler Field with the Green Bank Telescope at 1.15–1.73 GHz," *Astronomical Journal* 155, no. 2 (2018), 209, https://doi.org/10.3847/1538-3881/aabb03; and thirty-one sun-like stars: Jean-Luc Margot et al., "A Search for Technosignatures around 31 Sun-like Stars with the Green Bank Telescope at 1.15–1.73 GHz," *Astronomical Journal* 161, no. 2 (2021), 55, https://doi.org/10.3847/1538-3881/abcc77. For a search of 1,327 stars, see D.C. Price et al., "The Breakthrough Listen Search for Intelligent Life: Observations of 1327 Nearby Stars Over 1.10–3.45 GHz," *Astronomical Journal* 159, no. 3 (2020), 86, https://doi.org/10.3847/1538-3881/ab65f1; as well as a search for laser emissions from 5,600 candidate stars: Nathaniel K. Tellis and Geoffrey W. Marcy, "A Search for Laser Emission with

Megawatt Thresholds from 5600 FGKM Stars," *Astronomical Journal* 153, no. 6 (2017), 251, https://doi.org/10.3847/1538-3881/aa6d12. The latter note that potentially this includes approximately 2,000 habitable Earth-like planets, leading them to conclude that maybe Fermi was onto something after all. On the larger scale, for example, see listening in to the Andromeda (M31) and Triangulum (M33) galaxies: Robert H. Gray and Kunal Mooley, "A VLA Search for Radio Signals from M31 and M33," *Astronomical Journal* 153, no. 3 (2017), 110, https://doi.org/10.3847/1538-3881/153/3/110.

24. F. J. Dyson, "Search for Artificial Stellar Sources of Infrared Radiation," *Science* 131, no. 3414 (1960): 1667–68.
25. Roger L. Griffith, Jason T. Wright, Jessica Maldonado, Matthew S. Povich, Steinn Sigurđsson, and Brendan Mullan, "The Ĝ Infrared Search for Extraterrestrial Civilizations with Large Energy Supplies. III. The Reddest Extended Sources in WISE," *Astrophysical Journal, Supplement Series* 217, no. 2 (2015), 25, https://doi.org/10.1088/0067-0049/217/2/25; also Erik Zackrisson, Per Calissendorff, Saghar Asadi, and Anders Nyholm, "Extragalactic SETI: The Tully–Fisher Relation as a Probe of Dysonian Astroengineering in Disk Galaxies," *Astrophysical Journal* 810, no. 1 (2015), 23, https://doi.org/10.1088/0004-637X/810/1/23.
26. Somewhat more pessimistic than Jianpo Guo, Fenghui Zhang, Xuefei Chen, and Zhanwen Han, "Probability Distribution of Terrestrial Planets in Habitable Zones around Host Stars," *Astrophysics and Space Science* 323, no. 4 (2009): 367–73.
27. A similar search for Dyson spheres in our immediate neighborhood (around 1,000 light years) also drew a blank; see Richard A. Carrigan Jr., "*IRAS*-Based Whole-Sky Upper Limit on Dyson Spheres," *Astrophysical Journal* 698 (2009): 2075–86.
28. See, for example, T. L. Campante et al., "An Ancient Extrasolar System with Five Sub-Earth-Size Planets," *Astrophysical Journal* 799, no. 2 (2015), 170, https://doi.org/10.1088/0004-637X/799/2/170.
29. See, for example, Darach Watson, Lise Christensen, Kirsten Kraiberg Knudsen, Johan Richard, Anna Gallazzi, and Michał Jerzy Michałowski, "A Dusty, Normal Galaxy in the Epoch of Reionization," *Nature* 519, no. 7543 (2015): 327–30.

30. Charles H. Lineweaver, “An Estimate of the Age Distribution of Terrestrial Planets in the Universe: Quantifying Metallicity as a Selection Effect,” *Icarus* 151 (2001): 307–13. In a subsequent analysis, Peter Behroozi and Molly S. Peeples, “On the History and Future of Cosmic Planet Formation,” *Monthly Notices of the Royal Astronomical Society* 454, no. 2 (2015): 1811–17, suggest that in the galaxy 80 percent of planets formed before the formation of the solar system.
31. See, for example, J. Fritz et al., “Earth-like Habitats in Planetary Systems,” *Planetary and Space Science* 98 (2014): 254–67. For a list of potential candidate exoplanets, see Dirk Schulze-Makuch et al., “A Two-Tiered Approach to Assessing the Habitability of Exoplanets,” *Astrobiology* 20, no. 10 (2020): 1394–404.
32. Peter D. Ward and Donald Brownlee, *Rare Earth: Why Complex Life Is Uncommon in the Universe* (New York: Copernicus Books, 2000). As a sort of counterblast that other planets may be more habitable than the Earth, see René Heller and John Armstrong, “Superhabitable Worlds,” *Astrobiology* 14, no. 1 (2014): 50–66.
33. Dirk Schulze-Makuch and Ian A. Crawford, “Was There an Early Habitability Window for Earth’s Moon?” *Astrobiology* 18, no. 8 (2018): 985–88. Overall the potential for carbonaceous asteroids appears low; see O. Abramov and S.-J. Mojzsis, “Abodes for Life in Carbonaceous Asteroids?” *Icarus* 213, no. 1 (2011): 273–79.
34. In the case of Mars there is a regular series of teasers, such as claims for minute quantities of methane; see, for example, Christopher R. Webster et al., “Background Levels of Methane in Mars’ Atmosphere Show Strong Seasonal Variations,” *Science* 360, no. 6393 (2018): 1093–96. More recently, however, the evidence for both methane and other organics such as ethylene has, so to speak, evaporated; see, respectively, Oleg Korablev et al., “No Detection of Methane on Mars from Early ExoMars Trace Gas Orbiter Observations,” *Nature* 568 (2019): 517–20; Elise W. Knutsen et al., “Comprehensive Investigation of Mars Methane and Organics with ExoMars/NOMAD,” *Icarus* 357 (2021), 114266, https://doi.org/10.1016/j.icarus.2020.114266. Curiously, the arguments for Venusian life might be more compelling. Not on its torrid surface, of course, but at an altitude of approximately fifty kilometers the conditions are

more or less equivalent to earth surface pressure and temperature. The extreme acidity may be less of a challenge given the low pH tolerance of some terrestrial extremophiles, but the near absence of water may be fatal, given that terrestrial counterparts cannot survive when the so-called water activity falls below a critical value. Similarly, the excitement about the apparent detection of phosphine (PH_3) in the atmosphere, a potential signature of biotic activity, is now looking much more likely to be sulphur dioxide (SO_2); see, for example, Jane S. Greaves et al., "Phosphine Gas in the Cloud Decks of Venus," *Nature Astronomy* 5 (2021): 655–64; but then Alex B. Akins, Andrew P. Lincowski, Victoria S. Meadows, and Paul G. Steffes, "Complications in the ALMA Detection of Phosphine at Venus," *Astrophysical Journal Letters* 907, no. 2 (2021), L27, https://doi.org/10.3847/2041-8213/abd56a; and Andrew P. Lincowski et al., "Claimed Detection of PH_3 in the Clouds of Venus Is Consistent with Mesospheric SO_2," *Astrophysical Journal Letters* 908, no. 2 (2021), L44, https://doi.org/10.3847/2041-8213/abde47; and M. A. Thompson, "The Statistical Reliability of 267-GHz JCMT Observations of Venus: No Significant Evidence for Phosphine Absorption," *Monthly Notices of the Royal Astronomical Society: Letters* 501, no. 1 (2021): L18–L22, https://doi.org/10.1093/mnrasl/slaa187. More generally, regarding the possibility of Venusian life, see, for example, Sanjay S. Limaye, Rakesh Mogul, David J. Smith, Arif H. Ansari, Grzegorz P. Słowik, and Parag Vaishampayan, "Venus' Spectral Signatures and the Potential for Life in the Clouds," *Astrobiology* 18, no. (2018): 1181–98.

35. See, for example, Brandon Carter, "Five- or Six-Step Scenario for Evolution?" *International Journal of Astrobiology* 7, no. 2 (2008): 177–82. For an update, see Andrew E. Snyder-Beattie, Anders Sandberg, K. Eric Drexler, and Michael B. Bonsall, "The Timing of Evolutionary Transitions Suggests Intelligent Life Is Rare," *Astrobiology* 21, no. 3 (2021): 265–78.
36. Robert J. Stern, "Is Plate Tectonics Needed to Evolve Technological Species on Exoplanets?" *Geoscience Frontiers* 7, no. 4 (2015): 573–80.
37. See, for example, David C. Catling, Christopher R. Glein, Kevin J. Zahnle, and Christopher P. McKay, "Why O_2 Is Required by Complex Life on Habitable Planets and the Concept of Planetary 'Oxygenation Time,'" *Astrobiology* 5, no. 3 (2005): 415–38.

38. See, for example, William Bains and Dirk Schulze-Makuch, "Mechanisms of Evolutionary Innovation Point to Genetic Control Logic as the Key Difference between Prokaryotes and Eukaryotes," *Journal of Molecular Biology* 81, no. 1–2 (2015): 34–53; and William Bains and Dirk Schulze-Makuch, "The Cosmic Zoo: The (Near) Inevitability of the Evolution of Complex, Macroscopic Life," *Life (Basel)* 6, no. 3 (2016), 25, https://doi.org/10.3390/life6030025.
39. Aditya Chopra and Charles H. Lineweaver, "The Case for a Gaian Bottleneck: The Biology of Habitability," *Astrobiology* 16, no. 1 (2016): 7–22.
40. See, for example, J. T. Wright, B. Mullan, S. Sigurdsson, and M. S. Povich, "The Ĝ Infrared Search for Extraterrestrial Civilizations with Large Energy Supplies. I. Background and Justification," *Astrophysical Journal* 792, no. 1 (2014), 26, https://doi.org/10.1088/0004-637X/792/1/26; they suggest a time span of ca. 10^8–10^9 years but also indicate the diaspora might be considerably faster. See also the analysis by Nikos Prantzos, "A Joint Analysis of the Drake Equation and the Fermi Paradox," Special Issue 3, "The Drake Equation," *International Journal of Astrobiology* 12 (2013): 246–53, that if extraterrestrial civilizations are not that rare then colonization is highly probable.
41. Stuart Armstrong and Anders Sandberg, "Eternity in Six Hours: Intergalactic Spreading of Intelligent Life and Sharpening the Fermi Paradox," *Acta Astronautica* 89 (2013): 1–13.
42. Simon Conway Morris and Jean-Bernard Caron, "A Primitive Fish from the Cambrian of North America," *Nature* 512, no. 7515 (2014): 419–22.
43. And here we give at least a nod to the possibility of noncarbaquist life forms; see, for example, Dirk Schulze-Makuch and Louis N. Irwin, "The Prospect of Alien Life in Exotic Forms on Other Worlds," *Naturwissenschaften* 93, no. 4 (2006): 155–72. However, when it comes to the most-cited alternative, silicon, then the prospects, certainly in association with water, are pretty dim, and the best hope might be sulfuric acid (and so maybe a clue to Venusian life); see Janusz Jurand Petkowski, William Bains, and Sara Seager, "On the Potential of Silicon as a Building Block for Life," *Life (Basel)* 10, no. 6 (2020), 84, https://doi.org/10.3390/life10060084.

44. Nick Bostrom, "Where Are They: Why I Hope the Search for Extraterrestrial Life Finds Nothing," *MIT Technology Review*, May/June 2008, 72–78, p. 72.
45. See the influential article by James Annis, "An Astrophysical Explanation for the 'Great Silence,'" *Journal of the British Interplanetary Society* 52, no. 1 (1999): 19–22.
46. See, for example, James E. Horvath and Douglas Galante, "Effects of High-Energy Astrophysical Events on Habitable Planets," *International Journal of Astrobiology* 11, no. 4 (2012): 279–86; and Dimitra Atri, Adrian L. Melott, and Andrew Karam, *International Journal of Astrobiology* 13, no. 3 (2014): 224–28.
47. Milan M. Ćirković and Branislav Vukotić, "Astrobiological Phase Transition: Towards Resolution of Fermi's Paradox," *Origins of Life and Evolution of Biospheres* 38, no. 6 (2008): 535–47.
48. At least as far as galaxies like the Milky Way are concerned, gamma ray bursts are largely the hallmark of low luminosity, low stellar mass, and low metallicity galaxies. K. Z. Stanek et al., "Protecting Life in the Milky Way: Metals Keep the GRBs Away," *Acta Astronomica* 56 (2006): 333–45; also Raul Jimenez and Tsvi Piran, "Reconciling the Gamma-Ray Burst Rate and Star Formation Histories," *Astrophysical Journal* 773, no. 2 (2013), 216, https://doi.org/10.1088/0004-637X/773/2/126; and Michael G. Gowanlock, "Astrobiological Effects of Gamma-Ray Bursts in the Milky Way Galaxy," *Astrophysical Journal* 832, no. 1 (2016), 38, https://doi.org/10.3847/0004-637X/832/1/38.
49. Simon Conway Morris, "Three Explanations for Extraterrestrials: Sensible, Unlikely, Mad," *International Journal of Astrobiology* 17, no. 4 (2018): 287–93.
50. For one such gloomy prognosis, see Robert Klee, "Human Expunction," *International Journal of Astrobiology* 16, no. 4 (2017): 379–88; and a spirited response by Milan M. Ćirković, "The Reports of Expunction Are Grossly Exaggerated: A Reply to Robert Klee," *International Journal of Astrobiology* 18, no. 1 (2019): 14–17, along with a robust reply by Klee, "The Importance of Being Sufficiently Realistic: A Reply to Milan Ćirković," 10–13.

51. See, for example, John A. Ball, "The Zoo Hypothesis," *Icarus* 19, no. 3 (1973): 347–49, as well as various critiques such as the ones by Duncan H. Forgan: "Spatio-Temporal Constraints on the Zoo Hypothesis, and the Breakdown of Total Hegemony," *International Journal of Astrobiology* 10, no. 4 (2011): 341–47, and "The Galactic Club or Galactic Cliques? Exploring the Limits of Interstellar Hegemony and the Zoo Hypothesis," *International Journal of Astrobiology* 16, no. 4 (2017): 349–54. For a variant of the zoo hypothesis, see the "interdict hypothesis" of Martyn J. Fogg, "Temporal Aspects of the Interaction among the First Galactic Civilizations: The Interdict 'Hypothesis,'" *Icarus* 69, no. 2 (1987): 370–84.
52. See, for example, the paper by David M. Kipping and Alex Teachey, "A Cloaking Device for Transiting Planets," *Monthly Notices of the Royal Astronomical Society* 459, no. 2 (2016): 1233–41.
53. Anders Sandberg, Stuart Armstrong, and Milan M. Ćirković, "That Is Not Dead Which Can Eternal Lie: The Aestivation Hypothesis for Resolving Fermi's Paradox," *Journal of the British Interplanetary Society* 69, no. 11 (2016): 406–15. For a strong critique, see Charles H. Bennett, Robin Hanson, and C. Jess Riedel, "Comment on 'The Aestivation Hypothesis for Resolving Fermi's Paradox,'" *Foundations of Physics* 49, no. 8 (2019): 820–29.
54. Since I wrote *Life's Solution: Inevitable Humans in a Lonely Universe* (Cambridge: Cambridge University Press, 2003), the question of the potential importance of evolutionary convergence in the context of extraterrestrials has gained considerable traction. For recent essays see, for example, by Claudio L. Flores Martinez, "SETI in the Light of Cosmic Convergent Evolution," *Acta Astronautica* 104, no. 1 (2014): 341–49; and Milan M. Ćirković, "Woodpeckers and Diamonds: Some Aspects of Evolutionary Convergence in Astrobiology," *Astrobiology* 18, no. 5 (2018): 491–502. For a parallel argument that complex life is also very much on the cards, see Bains and Schulze-Makuch, "Cosmic Zoo," 25.
55. Stephen Baxter, "The Planetarium Hypothesis: A Resolution of the Fermi Paradox," *Journal of the British Interplanetary Society* 54, no. 5–6 (2001): 210–16.

56. Published in Stephen Baxter, *Phase Space* (London: Voyager, 2003), 380–408.
57. Whether we could be so-called Sims is examined by Nick Bostrom, "Are We Living in a Computer Simulation?" *Philosophical Quarterly* 53, no. 211 (2003): 243–55; Brian Weatherson, "Are You a Sim?" *Philosophical Quarterly* 53, no. 212 (2003): 425–31; and Claus Beisbart, "Are We Sims? How Computer Simulations Represent and What This Means for the Simulation Argument," *Monist* 97, no. 3 (2014): 399–417.
58. See, for example, Silas R. Beane, Zohreh Davoudi, and Martin J. Savage, "Constraints on the Universe as a Numerical Simulation," *European Physical Journal, Series A: Hadrons and Nuclei* 50 (2014), 148, https://doi.org/10.1140/epja/i2014-14148-0.
59. See, for example, V. T. Dokuchaev, "Is There Life Inside Black Holes?" *Classical and Quantum Gravity* 28, no. 23 (2011), 235015, https://doi.org/10.1088/0264-9381/28/23/235015; and Rohan M. Ganapathy and Mohammed Shazin Shoukath, "Black Holes: Attractors for Intelligence?" *Proceedings of the International Astronautical Congress* (IAC) (Naples, Italy) 2, (2013), 1607–18.
60. John M. Smart, "The Transcension Hypothesis: Sufficiently Advanced Civilizations Invariably Leave Our Universe, and Implications for METI and SETI," *Acta Astronautica* 78 (2012): 55–68.
61. See, for example, Bernard Haisch, "Is the Universe a Vast, Consciousness-Created Virtual Reality Simulation?" *Cosmos and History* 10, no. 1 (2014): 48–60.
62. To refer to a perhaps less familiar example, take Brian Aldiss's account in his *The Twinkling of an Eye, Or, My Life as an Englishman* (London: Little, Brown, 1998) of a haunting in his home in East Dereham, one that persisted long after his family had left and the building became offices.
63. See the brief account in Peter Ackroyd, *The English Ghost* (London: Chatto and Windus, 2011), 112. This account has been retold many times; see, for example, Aubrey L. Parke, "The Folklore of Sixpenny Handley, Dorset, Part I," *Folklore* 74, no. 3 (1963): 481–87.
64. See, for example, Ackroyd's account of events in the early nineteenth century in the Tower of London (*English Ghost*, 20–22).

65. Charles Fort, *The Book of the Damned* (New York: Boni and Liverright, 1919).
66. Although poltergeists centering on adolescents is strongly recurrent, with concomitant phenomena such as objects flying all over the place and badly damaging inanimate things yet curiously seldom harming the baffled observers.
67. Jeffery J. Kripal, *Secret Body: Erotic and Esoteric Currents in the History of Religions* (Chicago: University of Chicago Press, 2017), 206; and so Sir Arthur Eddington in his *Science and the Unseen World: Swarthmore Lecture, 1929* (London: George Allen & Unwin; 1929) remarks, "Consciousness is not wholly, nor even primarily a device for receiving self-impressions. . . . We may the more boldly insist that there is another outlook than the scientific one, because in practice a more transcendental outlook is almost universally admitted." Eddington proceeds then to outline "the poetry of existence" (28–29).
68. Simon Conway Morris, *The Runes of Evolution* (Conshohocken, PA: Templeton Press, 2015), 301–2; and drawing on the absorbing biography by Robert Kanigel, *The Man Who Knew Infinity: A Life of the Genius Ramanujan* (New York: Charles Scribner's Sons, 1991).
69. See, for example, the account by Robert Louis Stevenson in chapter 8 of his *Across the Plains with Other Memories and Essays* (London: Chatto & Windus, 1918) of how his celebrated Dr. Jekyll and Mr. Hyde (and other stories) were arrived at.
70. See Theodora Gay Scutt, *Cuckoo in the Powys Nest: A Memoir* (Harleston, UK: Brynmill Press, 2000), 208–9; see also p. 243 for another very odd story.
71. Kripal, *Secret Body*, 332n63.
72. For an overview of this apparently paradoxical reality, see Anthony Peake, *A Life of Philip K. Dick: The Man Who Remembered the Future* (London: Arcturus, 2013).
73. See her chapter "But What Did He Really Mean?" in Verlyn Flieger, *There Would Always Be a Fairy Tale: More Essays on Tolkien* (Ashland, OH: Kent State University Press, 2017).

CODA

1. Charles Darwin, *The Descent of Man, and Selection in Relation to Sex*, 2nd ed. (London: Murray, 1889), 126.
2. Darwin Correspondence Project, "To William Graham, 3 July 1881," letter 13230, https://www.darwinproject.ac.uk/letter/DCP-LETT-13230.xml.
3. Darwin Correspondence Project, "To Asa Gray, 22 May 1860," letter 2814, https://www.darwinproject.ac.uk/letter/DCP-LETT-2814.xml.
4. See John of Salisbury, *The Metalogicon: A Twelfth-Century Defense of the Verbal and Logical Arts of the Trivium*, trans. Daniel D. McGarry (Berkeley: University of California Press, 1955), book III, 167.
5. See, for example, Jeffery J. Kripal, *Secret Body: Erotic and Esoteric Currents in the History of Religions* (Chicago: University of Chicago Press, 2017).
6. Emphasis in the original (286), in his essay "Possible Worlds," in J. B. S. Haldane, *Possible Worlds and Other Papers* (London: Chatto & Windus, 1928), 260–68.

ABOUT THE AUTHOR

Photo by Steve Bond.

Simon Conway Morris is the Emeritus Professor of Evolutionary Palaeobiology at the University of Cambridge. Dr. Conway Morris is well known for his work on the early evolution of metazoans (popularly referred to as the "Cambrian Explosion") and his extensive studies on convergent evolution. He is the author of more than 100 scientific articles and is the author or editor of seven books. These include *The Crucible of Creation: The Burgess Shale and the Rise of Animals* (Oxford University Press, 1998), *Life's Solution: Inevitable Humans in a Lonely Universe* (Cambridge University Press, 2003), and *The Runes of Evolution: How the Universe Became Self-Aware* (Templeton Press, 2015). Dr. Conway Morris has received the Walcott Medal from the National Academy of Sciences, the Charles Schuchert Award from the Paleontological Society, and the Lyell Medal from the Geological Society of London. He was elected a Fellow of the Royal Society in 1990. He has spoken extensively at the intersection of science and religion, including giving the Gifford Lectures in 2007 at the University of Edinburgh.

INDEX